MOLECULAR DESIGN OF LIFE

MOLECULAR DESIGN OF LIFE

Lubert Stryer

STANFORD UNIVERSITY

W. H. FREEMAN AND COMPANY / NEW YORK

Molecular Design of Life consists of Chapters 1–8, Appendixes A–D, and the corresponding Answers to Problems of Lubert Stryer's *Biochemistry*, Third Edition.

Library of Congress Cataloging-in-Publication Data

Stryer, Lubert.
  Molecular design of life / Lubert Stryer.
    p. cm.
  Includes index.
  ISBN 0-7167-2049-3.—ISBN 0-7167-2050-7 (pbk.)
  1. Molecular biology. I. Title.
QH506.S875 1989   88-32219
574.8′8—dc19   CIP

Copyright © 1988, 1989 by Lubert Stryer

No part of this book may be reproduced by any mechanical, photographic, or electronic process, or in the form of a phonographic recording, nor may it be stored in a retrieval system, transmitted, or otherwise copied for public or private use, without written permission from the publisher.

Printed in the United States of America

1 2 3 4 5 6 7 8 9 0 RRD 7 6 5 4 3 2 1 0 8 9

To my teachers

    Paul F. Brandwein
    Daniel L. Harris
    Douglas E. Smith
    Elkan R. Blout
    Edward M. Purcell

Contents

List of Topics ix
Preface xiii

CHAPTER 1. Prelude 3
2. Protein Structure and Function 15
3. Exploring Proteins 43
4. DNA and RNA: Molecules of Heredity 71
5. Flow of Genetic Information 91
6. Exploring Genes: Analyzing, Constructing, and Cloning DNA 117
7. Oxygen-transporting Proteins: Myoglobin and Hemoglobin 143
8. Introduction to Enzymes 177

Appendixes 201
Answers to Problems 207
Index 211

List of Topics

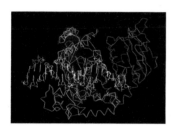

CHAPTER 1 Prelude 3

Molecular models depict three dimensional structure 4
Space, time, and energy 5
Reversible interactions of biomolecules are mediated by three kinds of noncovalent bonds 7
The biologically important properties of water are its polarity and cohesiveness 9
Water solvates polar molecules and weakens ionic and hydrogen bonds 9
Hydrophobic interactions: nonpolar groups tend to associate in water 10
Design of this book 11

CHAPTER 2 Protein Structure and Function 15

Proteins are built from a repertoire of twenty amino acids 16
Amino acids are linked by peptide bonds to form polypeptide chains 22
Proteins have unique amino acid sequences that are specified by genes 23
Protein modification and cleavage confer new capabilities 24
The peptide unit is rigid and planar 25
Polypeptide chains can fold into regular structures: the α helix and β pleated sheets 25
Polypeptide chains can reverse direction by making β-turns 28
Proteins are rich in hydrogen-bonding potentiality 28
Water-soluble proteins fold into compact structures with nonpolar cores 29
Levels of structure in protein architecture 31
Amino acid sequence specifies three-dimensional structure 32
Proteins fold by the association of α-helical and β-strand segments 34
Prediction of conformation from amino acid sequence 35
Essence of protein action: specific binding and transmission of conformational changes 37
Appendix: Acid-base concepts 41

CHAPTER 3 Exploring Proteins 43

Proteins can be separated by gel electrophoresis and displayed 44
Proteins can be purified according to size, charge, and binding affinity 46
Ultracentrifugation is valuable for separating biomolecules and determining molecular weights 49
Amino acid sequences can be determined by automated Edman degradation 50
Proteins can be specifically cleaved into small peptides to facilitate analysis 55

Recombinant DNA technology has revolutionized protein sequencing   57
Amino acid sequences provide many kinds of insights   58
X-ray crystallography reveals three-dimensional structure in atomic detail   59
Proteins can be quantitated and localized by highly specific antibodies   62
Peptides can be synthesized by automated solid-phase methods   64

CHAPTER 4   DNA and RNA: Molecules of Heredity   71

DNA consists of four kinds of bases joined to a sugar-phosphate backbone   72
Transformation of pneumococci by DNA revealed that genes are made of DNA   73
The Watson-Crick DNA double helix   76
The complementary chains act as templates for each other in DNA replication   78
DNA replication is semiconservative   79
The double helix can be reversibly melted   80
DNA molecules are very long   82
Some DNA molecules are circular and supercoiled   83
DNA is replicated by polymerases that take instructions from templates   84
Some viruses have single-stranded DNA during part of their life cycle   85
The genes of some viruses are made of RNA   86
RNA tumor viruses replicate through double-helical DNA intermediates   87

CHAPTER 5   Flow of Genetic Information   91

Several kinds of RNA play key roles in gene expression   92
Formulation of the concept of messenger RNA   93
Experimental evidence for messenger RNA, the informational intermediate in protein synthesis   94
Hybridization studies showed that messenger RNA is complementary to its DNA template   95
Ribosomal RNA and transfer RNA are also synthesized on DNA templates   96
All cellular RNA is synthesized by RNA polymerases   96
RNA polymerase takes instructions from a DNA template   97
Transcription begins near promoter sites and ends at terminator sites   98
Transfer RNA is the adaptor molecule in protein synthesis   99
Amino acids are coded by groups of three bases starting from a fixed point   99
Deciphering the genetic code: synthetic RNA can serve as messenger   101
Trinucleotides promote the binding of specific transfer RNA molecules to ribosomes   103
Copolymers with a defined sequence were also instrumental in breaking the code   104
Major features of the genetic code   106

Start and stop signals for protein synthesis   107
The genetic code is nearly universal   108
The sequences of genes and their encoded proteins are colinear   109
Most eucaryotic genes are mosaics of introns and exons   110
Many exons encode protein domains   111
RNA probably came before DNA and proteins in evolution   113

CHAPTER 6   Exploring Genes: Analyzing, Constructing, and Cloning DNA   117

Restriction enzymes split DNA into specific fragments   118
Restriction fragments can be separated by gel electrophoresis and visualized   119
DNA can be sequenced by specific chemical cleavage (Maxam-Gilbert method)   120
DNA can be sequenced by controlled interruption of replication (Sanger dideoxy method)   121
DNA probes and genes can be synthesized by automated solid-phase methods   123
New genomes can be constructed, cloned, and expressed   124
Restriction enzymes and DNA ligase are key tools in forming recombinant DNA molecules   126
Plasmids and lambda phage are choice vectors for DNA cloning in bacteria   127
Specific genes can be cloned from a digest of genomic DNA   130
Complementary DNA (cDNA) prepared from mRNA can be expressed in host cells   132
New genes inserted into eucaryotic cells can be efficiently expressed   133
Tumor-inducing (Ti) plasmids can be used to bring new genes into plant cells   135
Novel proteins can be engineered by site-specific mutagenesis   136
Recombinant DNA technology has opened new vistas   137

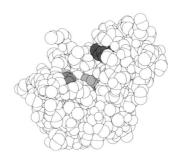

CHAPTER 7   Oxygen-transporting Proteins: Myoglobin and Hemoglobin   143

Oxygen binds to a heme prosthetic group   144
Myoglobin was the first protein to be seen at atomic resolution   144
Myoglobin has a compact structure and a high content of alpha helices   145
Oxygen binds within a crevice in myoglobin   146

A hindered heme environment is essential for reversible oxygenation   148
Carbon monoxide binding is diminished by the presence of the distal histidine   149
The central exon of myoglobin encodes a functional heme-binding unit   149
Hemoglobin consists of four polypeptide chains   150
X-ray analysis of hemoglobin   151
The alpha and beta chains of hemoglobin closely resemble myoglobin   152
Critical residues in the amino acid sequence   153
Hemoglobin is an allosteric protein   154
Oxygen binds cooperatively to hemoglobin   154
The cooperative binding of oxygen by hemoglobin enhances oxygen transport   156
H^+ and CO_2 promote the release of O_2 (the Bohr effect)   156
BPG lowers the oxygen affinity of hemoglobin   156
Fetal hemoglobin has a higher oxygen affinity than maternal hemoglobin   157
Subunit interactions are required for allosteric effects   157
The quaternary structure of hemoglobin changes markedly on oxygenation   158
Deoxyhemoglobin is constrained by salt links between different chains   159
Oxygenation moves the iron atom into the plane of the porphyrin   159
Movement of the iron atom is transmitted to other subunits by the proximal histidine   160
Mechanism of the cooperative binding of oxygen   160
BPG decreases oxygen affinity by cross-linking deoxyhemoglobin   161
CO_2 binds to the terminal amino groups of hemoglobin and lowers its oxygen affinity   162
Mechanism of the Bohr effect   162
Communication within a protein molecule   163
Sickle-shaped red blood cells in a case of severe anemia   163
Sickle-cell anemia is a genetically transmitted, chronic, hemolytic disease   164
The solubility of deoxygenated sickle hemoglobin is abnormally low   164
Hemoglobin S has an abnormal electrophoretic mobility   165
A single amino acid in the beta chain is altered in sickle-cell hemoglobin   165
Sickle hemoglobin has sticky patches on its surface   166
Deoxyhemoglobin S forms long helical fibers   167
High incidence of the sickle gene is due to the protection conferred against malaria   168
Fetal DNA can be analyzed for the presence of the sickle-cell gene   169
Molecular pathology of hemoglobin   170
Thalassemias are genetic disorders of hemoglobin synthesis   171
Impact of the discovery of molecular diseases   171

CHAPTER 8   Introduction to Enzymes   177

Enzymes have immense catalytic power   177
Enzymes are highly specific   178
The catalytic activities of many enzymes are regulated   179
Enzymes transform different forms of energy   180
Free energy is the most useful thermodynamic function in biochemistry   180
Standard free-energy change of a reaction and its relation to the equilibrium constant   182
Enzymes cannot alter reaction equilibria   183
Enzymes accelerate reactions by stabilizing transition states   184
Formation of an enzyme-substrate complex is the first step in enzymatic catalysis   184
Some key features of active sites   185
The Michaelis-Menten model accounts for the kinetic properties of many enzymes   187
V_{max} and K_M can be determined by varying the substrate concentration   189
Significance of K_M and V_{max} values   190
Kinetic perfection in enzymatic catalysis: the k_{cat}/K_M criterion   191
Enzymes can be inhibited by specific molecules   192
Allosteric enzymes do not obey Michaelis-Menten kinetics   193
Competitive and noncompetitive inhibition are kinetically distinguishable   193
Ethanol is used therapeutically as a competitive inhibitor to treat ethylene glycol poisoning   195
Penicillin irreversibly inactivates a key enzyme in bacterial cell-wall synthesis   195

Preface

Our view of the molecular basis of life has been profoundly altered by the recombinant DNA revolution. The genome is now an open book—any passage can be read. The cloning and sequencing of millions of bases of DNA have greatly enriched our understanding of genes and proteins. The intricate interplay of genotype and phenotype is now being unraveled at the molecular level. One of the fruits of this harvest is insight into how the genome is organized and its expression is controlled. The molecular circuitry of growth and development is coming into view. The reading of the genome is also providing a wealth of amino acid sequence information that illuminates the entire protein landscape. Scarce proteins can be produced in abundance by cells harboring new genes. Moreover, precisely designed novel proteins can be generated by directed mutagenesis to elucidate how proteins fold, catalyze reactions, transduce signals, transport ions, and interconvert different forms of free energy.

These remarkable advances stimulated me to reshape the architecture of my textbook of biochemistry. The third edition of *Biochemistry*, which was published earlier this year, is built around a new framework for the exposition of fundamental themes and principles of the molecular basis of life. *Biochemistry* begins with a new section, entitled *Molecular Design of Life*, that provides an overview of the central molecules of life—DNA, RNA, and proteins—and their interplay. Recombinant DNA technology and other experimental methods for exploring proteins and genes are also presented in this part of my textbook.

After the publication of *Biochemistry*, I began to reflect on the teaching of biology. It occurred to me that publication of the first eight chapters of *Biochemistry* as a separate volume would give beginning biology students and other interested readers ready access to the heart of biochemistry and molecular biology—the flow of information from gene to protein. My goal has been to make these concepts comprehensible and to share their beautiful imagery. The deep interest and commitment of my publisher has led to the rapid publication of this volume.

The planning of the third edition of *Biochemistry* unexpectedly took place in terrain quite different from Yale, Stanford, and Aspen, where the first two editions took form. In December 1985, my family and I went to Nepal to trek in the Everest region. After two rewarding days in Katmandu, we were on the verge of boarding the plane to Lukla, only to be turned back with the disappointing news that the landing strip was closed because of snow. We then headed for the Annapurna region in four-wheel drive vehicles but had to return a day later because the road vanished in the heavy rain. The inaccessibility of the high Himalayas led to an abrupt change of itinerary. We flew to Bangkok and arrived in 95-degree heat, carrying our parkas and arctic sleeping bags. Instead of hiking at 12,000 feet, we found ourselves at a hotel pool at sea level. This dislocation led to a totally unforeseen benefit. I was able to unhurriedly reflect on the remarkable development of biochemistry since I wrote my last edition. I had the leisure to think and dream and plan. Best of all, I was able to share my thoughts with my son Daniel, who was then a senior majoring in human biology, and gain from his insights. *Molecular Design of Life* is the outcome of our discussions.

I am grateful to Alexander Glazer, Daniel Koshland, Jr., and Alexander Rich for having encouraged me to write the third edition of my textbook. Alexander Glazer, Richard Gumport, Roger Koeppe, James Rawn, Carl Rhodes, and Peter Rubenstein read the entire manuscript. I benefited greatly from their scholarly and perceptive criticism. The contributors of many striking and informative illustrations are acknowledged in the figure legends. I was able to concentrate on the writing of the book because my office was in the capable hands of Joanne Tisch. She played a critical role in preparing the manuscript and reading the proofs. Her sensitivity, intelligence, and good spirits lightened my load. Andrew Kudlacik edited the manuscript with a fine sense of style and meaning. Mike Suh skillfully integrated word and picture in the design of each page. Susan Moran kept a watchful and discerning eye over many thousands of pages of manuscript, figures, and proofs. I also wish to thank Tom Cardamone and Shirley Baty for many outstanding drawings.

I am grateful to my family for their sustained support of the endeavor, which was more arduous than anticipated. My sons, Michael and Daniel, now embarked on their own careers, cheered me from afar. My wife, Andrea, provided criticism, advice, and encouragement in just the right proportions. I have been nurtured, too, by many who have reached out to express their warmth and interest in continuing this dialogue of biochemistry. I feel very fortunate and privileged to partake in this process at such a wonderful time.

Lubert Stryer
OCTOBER 1988

MOLECULAR DESIGN OF LIFE

On the facing page: Proteins and nucleic acids are the central molecules of life. A complex of DNA bound to the protein that replicates it is shown here. DNA polymerase is shown in blue, one strand of the double-helical DNA template in green, and the other in red. [Courtesy of Dr. Thomas Steitz.]

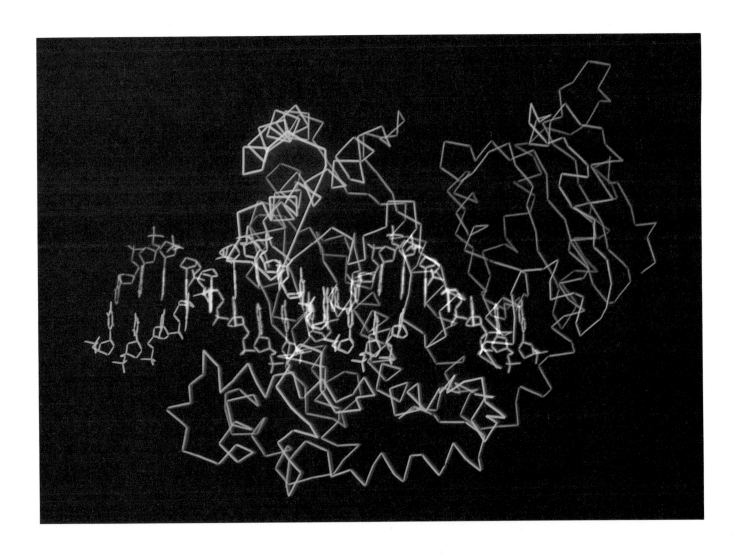

CHAPTER 1

Prelude

Biochemistry is the study of the molecular basis of life. There is much excitement and activity in biochemistry today for several reasons.

First, the chemical bases of many central processes are now understood. The discovery of the double-helical structure of deoxyribonucleic acid (DNA), the elucidation of the flow of information from gene to protein, the determination of the three-dimensional structure and mechanism of action of many protein molecules, the unraveling of central metabolic pathways and energy-conversion mechanisms, and the development of recombinant DNA technology are some of the outstanding achievements of biochemistry.

Second, it is now known that common molecular patterns and principles underlie the diverse expressions of life. Organisms as different as the bacterium *Escherichia coli* and human beings use the same building blocks to construct macromolecules. The flow of genetic information from DNA to ribonucleic acid (RNA) to protein is essentially the same in all organisms. Adenosine triphosphate (ATP), the universal currency of energy in biological systems, is generated in similar ways by all forms of life.

Third, biochemistry is profoundly influencing medicine. The molecular mechanisms of many diseases, such as sickle-cell anemia and numerous inborn errors of metabolism, have been elucidated. Assays for enzyme activity are indispensable in clinical diagnosis. For example, the levels of certain enzymes in serum reveal whether a patient has recently had a myocardial infarction. DNA probes are coming into play in the diagnosis of genetic disorders, infectious diseases, and cancers. Genetically engineered strains of bacteria containing recombinant DNA are producing valuable proteins such as insulin and growth hormone. Furthermore, biochemistry is a basis for the rational design of new drugs. Agriculture, too, is likely to benefit from recombinant DNA technology, which can produce designed changes in the genetic endowment of organisms.

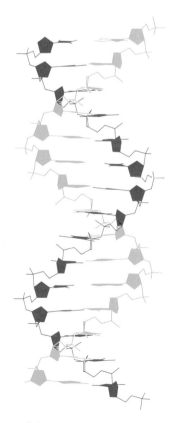

Figure 1-1
Model of the DNA double helix. The diameter of the helix is about 20 Å.

Part I
MOLECULAR DESIGN OF LIFE

Fourth, the rapid development of powerful biochemical concepts and techniques in recent years has enabled investigators to tackle some of the most challenging and fundamental problems in biology and medicine. How does a fertilized egg give rise to cells as different as those in muscle, the brain, and the liver? How do cells find each other in forming a complex organ? How is the growth of cells controlled? What are the causes of cancer? What is the mechanism of memory? What is the molecular basis of schizophrenia?

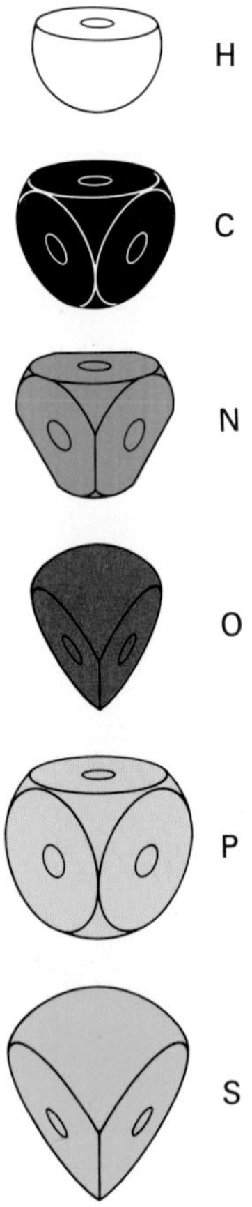

MOLECULAR MODELS DEPICT THREE-DIMENSIONAL STRUCTURE

The interplay between the three-dimensional structure of biomolecules and their biological function is the unifying motif of this book. Three types of atomic models will be used to depict molecular architecture: space-filling, ball-and-stick, and skeletal. The *space-filling models* are the most realistic. The size and configuration of an atom in a space-filling model are determined by its bonding properties and van der Waals radius (Figure 1-2). The colors of the model atoms are set by convention:

> Hydrogen, white   Oxygen, red
> Carbon, black   Phosphorus, yellow
> Nitrogen, blue   Sulfur, yellow

Space-filling models of several simple molecules are illustrated in Figure 1-3.

Figure 1-2
Space-filling models of hydrogen, carbon, nitrogen, oxygen, phosphorus, and sulfur atoms.

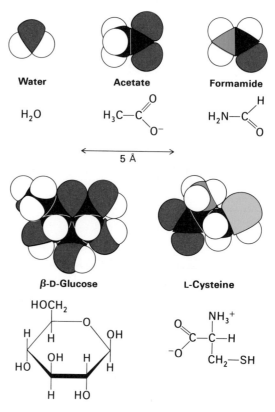

Figure 1-3
Space-filling models of water, acetate, formamide, glucose, and cysteine.

Ball-and-stick models are not as realistic as space-filling models because the atoms are depicted as spheres of radius smaller than the van der Waals radius. However, the bonding arrangement is easier to see because the bonds are explicitly represented by sticks. In an illustration, the taper of a stick tells whether the direction of the bond is in front of the plane of the page or behind it. More of a complex structure can be seen in a ball-and-stick model than in a space-filling model. An even simpler image is achieved with *skeletal models*, which show only the molecular framework. In these models, atoms are not shown explicitly. Rather, their positions are implied by the junctions and ends of bonds. Skeletal models are frequently used to depict large biological macromolecules, such as protein molecules having several thousand atoms. Space-filling, ball-and-stick, and skeletal models of ATP are compared in Figure 1-4.

SPACE, TIME, AND ENERGY

In considering molecular structure, it is important to have a sense of scale (Figure 1-5). The angstrom (Å) unit, which is equal to 10^{-10} meter (m) or 0.1 nanometer (nm), is customarily used as the measure of length at the atomic level. The length of a C—C bond, for example, is 1.54 Å. Small biomolecules, such as sugars and amino acids, are typically several angstroms long. Biological macromolecules, such as proteins, are at least tenfold larger. For example, hemoglobin, the oxygen-carrying protein in red blood cells, has a diameter of 65 Å. Another tenfold increase in size brings us to assemblies of macromolecules. Ribosomes, the protein-synthesizing machinery of the cell, have diameters of about 300 Å. The range from 100 Å (10 nm) to 1000 Å (100 nm) also encompasses most viruses. Cells are typically a hundred times as large, in the range of micrometers (μm). For example, a red blood cell is 7 μm (7×10^4 Å) long. It is important to note that the limit of resolution of the light microscope is about 2000 Å (0.2 μm), which corresponds to the size of many subcellular organelles. Mitochondria, the major generators of ATP in aerobic cells, can just be resolved by the light microscope. Most of our knowledge of biological structure in the range from 1 Å (0.1 nm) to 10^4 Å (1 μm) has come from electron microscopy and x-ray diffraction.

The molecules of life are constantly in flux. Chemical reactions in biological systems are catalyzed by enzymes, which typically convert

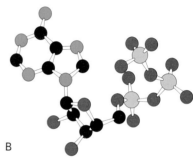
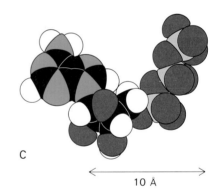

Figure 1-4
Comparison of (A) skeletal, (B) ball-and-stick, and (C) space-filling models of ATP. Hydrogen atoms are not shown in models A and B.

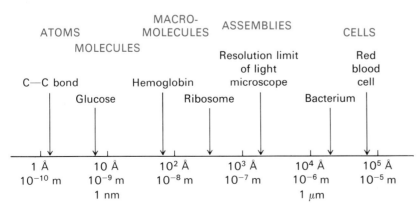

Figure 1-5
Dimensions of some biomolecules, assemblies, and cells.

substrate into product in milliseconds (ms, 10^{-3} s). Some enzymes act even more rapidly, in times as short as a few microseconds (μs, 10^{-6} s). Many conformational changes in biological macromolecules also are rapid. For example, the unwinding of the DNA double helix, which is essential for its replication and expression, is a microsecond event. The rotation of one domain of a protein with respect to another can take place in only nanoseconds (ns, 10^{-9} s). Many noncovalent interactions between groups in macromolecules are formed and broken in nanoseconds. Even more rapid processes can be probed with very short light pulses from lasers. It is remarkable that the primary event in vision—a change in structure of the light-absorbing group—occurs within a few picoseconds (ps, 10^{-12} s) after the absorption of a photon (Figure 1-6). From such brevity to the scale of evolutionary time, biological systems span a broad range. Life on earth arose some 3.5×10^9 years ago, or 1.1×10^{17} s ago.

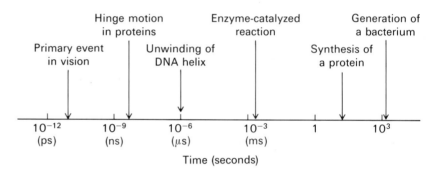

Figure 1-6
Typical rates of some processes in biological systems.

We shall be concerned with energy changes in molecular events (Figure 1-7). The ultimate source of energy for life is the sun. The energy of a green photon, for example, is 57 kilocalories per mole (kcal/mol). ATP, the universal currency of energy, has a usable energy content of about 12 kcal/mol. In contrast, the average energy of each vibrational degree of freedom in a molecule is much smaller, 0.6 kcal/mol at 25°C. This amount of energy is much less than that needed to dissociate covalent bonds (e.g., 83 kcal/mol for a C—C bond). Hence, the covalent framework of biomolecules is stable in the absence of enzymes and inputs of energy. On the other hand, noncovalent bonds in biological systems typically have an energy of only a few kilocalories per mole, so that thermal energy is enough to make and break them. An alternative unit of energy is the joule, which is equal to 0.239 calorie.

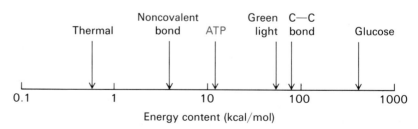

Figure 1-7
Some biologically important energies.

REVERSIBLE INTERACTIONS OF BIOMOLECULES ARE MEDIATED BY THREE KINDS OF NONCOVALENT BONDS

Reversible molecular interactions are at the heart of the dance of life. Weak, noncovalent forces play key roles in the faithful replication of DNA, the folding of proteins into intricate three-dimensional forms, the specific recognition of substrates by enzymes, and the detection of signal molecules. Indeed, all biological structures and processes depend on the interplay of noncovalent interactions as well as covalent ones. The three fundamental noncovalent bonds are *electrostatic bonds, hydrogen bonds,* and *van der Waals bonds.* They differ in geometry, strength, and specificity. Furthermore, these bonds are profoundly affected in different ways by the presence of water. Let us consider the characteristics of each:

1. *Electrostatic bonds.* A charged group on a substrate can attract an oppositely charged group on an enzyme. The force of such an *electrostatic attraction* is given by Coulomb's law:

$$F = \frac{q_1 q_2}{r^2 D}$$

in which q_1 and q_2 are the charges of the two groups, r is the distance between them, and D is the dielectric constant of the medium. The attraction is strongest in a vacuum (where D is 1) and is weakest in a medium such as water (where D is 80). This kind of attraction is also called an ionic bond, salt linkage, salt bridge, or ion pair. The distance between oppositely charged groups in an optimal electrostatic attraction is 2.8 Å.

2. *Hydrogen bonds* can be formed between uncharged molecules as well as charged ones. In a hydrogen bond, *a hydrogen atom is shared by two other atoms.* The atom to which the hydrogen is more tightly linked is called the hydrogen donor, whereas the other atom is the hydrogen acceptor. The acceptor has a partial negative charge that attracts the hydrogen atom. In fact, a hydrogen bond can be considered an intermediate in the transfer of a proton from an acid to a base. It is reminiscent of a ménage à trois.

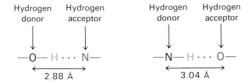

The donor in a hydrogen bond in biological systems is an oxygen or nitrogen atom that has a covalently attached hydrogen atom. The acceptor is either oxygen or nitrogen. The kinds of hydrogen bonds formed and their bond lengths are given in Table 1-1. The bond energies range from about 3 to 7 kcal/mol. Hydrogen bonds are stronger than van der Waals bonds but much weaker than covalent bonds. The length of a hydrogen bond is intermediate between that of a covalent

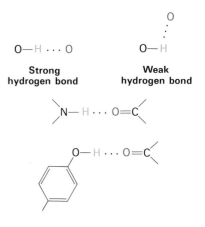

Table 1-1
Typical hydrogen-bond lengths

Bond	Length (Å)
O—H···O	2.70
O—H···O⁻	2.63
O—H···N	2.88
N—H···O	3.04
N⁺—H···O	2.93
N—H···N	3.10

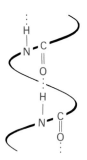

Figure 1-8
Schematic diagram of hydrogen bonding between an amide and a carbonyl group in an α-helix of a protein.

Table 1-2
Van der Waals contact radii of atoms (Å)

Atom	Radius
H	1.2
C	2.0
N	1.5
O	1.4
S	1.85
P	1.9

bond and a van der Waals bond. *An important feature of hydrogen bonds is that they are highly directional.* The strongest hydrogen bonds are those in which the donor, hydrogen, and acceptor atoms are colinear. The α-helix, a recurring motif in proteins, is stabilized by hydrogen bonds between amide (—NH) and carbonyl (—CO) groups (Figure 1-8). Another example of the importance of hydrogen bonding is the DNA double helix, which is held together by hydrogen bonds between bases on opposite strands (Figure 1-1).

3. *Van der Waals bonds,* a nonspecific attractive force, come into play when any two atoms are 3 to 4 Å apart. Though weaker and less specific than electrostatic and hydrogen bonds, van der Waals bonds are no less important in biological systems. The basis of a van der Waals bond is that the distribution of electronic charge around an atom changes with time. At any instant, the charge distribution is not perfectly symmetric. This transient asymmetry in the electronic charge around an atom encourages a similar asymmetry in the electron distribution around its neighboring atoms. The resulting attraction between a pair of atoms increases as they come closer, until they are separated by the van der Waals *contact distance* (Figure 1-9). At a shorter distance, very strong

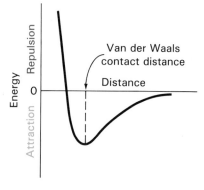

Figure 1-9
Energy of a van der Waals interaction as a function of the distance between two atoms.

repulsive forces become dominant because the outer electron clouds overlap. The contact distance between an oxygen and carbon atom, for example, is 3.4 Å, which is obtained by adding 1.4 and 2.0 Å, the contact radii (Table 1-2) of the O and C atoms.

The van der Waals bond energy of a pair of atoms is about 1 kcal/mol. It is considerably weaker than a hydrogen or electrostatic bond, which is in the range of 3 to 7 kcal/mol. A single van der Waals bond counts for very little because its strength is only a little more than the average thermal energy of molecules at room temperature (0.6 kcal/mol). Furthermore, the van der Waals force fades rapidly when the distance between a pair of atoms becomes even 1 Å greater than their contact distance. It becomes significant only when numerous atoms in one of a pair of molecules can simultaneously come close to many atoms of the other. This can happen only if the shapes of the molecules match. In other words, effective van der Waals interactions depend on *steric complementarity.* Though there is virtually no specificity in a single van der Waals interaction, *specificity arises when there is an opportunity to make a large number of van der Waals bonds simultaneously.* Repulsions between atoms closer than the van der Waals contact distance are as important as attractions for establishing specificity.

THE BIOLOGICALLY IMPORTANT PROPERTIES OF WATER ARE ITS POLARITY AND COHESIVENESS

Water profoundly influences all molecular interactions in biological systems. Two properties of water are especially important in this regard:

1. *Water is a polar molecule.* The shape of the molecule is triangular, not linear, and so there is an asymmetrical distribution of charge. The oxygen nucleus draws electrons away from the hydrogen nuclei, which leaves the region around those nuclei with a net positive charge. The water molecule is thus an electrically polar structure.

2. *Water molecules have a high affinity for each other.* A positively charged region in one water molecule tends to orient itself toward a negatively charged region in one of its neighbors. Ice has a highly regular crystalline structure in which all potential hydrogen bonds are made (Figure 1-10). Liquid water has a partly ordered structure in which hydrogen-bonded clusters of molecules are continually forming and breaking up. Each molecule is hydrogen bonded to an average of 3.4 neighbors in liquid water, compared with 4 in ice. *Water is highly cohesive.*

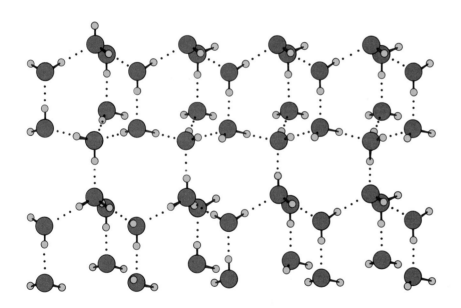

Figure 1-10
Structure of a form of ice. [After L. Pauling and P. Pauling. *Chemistry* (W. H. Freeman, 1975), p. 289.]

WATER SOLVATES POLAR MOLECULES AND WEAKENS IONIC AND HYDROGEN BONDS

The polarity and hydrogen-bonding capability of water make it a highly interacting molecule. Water is an excellent solvent for polar molecules. The reason is that water greatly weakens electrostatic forces and hydrogen bonding between polar molecules by competing for their attractions. For example, consider the effect of water on hydrogen bonding between a carbonyl and an amide group (Figure 1-11). The hydrogen atoms of water can replace the amide hydrogen group as hydrogen-bond donors, and the oxygen atom of water can replace the carbonyl oxygen as the acceptor. Hence, a strong hydrogen bond between a CO and an NH group forms only if water is excluded.

Water diminishes the strength of electrostatic attractions by a factor of 80, the dielectric constant of water, compared with the same interactions in a vacuum. Water has an unusually high dielectric constant (Table 1-3) because of its polarity and capacity to form oriented solvent shells around ions (Figure 1-12). These oriented solvent shells produce electric fields of their own, which oppose the fields produced by the ions. Consequently, electrostatic attractions between ions are markedly weakened by the presence of water.

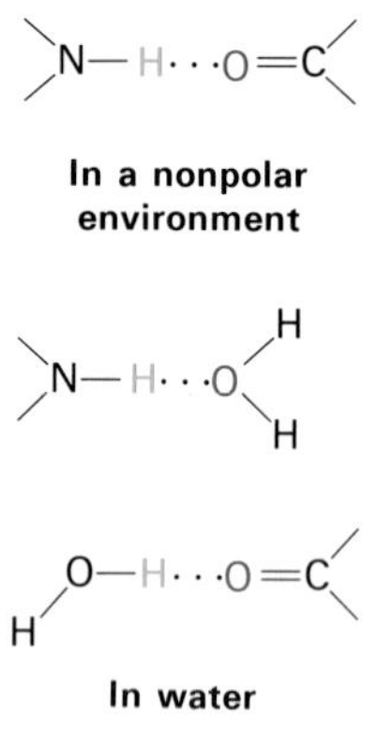

Figure 1-11
Water competes for hydrogen bonds.

Table 1-3
Dielectric constants of some solvents at 20°C

Substance	Dielectric constant
Hexane	1.9
Benzene	2.3
Diethyl ether	4.3
Chloroform	5.1
Acetone	21.4
Ethanol	24
Methanol	33
Water	80
Hydrogen cyanide	116

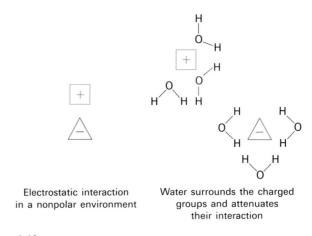

Figure 1-12
Water attenuates electrostatic attractions between charged groups.

The existence of life on earth depends critically on the capacity of water to dissolve a remarkable array of polar molecules that serve as fuels, building blocks, catalysts, and information carriers. High concentrations of these molecules can coexist in water, where they are free to diffuse and find each other. However, the excellence of water as a solvent poses a problem, for it also weakens interactions between polar molecules. Biological systems have solved this problem by creating water-free microenvironments where polar interactions have maximal strength. We shall see many examples of the critical importance of these specially constructed niches in protein molecules.

HYDROPHOBIC ATTRACTIONS: NONPOLAR GROUPS TEND TO ASSOCIATE IN WATER

The sight of dispersed oil droplets coming together in water to form a single large oil drop is a familiar one. An analogous process occurs at the atomic level: *nonpolar molecules or groups tend to cluster together in water*. These associations are called *hydrophobic attractions*. In a figurative sense, water tends to squeeze nonpolar molecules together.

Let us examine the basis of hydrophobic attractions, which are a major driving force in the folding of macromolecules, the binding of substrates to enzymes, and the formation of membranes that define the boundaries of cells and their internal compartments. Consider the introduction of a single nonpolar molecule, such as hexane, into some water. A cavity in the water is created, which temporarily disrupts some hydrogen bonds between water molecules. The displaced water molecules then reorient themselves to form a maximum number of new hydrogen bonds. This is accomplished at a price: the number of ways of forming hydrogen bonds in the cage of water around the hexane molecule is much fewer than in pure water. The water molecules around the hexane molecule are much more ordered than elsewhere in the solution. Now consider the arrangement of two hexane molecules in water. Do they sit in two small cavities (Figure 1-13A) or in a single larger one (Figure 1-13B)? The experimental fact is that the two hexane molecules come together and occupy a single large cavity. This association releases some of the more ordered water molecules around the separated hexanes. In fact, the basis of a hydrophobic attraction is this enhanced freedom of released water molecules. *Nonpolar solute molecules are driven together in water not primarily because they have a high affinity for each other but because water bonds strongly to itself.*

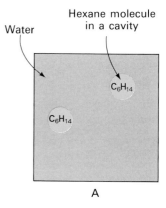

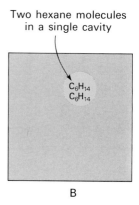

Figure 1-13
A schematic representation of two molecules of hexane in a small volume of water: (A) the hexane molecules occupy different cavities in the water structure, or (B) they occupy the same cavity, which is energetically more favored.

DESIGN OF THIS BOOK

This book has six parts, each having a major theme.

    I: Molecular Design of Life
   II: Protein Conformation, Dynamics, and Function
  III: Generation and Storage of Metabolic Energy
  IV: Biosynthesis of Macromolecular Precursors
   V: Genetic Information
  VI: Molecular Physiology

Part I is an overview of the central molecules of life—DNA, RNA, and proteins—and their interplay. We begin with proteins, which are unique in being able to recognize and bind a remarkably diverse array of molecules. Proteins determine the pattern of chemical transformations in biological systems by catalyzing nearly all of the necessary chemical reactions. We then turn to DNA, the repository of genetic information in all cells. The discovery of the DNA double helix led immediately to an understanding of how DNA replicates. The following chapter deals with the flow of genetic information from DNA to RNA to protein. The first step, called transcription, is the synthesis of RNA, and the second, called translation, is the synthesis of proteins according to instructions given by templates of messenger RNA. The genetic code, which specifies the relation between the sequence of four kinds of bases in DNA and RNA and the twenty kinds of amino acids in proteins, is beautiful in its simplicity. Three bases constitute a codon, the unit that specifies an amino acid. Translation is carried out by the coordinated interplay of more than a hundred kinds of protein and RNA molecules in an organized assembly called the ribosome. Experimental methods for exploring proteins and genes are also presented in Part I. Recombinant DNA technology is introduced here and some examples of its power and generality in analyzing and altering both genes and proteins are given.

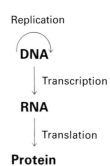

Figure 1-14
Flow of genetic information.

The interplay of three-dimensional structure and biological activity as exemplified by proteins is the major theme of Part II. The structure and function of myoglobin and hemoglobin, the oxygen-carrying proteins in vertebrates, are presented in detail because these proteins illustrate many general principles. Hemoglobin is especially interesting because its binding of oxygen is regulated by specific molecules in its environment. The molecular pathology of hemoglobin, particularly sickle-cell anemia, is also presented. We then turn to enzymes and consider how they recognize substrates and enhance reaction rates by factors of a million or more. The enzymes lysozyme, carboxypeptidase A, and chymotrypsin are examined in detail because the study of them has elucidated many general principles of catalysis. The regulation of enzymatic activity by specific control proteins and other signal molecules is considered next. Rather different facets of the theme of conformation emerge in the chapter on collagen and elastin, two connective-tissue proteins. The final chapter in Part II is an introduction to biological membranes, which are organized assemblies of lipids and proteins. Membranes serve to create compartments and control the flow of matter and information between them.

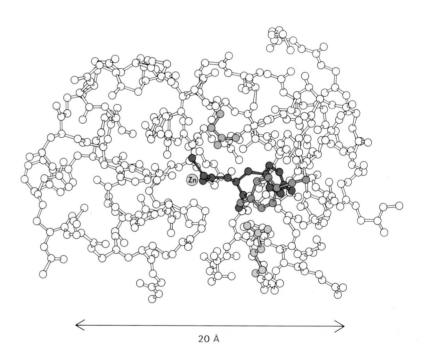

Figure 1-15
Structure of an enzyme-substrate complex. Glycyltyrosine (shown in red) is bound to carboxypeptidase A, a digestive enzyme. Only a quarter of the enzyme is shown. [After W. N. Lipscomb. *Proc. Robert A. Welch Found. Conf. Chem. Res.* 15(1971):141.]

Part III deals with the generation and storage of metabolic energy. First, the overall strategy of metabolism is presented. Cells convert energy from fuel molecules into ATP. In turn, ATP drives most energy-requiring processes in cells. In addition, reducing power in the form of nicotinamide adenine dinucleotide phosphate (NADPH) is generated for use in biosyntheses. The metabolic pathways that carry out these reactions are then presented in detail. For example, the generation of ATP from glucose requires a sequence of three series of reactions—glycolysis, the citric acid cycle, and oxidative phosphorylation. The last two are also common to the generation of ATP from the oxidation of fats and some amino acids, the other major fuels. We see here an illustration of molecular economy. Two storage forms of fuel molecules, glycogen and triacylglycerols (neutral fats), are also discussed in Part III. The concluding topic of this part of the book is photosynthesis, in which the primary event is the light-activated transfer of an electron from one substance to another against a chemical potential gradient. As in oxidative phosphorylation, electron flow leads to the pumping of protons across a membrane, which in turn drives the synthesis of ATP. In essence, life is powered by proton batteries that are ultimately energized by the sun.

Part IV deals with the biosynthesis of macromolecular precursors, starting with the synthesis of membrane lipids and steroids. The pathway for the synthesis of cholesterol, a 27-carbon steroid, is of particular interest because all of its carbon atoms come from a 2-carbon precursor. The reactions leading to the synthesis of selected amino acids and the heme group are then discussed. The control mechanisms in these pathways are of general significance. The biosynthesis of nucleotides, the activated precursors of DNA and RNA, is then considered. The final chapter in this part deals with the integration of metabolism. How are energy-yielding and energy-consuming reactions coordinated to meet the needs of an organism?

The transmission and expression of genetic information constitute the central theme of Part V. The genetic role and structure of DNA were introduced in Part I, as was the flow of genetic information. We now resume our consideration of this theme, enriched with a knowledge of proteins and metabolic transformations. The mechanism of DNA replication and DNA repair are discussed first. An intriguing aspect of DNA replication is its very high accuracy. The processes of genetic recombination and transposition, which produce new combinations of DNA, are then presented. We turn next to transcription and to the processing of nascent transcripts to form functional RNA mole-

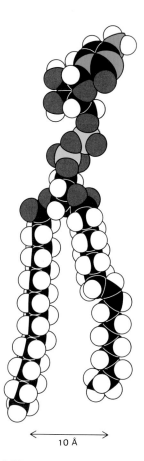

Figure 1-16
Model of CDP-diacylglycerol, an activated intermediate in the synthesis of some membrane lipids.

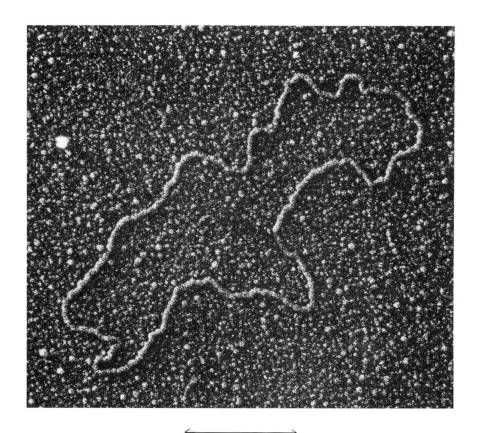

Figure 1-17
Electron micrograph of a DNA molecule. [Courtesy of Dr. Thomas Broker.]

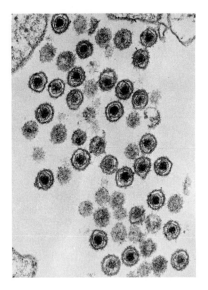

Figure 1-18
Electron micrograph of Rous sarcoma virus. This RNA virus can produce cancer in susceptible hosts. [Courtesy of Dr. Samuel Dales.]

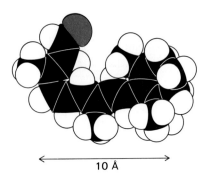

Figure 1-19
Model of 11-*cis*-retinal, the light-absorbing group in rhodopsin. The isomerization of this chromophore by light is the first event in visual excitation.

cules. The mechanism of protein synthesis, in which tRNAs, mRNAs, and ribosomes interact, comes next. We then consider how proteins are specifically targeted to many different destinations. The next chapter deals with the control of gene expression in bacteria. The focus here is on the lactose and tryptophan operons of *E. coli*, which are now understood in detail. This is followed by a discussion of the regulation of gene expression in higher organisms—molecules controlling the development of multicellular organisms are now being identified. Virus multiplication and assembly are considered next. Viral assembly exemplifies some general principles of how biological macromolecules form highly ordered structures from a few kinds of building blocks. Viruses have also provided insight into the molecular basis of cancer by revealing the existence of oncogenes, which are altered forms of genes that control cell growth.

Part VI, entitled "Molecular Physiology," is a transition from biochemistry to physiology. Many of the concepts that were developed earlier in this book are used here, because physiology involves the interplay of genetic information, conformation, and metabolism. We start with the molecular basis of the immune response. How does an organism detect a foreign substance? The next chapter deals with the problem of how the energy of chemical bonds is transformed into coordinated motion—myosin and actin, the major proteins in muscle, have a contractile role in most cells of higher organisms. The transport of ions, such as Na^+, K^+, and Ca^{2+}, and molecules is then considered. Molecular pumps in membranes control the transport of these ions to generate gradients that are at the heart of excitability. We then turn to the molecular basis of the action of hormones and growth factors. Families of receptors and signal-coupling proteins are being discovered, and recurring motifs of signal transduction are becoming evident. The final chapter deals with sensory processes and considers such questions as: How do bacteria detect nutrients in their environment and move toward them? How are action potentials propagated by nerve cells and transmitted across synapses? How is a retinal rod cell triggered by a single photon?

One of the most satisfying features of biochemistry is that it continually enriches our understanding of biological processes at all levels of organization.

CHAPTER 2

Protein Structure and Function

Proteins play crucial roles in virtually all biological processes. Their significance and the remarkable scope of their activity are exemplified in the following functions:

1. *Enzymatic catalysis.* Nearly all chemical reactions in biological systems are catalyzed by specific macromolecules called enzymes. Some of these reactions, such as the hydration of carbon dioxide, are quite simple. Others, such as the replication of an entire chromosome, are highly intricate. Enzymes exhibit enormous catalytic power. They usually increase reaction rates by at least a millionfold. Indeed, chemical transformations in vivo rarely proceed at perceptible rates in the absence of enzymes. Several thousand enzymes have been characterized, and many of them have been crystallized. The striking fact is that nearly all known enzymes are proteins. Thus, proteins play the unique role of determining the pattern of chemical transformations in biological systems.

2. *Transport and storage.* Many small molecules and ions are transported by specific proteins. For example, hemoglobin transports oxygen in erythrocytes, whereas myoglobin, a related protein, transports oxygen in muscle. Iron is carried in the plasma of blood by transferrin and is stored in the liver as a complex with ferritin, a different protein.

> *Protein—*
> A word coined by Jöns J. Berzelius in 1838 to emphasize the importance of this class of molecules. Derived from the Greek word *proteios*, which means "of the first rank."

Figure 2-1
Photomicrograph of a crystal of hexokinase, a key enzyme in the utilization of glucose. [Courtesy of Dr. Thomas Steitz and Dr. Mark Yeager.]

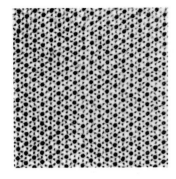

Figure 2-2
Electron micrograph of a cross section of insect flight muscle showing a hexagonal array of two kinds of protein filaments. [Courtesy of Dr. Michael Reedy.]

Figure 2-3
Electron micrograph of a fiber of collagen. [Courtesy of Dr. Jerome Gross and Dr. Romaine Bruns.]

3. *Coordinated motion.* Proteins are the major component of muscle. Muscle contraction is accomplished by the sliding motion of two kinds of protein filaments. On the microscopic scale, such coordinated motions as the movement of chromosomes in mitosis and the propulsion of sperm by their flagella also are produced by contractile assemblies consisting of proteins.

4. *Mechanical support.* The high tensile strength of skin and bone is due to the presence of collagen, a fibrous protein.

5. *Immune protection.* Antibodies are highly specific proteins that recognize and combine with such foreign substances as viruses, bacteria, and cells from other organisms. Proteins thus play a vital role in distinguishing between self and nonself.

6. *Generation and transmission of nerve impulses.* The response of nerve cells to specific stimuli is mediated by receptor proteins. For example, rhodopsin is the photoreceptor protein in retinal rod cells. Receptor proteins that can be triggered by specific small molecules, such as acetylcholine, are responsible for transmitting nerve impulses at synapses—that is, at junctions between nerve cells.

7. *Control of growth and differentiation.* Controlled sequential expression of genetic information is essential for the orderly growth and differentiation of cells. Only a small fraction of the genome of a cell is expressed at any one time. In bacteria, repressor proteins are important control elements that silence specific segments of the DNA of a cell. In higher organisms, growth and differentiation are controlled by growth factor proteins. For example, nerve growth factor guides the formation of neural networks. The activities of different cells in multicellular organisms are coordinated by hormones. Many of them, such as insulin and thyroid-stimulating hormone, are proteins. Indeed, proteins serve in all cells as sensors that control the flow of energy and matter.

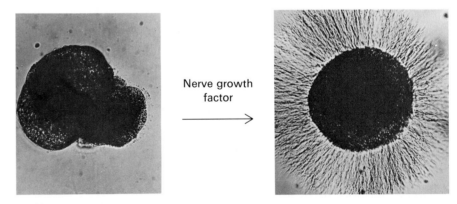

Figure 2-4
Photomicrograph of a ganglion showing the proliferation of nerves after addition of nerve growth factor, a complex of proteins. [Courtesy of Dr. Eric Shooter.]

PROTEINS ARE BUILT FROM A REPERTOIRE OF TWENTY AMINO ACIDS

Amino acids are the basic structural units of proteins. An α-amino acid consists of an amino group, a carboxyl group, a hydrogen atom, and a distinctive R group bonded to a carbon atom, which is called the α-carbon because it is adjacent to the carboxyl (acidic) group (Figure 2-5). An R group is referred to as a *side chain* for reasons that will be evident shortly.

Amino acids in solution at neutral pH are predominantly *dipolar ions* (or *zwitterions*) rather than un-ionized molecules. In the dipolar form of an amino acid, the amino group is protonated (—NH_3^+) and the carboxyl group is dissociated (—COO^-). The ionization state of an amino acid varies with pH (Figure 2-6). In acid solution (e.g., pH 1), the carboxyl group is un-ionized (—COOH) and the amino group is ionized (—NH_3^+). In alkaline solution (e.g., pH 11), the carboxyl group is ionized (—COO^-) and the amino group is un-ionized (—NH_2). For glycine, the pK of the carboxyl group is 2.3 and that of the amino group is 9.6. In other words, the midpoint of the first ionization is at pH 2.3, and that of the second is at pH 9.6. For a review of acid-base concepts and pH, see the Appendix to this chapter.

The tetrahedral array of four different groups about the α-carbon atom confers optical activity on amino acids. The two mirror-image forms are called the L-isomer and the D-isomer (Figure 2-7). *Only L-amino acids are constituents of proteins.* Hence, the designation of the optical isomer will be omitted and the L-isomer implied in discussions of proteins herein, unless otherwise noted.

Twenty kinds of side chains varying in *size, shape, charge, hydrogen-bonding capacity,* and *chemical reactivity* are commonly found in proteins. Indeed, all proteins in all species, from bacteria to humans, are constructed from the same set of twenty amino acids. This fundamental alphabet of proteins is at least two billion years old. The remarkable range of functions mediated by proteins results from the diversity and versatility of these twenty kinds of building blocks. We shall explore ways in which this alphabet is used to create the intricate three-dimensional structures that enable proteins to carry out so many biological processes.

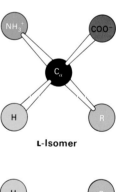
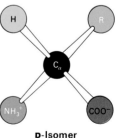

Figure 2-7
Absolute configurations of the L- and D-isomers of amino acids. R refers to the side chain.

Glycine (Gly, G) | Alanine (Ala, A) | Valine (Val, V) | Leucine (Leu, L) | Isoleucine (Ile, I)

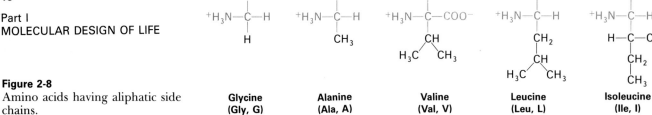

Figure 2-8
Amino acids having aliphatic side chains.

Let us look at this repertoire of amino acids. The simplest one is *glycine*, which has just a hydrogen atom as its side chain (Figure 2-8). *Alanine* comes next, with a methyl group as its side chain. Larger hydrocarbon side chains (three and four carbons long) are found in *valine*, *leucine*, and *isoleucine*. These larger aliphatic side chains are *hydrophobic*—that is, they have an aversion to water and like to cluster. As will be discussed later, the three-dimensional structure of water-soluble proteins is stabilized by the coming together of hydrophobic side chains to avoid contact with water. The different sizes and shapes of these hydrocarbon side chains (Figure 2-9) enable them to pack together to form compact structures with few holes.

Gly | Ala | Val | Leu | Ile

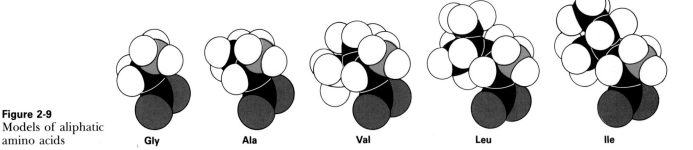

Figure 2-9
Models of aliphatic amino acids

Proline also has an aliphatic side chain but it differs from other members of the set of twenty in that its side chain is bonded to both the nitrogen and α-carbon atoms. The resulting cyclic structure (Figure 2-10) markedly influences protein architecture. Proline, often found in the bends of folded protein chains, is not averse to being exposed to water. Note that proline contains a secondary rather than a primary amino group, which makes it an *imino* acid.

Three amino acids with *aromatic side chains* are part of the fundamental repertoire (Figure 2-11). *Phenylalanine*, as its name indicates, contains a phenyl ring attached to a methylene (—CH_2—) group. *Tryptophan* has an indole ring joined to a methylene group; this side chain contains a nitrogen atom in addition to carbon and hydrogen atoms.

Proline
(Pro, P)

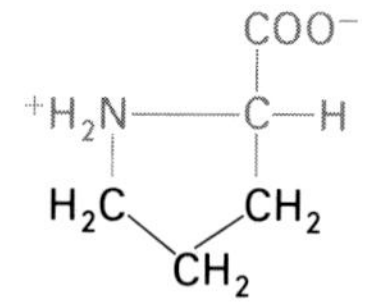

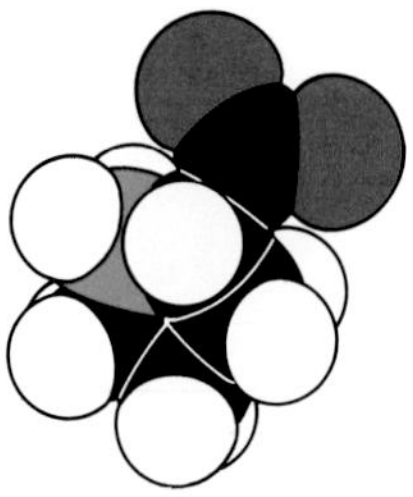

Figure 2-10
Proline differs from the other common amino acids in having a secondary amino group.

Phenylalanine (Phe, F) | Tyrosine (Tyr, Y) | Tryptophan (Trp, W)

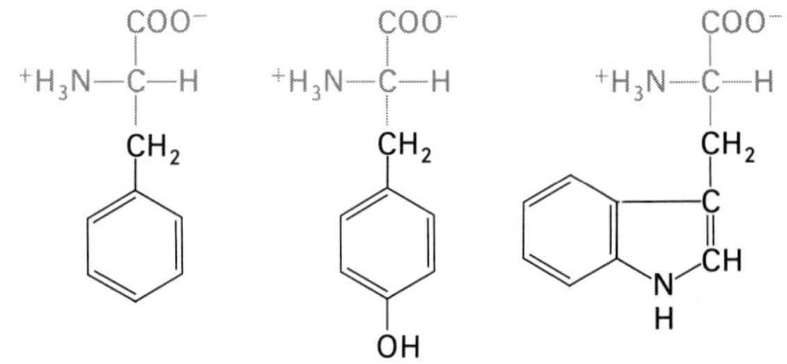

Figure 2-11
Phenylalanine, tyrosine, and tryptophan have aromatic side chains.

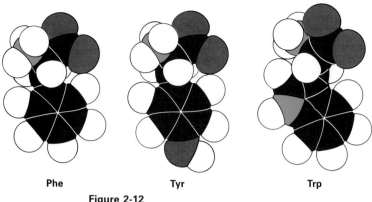

Figure 2-12
Models of the aromatic amino acids.

Phenylalanine and tryptophan are highly hydrophobic. The aromatic ring of *tyrosine* contains a hydroxyl group, which makes tyrosine less hydrophobic than phenylalanine. Moreover, this hydroxyl group is reactive, in contrast with the rather inert side chains of all the other amino acids discussed thus far. The aromatic rings of phenylalanine, tryptophan, and tyrosine contain delocalized pi-electron clouds that enable them to interact with other pi-systems and to transfer electrons.

A *sulfur atom* is present in the side chains of two amino acids (Figure 2-13). *Cysteine* contains a sulfhydryl group (—SH) and *methionine* contains a sulfur atom in a thioether linkage (—S—CH$_3$). Both of these sulfur-containing side chains are hydrophobic. The sulfhydryl group of cysteine is highly reactive. As will be discussed shortly, cysteine plays a special role in shaping some proteins by forming disulfide links.

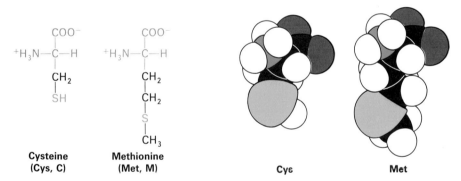

Figure 2-13
Cysteine and methionine have sulfur-containing side chains.

Figure 2-14
Models of cysteine and methionine.

Two amino acids, *serine* and *threonine*, contain aliphatic *hydroxyl groups* (Figure 2-15). Serine can be thought of as a hydroxylated version of alanine, and threonine as a hydroxylated version of valine. The hydroxyl groups on serine and threonine make them much more *hydrophilic* (water-loving) and *reactive* than alanine and valine. Threonine, like isoleucine, contains two centers of asymmetry. All other amino acids in the basic set of twenty, except for glycine, contain a single asymmetric center (the α-carbon atom). Glycine is unique in being optically inactive.

We turn now to amino acids with very polar side chains, which render them highly *hydrophilic*. *Lysine* and *arginine* are *positively charged* at neutral pH. *Histidine* can be uncharged or positively charged, depending on its local environment. Indeed, histidine is often found in the active

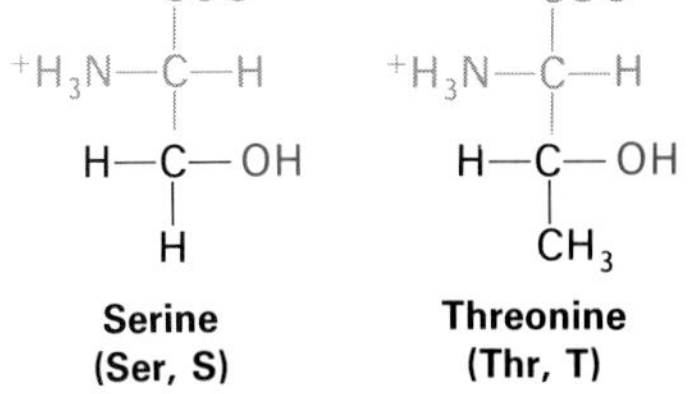

Figure 2-15
Serine and threonine have aliphatic hydroxyl side chains.

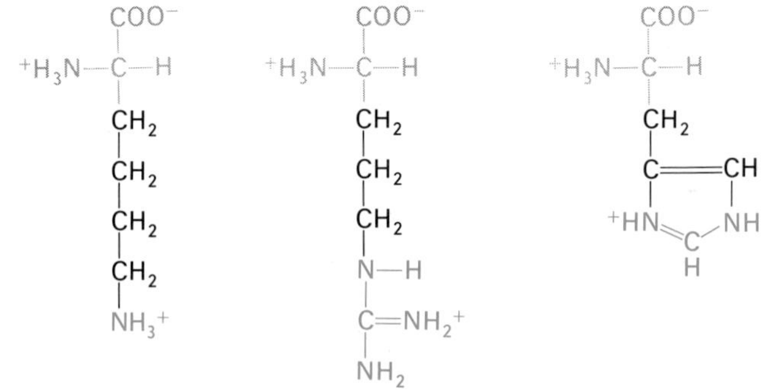

Part I
MOLECULAR DESIGN OF LIFE

Lysine (Lys, K) — **Arginine (Arg, R)** — **Histidine (His, H)**

Figure 2-16
Lysine, arginine, and histidine have basic side chains.

Arg

Figure 2-17
Model of arginine. The planar outer part of the side chain, consisting of three nitrogens bonded to a carbon atom, is called a guanidinium group.

sites of enzymes, where its imidazole ring can readily switch between these states to catalyze the making and breaking of bonds. These *basic amino acids* are depicted in Figure 2-16. The side chains of arginine and lysine are the longest ones in the set of twenty.

The repertoire of amino acids also contains two with *acidic side chains*, *aspartic acid* and *glutamic acid*. These amino acids are usually called *aspartate* and *glutamate* to emphasize that their side chains are nearly always negatively charged at physiological pH (Figure 2-18). Uncharged derivatives of glutamate and aspartate are *glutamine* and *asparagine*, which contain a terminal amide group in place of a carboxylate.

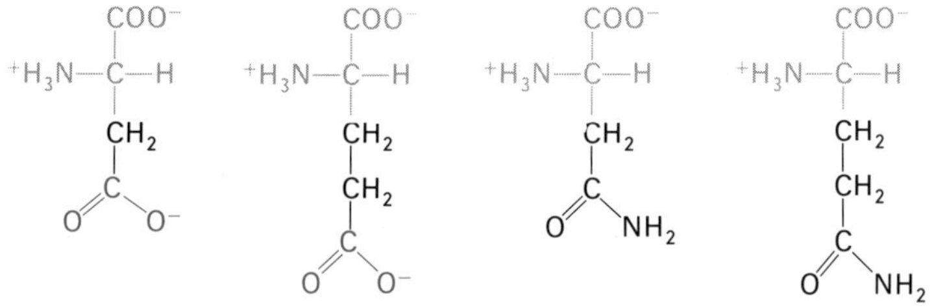

Aspartate (Asp, D) — **Glutamate (Glu, E)** — **Asparagine (Asn, N)** — **Glutamine (Gln, Q)**

Figure 2-18
Acidic amino acids (aspartate and glutamate) and their amide derivatives (asparagine and glutamine).

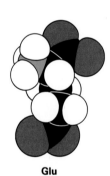

Glu

Figure 2-19
Model of glutamate.

Seven of the twenty amino acids have readily ionizable side chains. Equilibria and typical pK_a values for ionization of the side chains of arginine, lysine, histidine, aspartic and glutamic acids, cysteine, and tyrosine in proteins are given in Table 2-1. Two other groups in proteins, the terminal α-amino group and the terminal α-carboxyl group, can be ionized.

Amino acids are often designated by either a three-letter abbreviation or a one-letter symbol to facilitate concise communication (Table 2-2). The abbreviations for amino acids are the first three letters of their names, except for tryptophan (Trp), asparagine (Asn), glutamine (Gln), and isoleucine (Ile). The symbols for the small amino acids are the first letters of their names (e.g., G for glycine and L for leucine); the other symbols have been agreed upon by convention. These abbreviations and symbols are an integral part of the vocabulary of biochemists.

Table 2-1
pK values of ionizable groups in proteins

Group	Acid ⇌ base + H⁺	Typical pK*
Terminal carboxyl	—COOH ⇌ —COO⁻ + H⁺	3.1
Aspartic and glutamic acid	—COOH ⇌ —COO⁻ + H⁺	4.4
Histidine	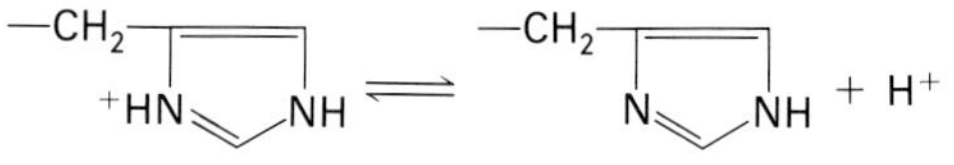 (imidazole ring protonation)	6.5
Terminal amino	—NH$_3^+$ ⇌ —NH$_2$ + H⁺	8.0
Cysteine	—SH ⇌ —S⁻ + H⁺	8.5
Tyrosine	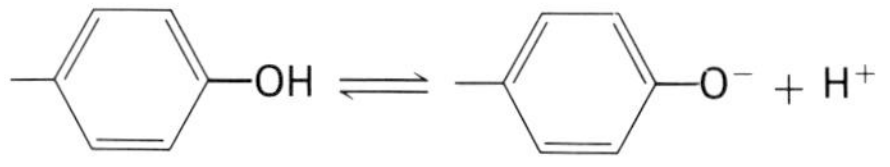 —C$_6$H$_4$—OH ⇌ —C$_6$H$_4$—O⁻ + H⁺	10.0
Lysine	—NH$_3^+$ ⇌ —NH$_2$ + H⁺	10.0
Arginine	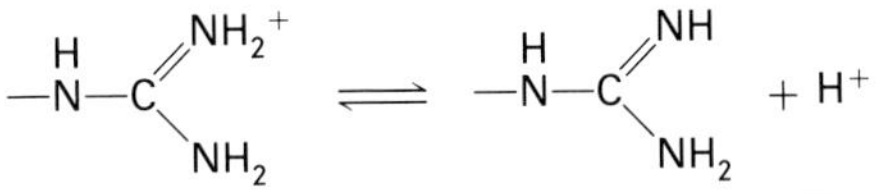 (guanidinium deprotonation)	12.0

*pK values depend on temperature, ionic strength, and the microenvironment of the ionizable group.

Table 2-2
Abbreviations for amino acids

Amino acid	Three-letter abbreviation	One-letter symbol
Alanine	Ala	A
Arginine	Arg	R
Asparagine	Asn	N
Aspartic acid	Asp	D
Asparagine or aspartic acid	Asx	B
Cysteine	Cys	C
Glutamine	Gln	Q
Glutamic acid	Glu	E
Glutamine or glutamic acid	Glx	Z
Glycine	Gly	G
Histidine	His	H
Isoleucine	Ile	I
Leucine	Leu	L
Lysine	Lys	K
Methionine	Met	M
Phenylalanine	Phe	F
Proline	Pro	P
Serine	Ser	S
Threonine	Thr	T
Tryptophan	Trp	W
Tyrosine	Tyr	Y
Valine	Val	V

Chapter 2
PROTEIN STRUCTURE
AND FUNCTION

AMINO ACIDS ARE LINKED BY PEPTIDE BONDS TO FORM POLYPEPTIDE CHAINS

In proteins, the α-carboxyl group of one amino acid is joined to the α-amino group of another amino acid by a *peptide bond* (also called an amide bond). The formation of a dipeptide from two amino acids by loss of a water molecule is shown in Figure 2-20. The equilibrium of this reaction lies on the side of hydrolysis rather than synthesis. Hence, the biosynthesis of peptide bonds requires an input of free energy, whereas their hydrolysis is thermodynamically downhill.

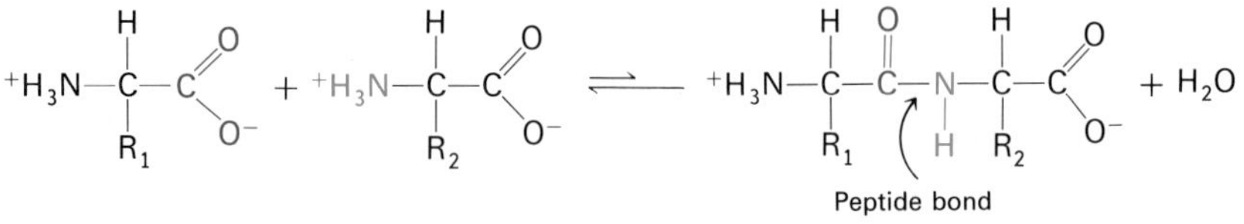

Figure 2-20
Formation of a peptide bond.

Many amino acids are joined by peptide bonds to form a *polypeptide chain*, which is unbranched (Figure 2-21). An amino acid unit in a polypeptide is called a *residue*. A polypeptide chain has direction because its building blocks have different ends—namely, the α-amino and the α-carboxyl groups. By convention, *the amino end is taken to be the beginning of a polypeptide chain*, and so the sequence of amino acids in a polypeptide chain is written starting with the amino-terminal residue. Thus, in the tripeptide Ala-Gly-Trp (AGW), alanine is the amino-terminal residue and tryptophan is the carboxyl-terminal residue. Note that Trp-Gly-Ala (WGA) is a different tripeptide.

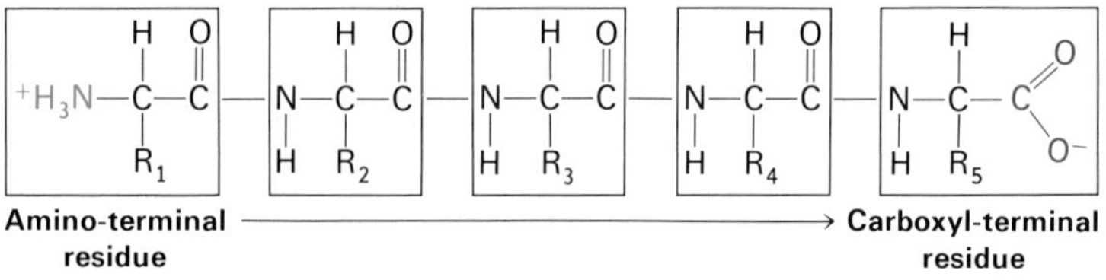

Figure 2-21
A pentapeptide. The constituent amino acid residues are outlined. The chain starts at the amino end.

A polypeptide chain consists of a regularly repeating part, called the *main chain,* and a variable part, comprising the distinctive *side chains* (Figure 2-22). The main chain is sometimes termed the backbone. Most natural polypeptide chains contain between 50 and 2000 amino acid residues. The mean molecular weight of an amino acid residue is about 110, and so the molecular weights of most polypeptide chains are between 5500 and 220,000. We can also refer to the mass of a protein, which is expressed in units of daltons; one *dalton* is equal to one atomic mass unit. A protein with a molecular weight of 50,000 has a mass of 50,000 daltons, or 50 kd (kilodaltons).

Figure 2-22
A polypeptide chain is made up of a regularly repeating *backbone* and distinctive *side chains* (R_1, R_2, R_3, shown in green).

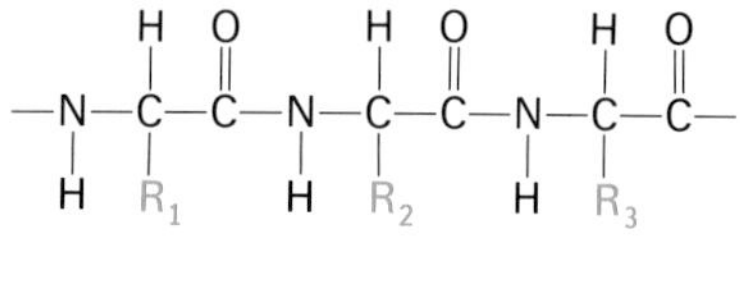

Dalton—
A unit of mass very nearly equal to that of a hydrogen atom (precisely equal to 1.0000 on the atomic mass scale). Named after John Dalton (1766–1844), who developed the atomic theory of matter.

Kilodalton (kd)—
A unit of mass equal to 1000 daltons.

Some proteins contain *disulfide bonds*. These cross-links between chains or between parts of a chain are formed by the oxidation of cysteine residues. The resulting disulfide is called *cystine* (Figure 2-23). Intracellular proteins usually lack disulfide bonds, whereas extracellular proteins often contain several. Nonsulfur cross-links derived from lysine side chains are present in some proteins. For example, collagen fibers in connective tissue are strengthened in this way, as are fibrin blood clots.

PROTEINS HAVE UNIQUE AMINO ACID SEQUENCES THAT ARE SPECIFIED BY GENES

In 1953, Frederick Sanger determined the amino acid sequence of insulin, a protein hormone (Figure 2-25). *This work is a landmark in biochemistry because it showed for the first time that a protein has a precisely defined amino acid sequence.* Moreover, it demonstrated that insulin consists only of L-amino acids in peptide linkage between α-amino and α-carboxyl groups. This accomplishment stimulated other scientists to carry out sequence studies of a wide variety of proteins. Indeed, the complete amino acid sequences of more than 2000 proteins are now known. The striking fact is that each protein has a unique, precisely defined amino acid sequence.

A series of incisive studies in the late 1950s and early 1960s revealed that the amino acid sequences of proteins are genetically determined. The sequence of nucleotides in DNA, the molecule of heredity, specifies a complementary sequence of nucleotides in RNA, which in turn specifies the amino acid sequence of a protein (p. 91). In particular, each of the twenty amino acids of the repertoire is encoded by one or more specific sequences of three nucleotides. Furthermore, proteins in all organisms are synthesized from their constituent amino acids by a common mechanism.

Amino acid sequences are important for several reasons. First, knowledge of the sequence of a protein is very helpful, indeed usually essential, in elucidating its mechanism of action (e.g., the catalytic mechanism of an enzyme). Second, analyses of relations between amino acid sequences and three-dimensional structures of proteins are uncovering the rules that govern the folding of polypeptide chains. The amino acid sequence is the link between the genetic message in DNA and the three-dimensional structure that performs a protein's biological function. Third, sequence determination is part of molecular pathology, an emerging area of medicine. Alterations in amino acid sequence can produce abnormal function and disease. Fatal disease, such as sickle-cell anemia, can result from a change in a single amino acid in

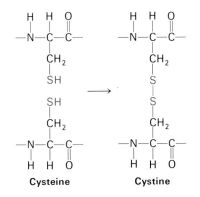

Figure 2-23
A disulfide bridge (—S—S—) is formed from the sulfhydryl groups (—SH) of two cysteine residues. The product is a *cystine* residue.

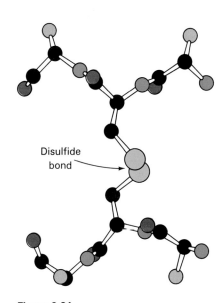

Figure 2-24
Model of a disulfide cross-link.

Figure 2-25
Amino acid sequence of bovine insulin.

a single protein. Fourth, the sequence of a protein reveals much about its evolutionary history. Proteins resemble one another in amino acid sequence only if they have a common ancestor. Consequently, molecular events in evolution can be traced from amino acid sequences; molecular paleontology is a flourishing area of research.

PROTEIN MODIFICATION AND CLEAVAGE CONFER NEW CAPABILITIES

The basic set of twenty amino acids can be modified after synthesis of a polypeptide chain to enhance its capabilities. For example, the amino-termini of many proteins are *acetylated*, which makes these proteins more resistant to degradation. In newly synthesized collagen, many proline residues become hydroxylated to form *hydroxyproline* (Figure 2-26). The added hydroxyl groups stabilize the collagen fiber. The bio-

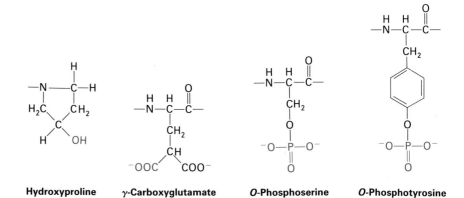

Figure 2-26
Some modified amino acid residues in proteins: hydroxyproline, γ-carboxyglutamate, *O*-phosphoserine, and *O*-phosphotyrosine. Groups added after the polypeptide chain is synthesized are shown in red.

logical significance of this modification is evident in scurvy, which results from insufficient hydroxylation of collagen because of a deficiency of vitamin C. Another specialized amino acid produced by a finishing touch is *γ-carboxyglutamate*. In vitamin K deficiency, insufficient carboxylation of glutamate in prothrombin, a clotting protein, can lead to hemorrhage. Most proteins, such as antibodies, that are secreted by cells acquire carbohydrate chains on specific asparagine residues. Many hormones, such as epinephrine (adrenaline), alter the activities of enzymes by stimulating the phosphorylation of the hydroxyl amino acids serine and threonine; *phosphoserine* and *phosphothreonine* are the most ubiquitous modified amino acids in proteins. Growth factors such as insulin act by triggering the phosphorylation of the hydroxyl group of tyrosine to form *phosphotyrosine*. The phosphate groups on these three modified amino acids can readily be removed, enabling them to act as reversible switches in regulating cellular processes. Indeed, some tumor viruses produce cancer by stimulating excessive phosphorylation of tyrosine residues on proteins that control cell proliferation.

Many proteins are cleaved and trimmed after synthesis. For example, digestive enzymes are synthesized as inactive precursors that can be stored safely in the pancreas. After being released into the intestine, these precursors become activated by peptide bond cleavage. In blood clotting, soluble fibrinogen is converted into insoluble fibrin by peptide bond cleavage. A number of polypeptide hormones that include adrenocorticotropic hormone arise from the splitting of a single large pre-

Figure 2-27
Model of phosphoserine.

cursor protein, a molecular cornucopia. The proteins of poliovirus, too, are produced by cleavage of a giant polyprotein precursor. We shall encounter many more examples of modification and cleavage as essential features of protein formation and function. Indeed, these finishing touches account for much of the versatility, precision, and elegance of protein action and regulation.

THE PEPTIDE UNIT IS RIGID AND PLANAR

A striking characteristic of proteins is that they have well-defined three-dimensional structures. A stretched-out or randomly arranged polypeptide chain is devoid of biological activity, as will be discussed shortly. *Function arises from conformation,* which is the three-dimensional arrangement of atoms in a structure. Amino acid sequences are important because they specify the conformation of proteins.

In the late 1930s, Linus Pauling and Robert Corey began x-ray crystallographic studies of the precise structure of amino acids and peptides. Their aim was to obtain a set of standard bond distances and bond angles for these building blocks and then use this information to predict the conformation of proteins. One of their important findings was that *the peptide unit is rigid and planar*. The hydrogen of the substituted amino group is nearly always *trans* (opposite) to the oxygen of the carbonyl group (Figure 2-28). There is no freedom of rotation about

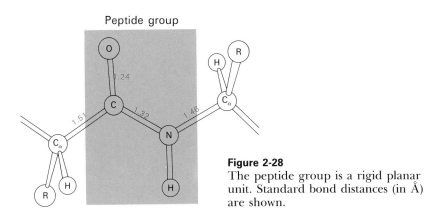

Figure 2-28
The peptide group is a rigid planar unit. Standard bond distances (in Å) are shown.

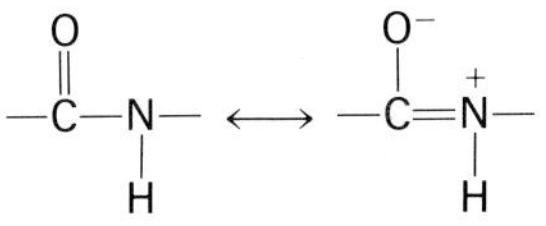

Figure 2-29
The peptide group is planar because the carbon–nitrogen bond has partial double-bond character.

the bond between the carbonyl carbon atom and the nitrogen atom of the peptide unit because this link has partial double-bond character (Figure 2-29). The length of this bond is 1.32 Å, which is between that of a C—N single bond (1.49 Å) and a C=N double bond (1.27 Å). In contrast, the link between the α-carbon atom and the carbonyl carbon atom is a pure single bond. The bond between the α-carbon atom and the peptide nitrogen atom also is a pure single bond. Consequently, *there is a large degree of rotational freedom about these bonds on either side of the rigid peptide unit* (Figure 2-30).

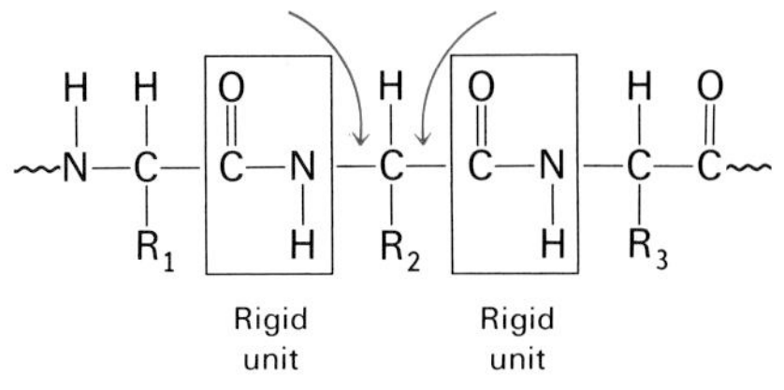

POLYPEPTIDE CHAINS CAN FOLD INTO REGULAR STRUCTURES: THE α HELIX AND β PLEATED SHEETS

Can a polypeptide chain fold into a regularly repeating structure? To answer this question, Pauling and Corey evaluated a variety of potential polypeptide conformations by building precise molecular models. They

Figure 2-30
There is considerable freedom of rotation about the bonds joining the peptide groups to the α-carbon atoms.

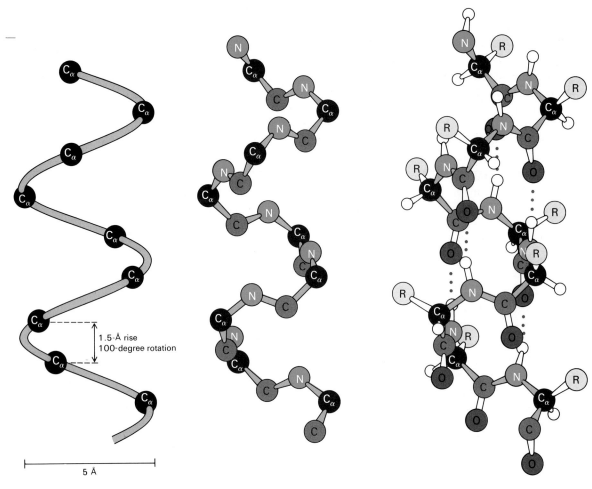

Figure 2-31
Models of a right-handed α helix: (A) only the α-carbon atoms are shown on a helical thread; (B) only the backbone nitrogen (N), α-carbon (C_α), and carbonyl carbon (C) atoms are shown; (C) entire helix. Hydrogen bonds (denoted in part C by red dots) between NH and CO groups stabilize the helix.

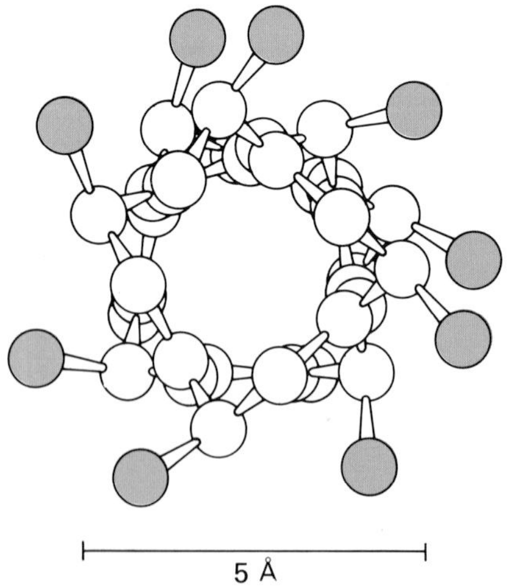

adhered closely to the experimentally observed bond angles and distances for amino acids and small peptides. In 1951, they proposed two periodic polypeptide structures, called the α helix (alpha helix) and the β pleated sheet (beta pleated sheet).

The α helix is a rodlike structure. The tightly coiled polypeptide main chain forms the inner part of the rod, and the side chains extend outward in a helical array (Figures 2-31 and 2-32). The α helix is stabilized by hydrogen bonds between the NH and CO groups of the main chain. The CO group of each amino acid is hydrogen bonded to the NH group of the amino acid that is situated four residues ahead in the linear sequence (Figure 2-33). Thus, *all the main-chain CO and NH groups*

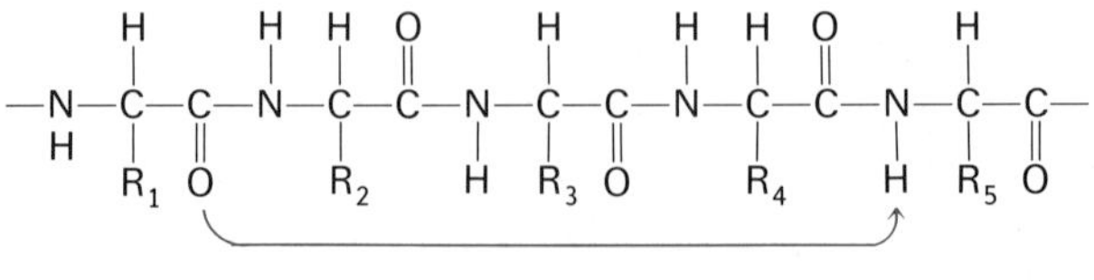

Figure 2-32
Cross-sectional view of an α helix. Note that the side chains (shown in green) are on the outside of the helix. The van der Waals radii of the atoms are larger than shown here; hence there is actually almost no free space inside the helix.

Figure 2-33
In the α helix, the CO group of residue n is hydrogen bonded to the NH group of residue $(n + 4)$.

are hydrogen bonded. Each residue is related to the next one by a translation of 1.5 Å along the helix axis and a rotation of 100°, which gives 3.6 amino acid residues per turn of helix. Thus, amino acids spaced three and four apart in the linear sequence are spatially quite close to one another in an α helix. In contrast, amino acids two apart in the linear sequence are situated on opposite sides of the helix and so are unlikely to make contact. The pitch of the α helix is 5.4 Å, the product of the translation (1.5 Å) and the number of residues per turn (3.6). The screw sense of a helix can be right-handed (clockwise) or left-handed (counterclockwise); the α helices found in proteins are right-handed.

The α helix content of proteins of known three-dimensional structure is highly variable. In some, such as myoglobin and hemoglobin, the α helix is the major structural motif. Other proteins, such as the digestive enzyme chymotrypsin, are virtually devoid of α helix. In most proteins, the single-stranded α helix discussed above is usually a rather short rod, typically less than 40 Å in length. However, the α helical theme is extended in some proteins to much longer rods, as long as 1000 Å (100 nm or 0.1 μm) or more. Two or more such α helices can entwine to form a cable. Such *α helical coiled coils* are found in keratin in hair, myosin and tropomyosin in muscle, epidermin in skin, and fibrin in blood clots. The helical cables in these proteins serve a mechanical role in forming stiff bundles of fibers.

The structure of the α helix was deduced by Pauling and Corey six years before it was actually to be seen in the x-ray reconstruction of the structure of myoglobin. *The elucidation of the structure of the α helix is a landmark in molecular biology because it demonstrated that the conformation of a polypeptide chain can be predicted if the properties of its components are rigorously and precisely known.*

In the same year, Pauling and Corey discovered another periodic structural motif, which they named the β pleated sheet (β because it was the second structure they elucidated, the α helix having been the first). The β pleated sheet differs markedly from the α helix in that it is a sheet rather than a rod. A polypeptide chain in the β pleated sheet is almost fully extended (Figure 2-34) rather than being tightly coiled as

Angstrom (Å)—
A unit of length equal to 10^{-10} meter.

$$1\text{ Å} = 10^{-10}\text{ m} = 10^{-8}\text{ cm}$$
$$= 10^{-4}\text{ μm} = 10^{-1}\text{ nm}$$

Named after Anders J. Ångström (1814–1874), a spectroscopist.

"When we consider that the fibrous proteins of the epidermis, the keratinous tissues, the chief muscle protein, myosin, and now the fibrinogen of the blood all spring from the same peculiar shape of molecule, and are therefore probably all adaptations of a single root idea, we seem to glimpse one of the great coordinating facts in the lineage of biological molecules."

K. BAILEY, W. T. ASTBURY, AND K. M. RUDALL
Nature, 1943

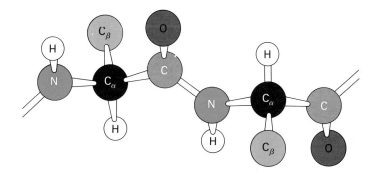

Figure 2-34
Conformation of a dipeptide unit in a β pleated sheet. The polypeptide chain is almost fully stretched out.

in the α helix. The axial distance between adjacent amino acids is 3.5 Å, in contrast with 1.5 Å for the α helix. Another difference is that the β pleated sheet is stabilized by hydrogen bonds between NH and CO groups in *different* polypeptide chains, whereas in the α helix the hydrogen bonds are between NH and CO groups in the *same* polypeptide chain. Adjacent chains in a β pleated sheet can run in the same direction (*parallel β sheet*) or in opposite directions (*antiparallel β sheet*) (Figure

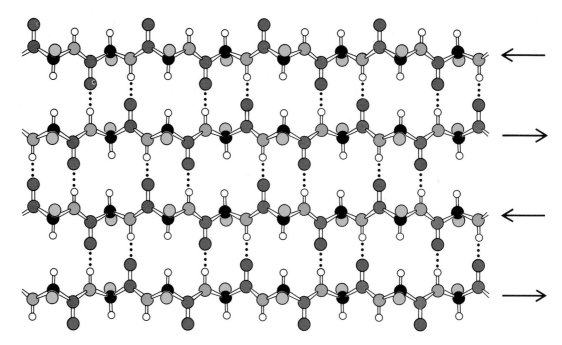

Figure 2-35
Antiparallel β pleated sheet. Adjacent strands run in opposite directions. Hydrogen bonds between NH and CO groups of adjacent strands stabilize the structure. The side chains (shown in green) are above and below the plane of the sheet.

2-35). For example, silk fibroin consists almost entirely of stacks of antiparallel β sheets. Such β-sheet regions are a recurring structural motif in many proteins. Structural units comprising from two to five parallel or antiparallel β strands are especially common.

The *collagen helix*, a third periodic structure, will be discussed in detail in Chapter 11. This specialized structure is responsible for the high tensile strength of collagen, the major protein of skin, bone, and tendon.

POLYPEPTIDE CHAINS CAN REVERSE DIRECTION BY MAKING β-TURNS

Most proteins have compact, globular shapes due to numerous reversals of the direction of their polypeptide chains. Analyses of the three-dimensional structures of numerous proteins have revealed that many of these chain reversals are accomplished by a common structural element called the *β-turn*. The essence of this hairpin turn is that the CO group of residue n of a polypeptide is hydrogen bonded to the NH group of residue $(n + 3)$ (Figure 2-36). Thus, a polypeptide chain can abruptly reverse its direction. β-Turns often connect antiparallel β strands; hence their name. β-Turns are also known as *reverse turns* or *hairpin bends*.

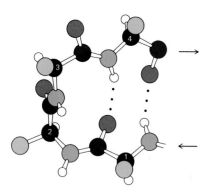

Figure 2-36
Structure of a β-turn. The CO group of residue 1 of the tetrapeptide shown here is hydrogen bonded to the NH group of residue 4, which results in a hairpin turn.

PROTEINS ARE RICH IN HYDROGEN-BONDING POTENTIALITY

What are the forces that determine the three-dimensional architecture of proteins? As Chapter 1 explained, all reversible molecular interactions in biological systems are mediated by three kinds of forces: *electro-*

static bonds, *hydrogen bonds*, and *van der Waals bonds*. We have already seen hydrogen bonds between main-chain NH and CO groups at work in forming α helices and β sheets. In fact, side chains of eleven of the twenty fundamental amino acids also can participate in hydrogen bonding. It is convenient to group these residues according to their hydrogen-bonding potentialities:

1. The side chains of tryptophan and arginine can serve as *hydrogen-bond donors only*.

2. Like the peptide group itself, the side chains of asparagine, glutamine, serine, and threonine can serve as *hydrogen-bond donors and acceptors*.

3. The hydrogen-bonding capabilities of lysine (and the terminal amino group), aspartic and glutamic acid (and the terminal carboxyl group), tyrosine, and histidine vary with pH. These groups can serve as both acceptors and donors over a certain range of pH, and as acceptors or donors (but not both) at other pH values, as shown for aspartate and glutamate in Figure 2-37. *The hydrogen-bonding modes of these ionizable residues are pH-dependent.*

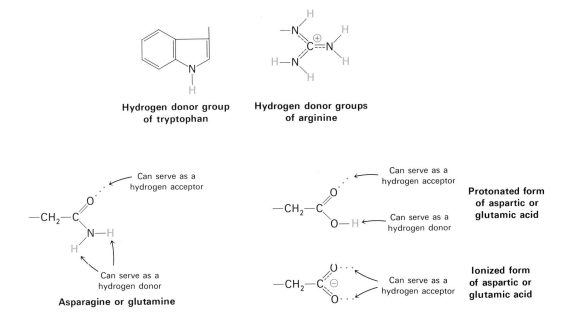

Figure 2-37
Hydrogen-bonding groups of several side chains in proteins.

WATER-SOLUBLE PROTEINS FOLD INTO COMPACT STRUCTURES WITH NONPOLAR CORES

Let us now see how these forces shape the structure of proteins. X-ray crystallographic studies have revealed the detailed three-dimensional structures of more than a hundred proteins. The experimental method of x-ray analysis and some examples of its results will be discussed in later chapters. We begin here with a preview of *myoglobin*, the first protein to be seen in atomic detail.

Myoglobin, the oxygen carrier in muscle, is a single polypeptide chain of 153 amino acids and has a mass of 18 kd. The capacity of myoglobin to bind oxygen depends on the presence of *heme*, a nonpolypeptide *prosthetic* (helper) *group* consisting of protoporphyrin

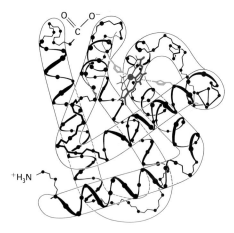

Figure 2-38
Model of myoglobin. Only the α-carbon atoms are shown. The heme group is shown in red and two adjacent histidines in green. [After R. E. Dickerson. In *The Proteins*, H. Neurath, ed., 2nd ed. (Academic Press, 1964), vol. 2, p. 634.]

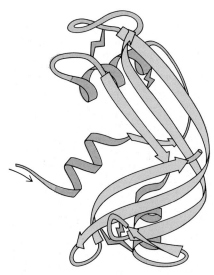

Figure 2-39
Ribbon model of ribonuclease S. Sections of α helix are shown in red, β-sheets in green, and disulfide bridges in yellow. [Courtesy of Dr. Jane Richardson.]

and a central iron atom. *Myoglobin is an extremely compact molecule with very little empty space inside* (Figure 2-38). Its overall dimensions are 45 × 35 × 25 Å, an order of magnitude less than if it were fully stretched out. *Myoglobin is built primarily of α helices*, of which there are eight. About 70% of the main chain is folded into α helices, and much of the rest of the chain forms turns between helices. Four of the turns contain proline, which disrupts α helices because of its rigid five-membered ring. The folding of the main chain of myoglobin, like that of other proteins, is complex and devoid of symmetry. However, a unifying principle emerges from the distribution of side chains. The striking fact is that *the interior consists almost entirely of nonpolar residues* such as leucine, valine, methionine, and phenylalanine. Polar residues such as aspartate, glutamate, lysine, and arginine are absent from the inside of myoglobin. The only polar residues inside are two histidines, which play critical roles in the binding of heme oxygen. The outside of myoglobin, on the other hand, consists of both polar and nonpolar residues.

This contrasting distribution of polar and nonpolar residues reveals a key facet of protein architecture. In an aqueous environment, protein folding is driven by the strong tendency of hydrophobic residues to be excluded from water. Recall that water is highly cohesive and that hydrophobic groups are thermodynamically more stable when clustered in the interior of the molecule than when extended into the aqueous surroundings (p. 10). *The polypeptide chain therefore folds spontaneously so that its hydrophobic side chains are buried and its polar, charged chains are on the surface.* The fate of the main chain accompanying the hydrophobic side chains is important, too. An unpaired peptide NH or CO markedly prefers water to a nonpolar milieu. The secret of burying a segment of main chain in a hydrophobic environment is to pair all the NH and CO groups by hydrogen bonding. This pairing is neatly accomplished in an α helix or β sheet. Van der Waals bonds between tightly packed hydrocarbon side chains also contribute to the stability of proteins. We can now understand why the repertoire of twenty amino acids contains so many aliphatic ones, differing subtly in size and shape. Nature can choose among them to fill the interior of a protein neatly and thereby maximize van der Waals interactions, which require intimate contact.

Ribonuclease S, a pancreatic enzyme that hydrolyzes RNA, exemplifies a rather different mode of protein folding. This single polypeptide chain of 124 residues is folded mainly into β-sheet strands, in contrast with myoglobin, which contains α helices and lacks β sheets. Ribonuclease, like myoglobin, contains a tightly packed, highly nonpolar interior. This enzyme is further stabilized by four disulfide bonds. The structure of a protein can be symbolized in highly schematic form by depicting β strands as broad arrows, α helices as helical ribbons, and connecting regions as strings. This representation is very useful for concisely representing relations of these elements in proteins, especially large ones, and for detecting structural motifs that recur in different proteins. A ribbon drawing of ribonuclease is shown in Figure 2-39.

Integral membrane proteins, those that traverse biological membranes, are designed differently from proteins that are soluble in aqueous solution. The permeability barrier of membranes is formed by lipids, which are highly hydrophobic. Thus, the part of a membrane protein that spans this region must have a hydrophobic exterior. As will be discussed in Chapter 12, the transmembrane portion of a membrane protein usually consists of bundles of α helices with nonpolar side chains (such as those of leucine and phenylalanine) facing out from the surface of the protein.

LEVELS OF STRUCTURE IN PROTEIN ARCHITECTURE

Four levels of structure are frequently cited in discussions of protein architecture. *Primary structure* is the amino acid sequence and the location of disulfides, if there are any. The primary structure is thus a complete description of the covalent connections of a protein. *Secondary structure* refers to the spatial arrangement of amino acid residues that are near one another in the linear sequence. Some of these steric relationships are of a regular kind, giving rise to a periodic structure. The α helix, β pleated sheet, and collagen helix are elements of secondary structure. *Tertiary structure* refers to the spatial arrangement of amino acid residues that are far apart in the linear sequence. The dividing line between secondary and tertiary structure is a matter of taste. Proteins containing more than one polypeptide chain exhibit an additional level of structural organization. Each polypeptide chain in such a protein is called a subunit. *Quaternary structure* refers to the spatial arrangement of such subunits and the nature of their contacts (Figure 2-40). The constituent chains of a multisubunit protein can be identical or different. For example, immunoglobulin G, the major antibody molecule in plasma, consists of two L chains and two H chains. The spherical shell of tomato bushy stunt virus, a plant pathogen, is formed from 180 identical coat protein molecules. The interfaces between subunits are often functionally significant. For example, in hemoglobin (consisting of four chains), the subunit interfaces participate in transmitting information between binding sites for O_2, CO_2, and H^+. In antibody molecules, the combining site for antigen is formed by segments of two different kinds of chains.

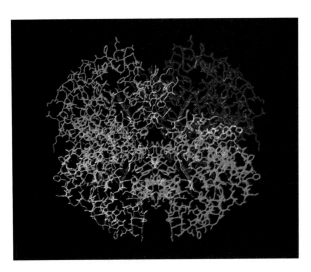

Figure 2-40
Three-dimensional structure of hemoglobin. The four subunits are shown in different colors. Each contains an oxygen-binding heme group (red).

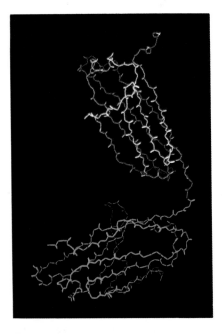

Figure 2-41
The light (L) chain of an antibody molecule consists of two distinct domains.

Recent studies of protein conformation, function, and evolution have revealed the importance of two additional levels of organization. *Supersecondary structure* refers to clusters of secondary structure. For example, a β strand separated from another β strand by an α helix is found in many proteins; this motif is called a βαβ unit. It is fruitful to regard supersecondary structures as intermediates between secondary and tertiary structure. Some polypeptide chains fold into two or more compact regions that may be joined by a flexible segment of polypeptide chain, rather like pearls on a string. These compact globular units, called *domains*, range in size from about 100 to 400 amino acid residues. For example, a 25-kd L chain of an antibody is folded into two domains (Figure 2-41). Indeed, these domains resemble one another, which suggests that they arose by duplication of a primordial gene. An important principle has emerged from analyses of genes and proteins in higher eucaryotes: *protein domains are often encoded by distinct parts of genes called exons* (p. 112). In our explorations of genes and proteins, exons and domains will often be at the focal point.

AMINO ACID SEQUENCE SPECIFIES THREE-DIMENSIONAL STRUCTURE

Insight into the relation between the amino acid sequence of a protein and its conformation came from the work of Christian Anfinsen on ribonuclease. As mentioned earlier, ribonuclease is a single polypeptide chain consisting of 124 amino acid residues (Figure 2-42). Its four disul-

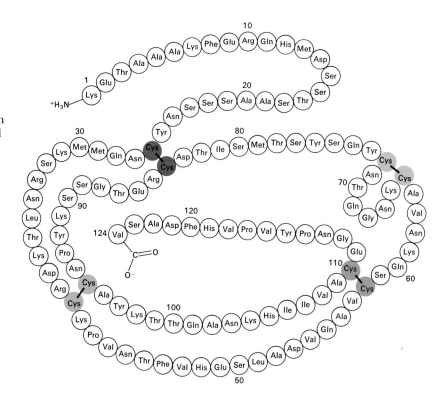

Figure 2-42
Amino acid sequence of bovine ribonuclease. The four disulfide bonds are shown in color. [After C. H. W. Hirs, S. Moore, and W. H. Stein. *J. Biol. Chem.* 235(1960):633.]

fide bonds can be cleaved reversibly by reducing them with a reagent such as *β-mercaptoethanol*, which forms mixed disulfides with cysteine side chains (Figure 2-43). In the presence of a large excess of β-mercaptoethanol, the mixed disulfides also are reduced, so that the final product is a protein in which the disulfides (cystines) are fully converted into sulfhydryls (cysteines). However, it was found that ribonuclease at 37°C and pH 7 cannot be readily reduced by β-mercaptoethanol unless the protein is partly unfolded by agents such as *urea* or *guanidine hydrochloride*. Although the mechanism of action of

$$H_2N-\underset{\underset{O}{\|}}{C}-NH_2$$
Urea

$$H_2N-\underset{\underset{NH_2}{\|}}{\overset{NH_2^+\ Cl^-}{C}}-NH_2$$
Guanidine hydrochloride

$$HO-CH_2-CH_2-SH$$
β-Mercaptoethanol

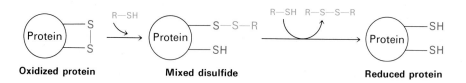

Figure 2-43
Reduction of the disulfide bonds in a protein by an excess of a sulfhydryl reagent such as β-mercaptoethanol.

these agents is not fully understood, it is evident that they disrupt noncovalent interactions. Most polypeptide chains devoid of cross-links assume a *random-coil conformation* in 8 M urea or 6 M guanidine HCl, as evidenced by physical properties such as viscosity and optical rotatory spectra. When ribonuclease was treated with β-mercaptoethanol in 8 M urea, the product was a fully reduced, randomly coiled polypeptide chain *devoid of enzymatic activity*. In other words, ribonuclease was *denatured* by this treatment (Figure 2-44).

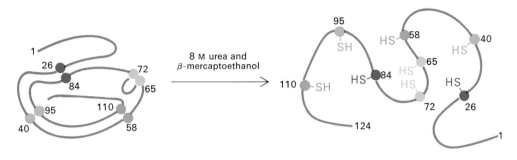

Figure 2-44
Reduction and denaturation of ribonuclease.

Anfinsen then made the critical observation that the denatured ribonuclease, freed of urea and β-mercaptoethanol by dialysis, slowly regained enzymatic activity. He immediately perceived the significance of this chance finding: the sulfhydryls of the denatured enzyme became oxidized by air and the enzyme spontaneously refolded into a catalytically active form. Detailed studies then showed that nearly all of the original enzymatic activity was regained if the sulfhydryls were oxidized under suitable conditions (Figure 2-45). All of the measured physical and chemical properties of the refolded enzyme were virtually identical with those of the native enzyme. These experiments showed that *the information needed to specify the complex three-dimensional structure of ribonuclease is contained in its amino acid sequence.* Subsequent studies of other proteins have established the generality of this central principle of molecular biology: *sequence specifies conformation.*

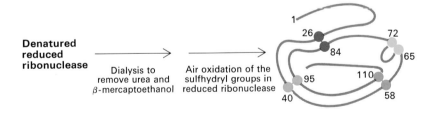

Figure 2-45
Renaturation of ribonuclease.

A quite different result was obtained when reduced ribonuclease was reoxidized while it was still in 8 M urea. This preparation was then dialyzed to remove the urea. Ribonuclease reoxidized in this way had only 1% of the enzymatic activity of the native protein. Why was the outcome of this experiment different from the one in which reduced ribonuclease was reoxidized in a solution free of urea? The reason is that wrong disulfide pairings were formed when the random-coil form of the reduced molecule was reoxidized. There are 105 different ways of pairing eight cysteines to form four disulfides; only one of these combinations is enzymatically active. The 104 wrong pairings have been picturesquely termed "scrambled" ribonuclease. Anfinsen then found that scrambled ribonuclease spontaneously converted into fully active, native ribonuclease when trace amounts of β-mercaptoethanol were added to an aqueous solution of the protein (Figure 2-46). The added

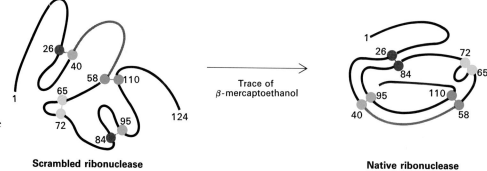

Figure 2-46
Formation of native ribonuclease from scrambled ribonuclease in the presence of a trace of β-mercaptoethanol.

β-mercaptoethanol catalyzed the rearrangement of disulfide pairings until the native structure was regained, which took about ten hours. This process was driven entirely by the decrease in free energy as the scrambled conformations were converted into the stable, native conformation of the enzyme. *Thus, the native form of ribonuclease appears to be the thermodynamically most stable structure.*

Anfinsen (1964) wrote:

> It struck me recently that one should really consider the sequence of a protein molecule, about to fold into a precise geometric form, as a line of melody written in canon form and so designed by Nature to fold back upon itself, creating harmonic chords of interaction consistent with biological function. One might carry the analogy further by suggesting that the kinds of chords formed in a protein with scrambled disulfide bridges, such as I mentioned earlier, are dissonant, but that, by giving an opportunity for rearrangement by the addition of mercaptoethanol, they modulate to give the pleasing harmonics of the native molecule. Whether or not some conclusion can be drawn about the greater thermodynamic stability of Mozart's over Schoenberg's music is something I will leave to the philosophers of the audience.

PROTEINS FOLD BY THE ASSOCIATION OF α-HELICAL AND β-STRAND SEGMENTS

How are the harmonic chords of interaction created in the conversion of an unfolded polypeptide chain into a folded protein? One possibility a priori is that all possible conformations are searched to find the energetically most favored form. How long would such a random search take? Consider a small protein with 100 residues. If each residue can assume three different positions, the total number of structures is 3^{100}, which is equal to 5×10^{47}. If it takes 10^{-13} seconds to convert one structure into another, the total search time would be $5 \times 10^{47} \times 10^{-13}$ seconds, which is equal to 5×10^{34} seconds, or 1.6×10^{27} years! Note that this length of time is a minimal estimate because the actual number of possible conformations per residue is greater than three and the time that it takes to change from one conformation into another is probably considerably longer than 10^{-13} seconds. Clearly, it would take much too long for even a small protein to fold properly by randomly trying out all possible conformations.

How, then, do proteins fold in a few seconds or minutes? The answer is not yet known in detail, but it seems likely that *small stretches of secondary structure serve as intermediates in the folding process.* According to this model, short segments (~15 residues) of an unfolded polypeptide chain flicker in and out of their native α-helical or β-sheet form. These transient structures find each other by diffusion and stabilize each other by forming a complex (Figure 2-47). For example, two α helices, two β strands, or an α helix and a β strand may come together. These αα, ββ, and αβ complexes, which are called *folding units,* then act as nuclei to attract and stabilize other flickering elements of secondary structure.

This model is supported by several lines of experimental evidence. The first is that the tendency of a polypeptide to adopt a regular secondary structure depends to a large degree on its amino acid composition. The formation of an α helix is favored by glutamate, methionine, alanine, and leucine residues, whereas β-sheet formation is enhanced by valine, isoleucine, and tyrosine residues. Second, the transition from a random coil to an α helix can occur in less than a microsecond. Thus, short segments of secondary structure can be formed very rapidly. Third, the postulated folding units (αα, ββ, and αβ complexes) are very similar to the supersecondary structural motifs discussed earlier (p. 31).

The folding of a polypeptide chain into its native structure is like solving a jigsaw puzzle. A complex puzzle (particularly one without straight borders) is best solved by recognizing component patterns, such as blue sky and green grass. The order in which they are found and assembled is unimportant. *A complex jigsaw puzzle can be solved in many different ways. Likewise, there are many different pathways for the folding of a polypeptide chain.* The analogs of blue sky and green grass are local structures that resemble parts of the final protein, such as the helix-turn-helix and beta-turn-beta supersecondary motifs. In the process of folding, incorrect structures are undoubtedly formed, but they are transient because they do not lead to subsequent productive interactions. In contrast, correctly assembled fragments of structure are likely to persist because they will be stabilized by stereospecific interactions with non-neighboring regions to form a compact structure resembling a native domain. The initial folding units will then undergo structural rearrangements to optimize these tertiary interactions. In short, it seems likely that local folding is followed by long-range interactions and then by local rearrangements to give the final folded state of the protein. The challenge now is to identify the sources of stability of native proteins and detect structural motifs that may begin the folding process and stabilize its intermediate forms.

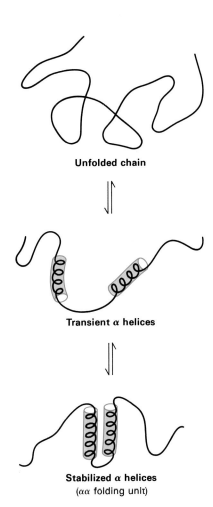

Figure 2-47
Postulated step in protein folding. Two segments of an unfolded polypeptide chain transiently become α helical. These helices are then stabilized by the formation of a complex between the two segments.

PREDICTION OF CONFORMATION FROM AMINO ACID SEQUENCE

Can the conformation of a protein be deduced from its amino acid sequence? Let us begin by considering the polypeptide backbone. Recall that the main chain can rotate on either side of each rigid peptide unit. The amount of rotation at the bond between the nitrogen and α-carbon atoms of the main chain is called phi (φ), and the rotation at the one between the α-carbon and carbonyl carbon atoms is called psi

Part I
MOLECULAR DESIGN OF LIFE

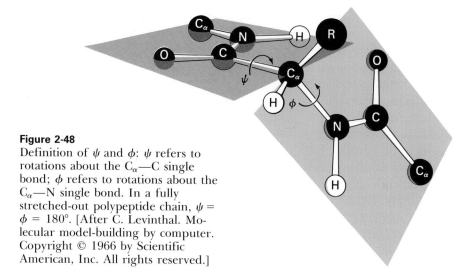

Figure 2-48
Definition of ψ and ϕ: ψ refers to rotations about the C_α—C single bond; ϕ refers to rotations about the C_α—N single bond. In a fully stretched-out polypeptide chain, $\psi = \phi = 180°$. [After C. Levinthal. Molecular model-building by computer. Copyright © 1966 by Scientific American, Inc. All rights reserved.]

(ψ) (Figure 2-48). The conformation of the main chain is completely defined when ϕ and ψ are specified for each residue in the chain. G. N. Ramachandran recognized that a residue in a polypeptide chain cannot have *any* pair of values of ϕ and ψ. Certain combinations are not accessible because of steric hindrance. Allowed ranges of ϕ and ψ can be predicted readily and visualized in steric contour diagrams called *Ramachandran plots*. Such a plot for poly-L-alanine shows three separate allowed ranges (Figure 2-49). In one of them lie the values that produce the antiparallel and parallel β sheets and the collagen helix; in another, those that produce the right-handed α helix; in the third, the left-handed α helix. Though sterically allowed, the left-handed α helix does not occur because it is energetically less favored than the right-handed one. For glycine, these three allowed regions are larger and a fourth

Figure 2-49
Ramachandran plot showing allowed values of ϕ and ψ for L-alanine residues (green regions). Additional conformations are accessible to glycine (yellow regions) because it has a very small side chain.

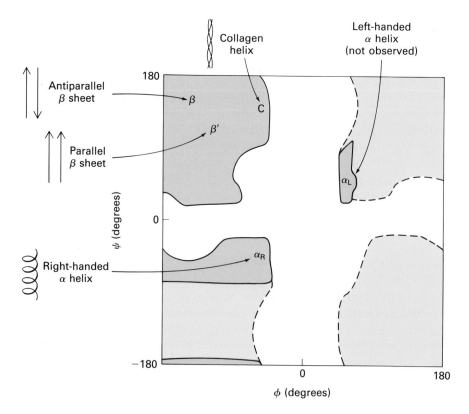

one appears (Figure 2-49) because a hydrogen atom causes less steric hindrance than a methyl group. Glycine enables the polypeptide backbone to make turns that would not be possible with another residue. In contrast, the five-membered ring of proline prevents rotation about the C_α—N bond, which markedly restricts the range of allowed conformations. The measured values of ϕ and ψ for more than 2500 residues in 13 accurately determined protein structures fit nicely into the regions predicted to be allowed.

Additional insight into protein conformation comes from the finding that amino acid residues have different frequencies of occurrence in α helices, β sheets, and reverse turns (Table 2-3). The formation of an α helix is favored by glutamate, methionine, and leucine. β-Sheet formation is enhanced by valine, isoleucine, and phenylalanine. Reverse turns, on the other hand, are promoted by proline, glycine, aspartate, asparagine, and serine. Studies of the conformation of synthetic polypeptides by Elkan Blout, Gerald Fasman, and Ephraim Katchalski have revealed some of the reasons for these preferences, as have analyses of known three-dimensional structures of proteins. For example, branching at the β-carbon atom (as in valine) tends to destabilize the α helix because of steric hindrance. Serine, aspartate, and asparagine tend to disrupt α helices because their side chains contain a hydrogen-bond donor and acceptors in close proximity to the main chain, where they compete for main-chain NH and CO groups.

Much effort has been devoted to predicting the secondary structure of proteins from amino acid sequence and a knowledge of the different tendencies of residues to occur in α helices, β sheets, and reverse turns. The predicted secondary structure agrees with the actual one for about 60% of the chain of proteins whose structures have been solved by x-ray methods. These are encouraging starts, but it is evident that much remains to be accomplished. A clue to the likely direction of fruitful effort is the finding that a pentapeptide sequence can be part of an α helix in one protein and of a β-sheet region in another protein. Hence, the local amino acid sequence is sometimes not enough to determine secondary structure. *The context in which a peptide segment folds may be crucial.* Powerful experimental approaches such as x-ray crystallography, nuclear magnetic resonance spectroscopy, and recombinant DNA cloning are now being used in concert to solve this fundamental problem.

ESSENCE OF PROTEIN ACTION: SPECIFIC BINDING AND TRANSMISSION OF CONFORMATIONAL CHANGES

The first step in the action of a protein is its binding of another molecule. *Proteins as a class of macromolecules are unique in being able to recognize and interact with highly diverse molecules.* For example, myoglobin tightly binds a heme group when its polypeptide chain is partly folded. The acquisition of heme enables myoglobin to carry out its biological function, which is to reversibly bind O_2. Proteins also combine with other proteins to produce highly ordered arrays, such as the contractile filaments in muscle. The binding of foreign molecules to antibody proteins is at the heart of the capacity of the immune system to distinguish between self and nonself. Furthermore, the expression of many genes is controlled by the binding of proteins that recognize specific DNA sequences. Proteins are able to interact specifically with such a wide range of molecules because they are highly proficient at forming *complementary surfaces and clefts* (Figure 2-50). The rich repertoire of side chains on

Table 2-3
Relative frequencies of occurrence of amino acid residues in the secondary structures of proteins

Amino acid	α helix	β sheet	β-turn
Ala	1.29	0.90	0.78
Cys	1.11	0.74	0.80
Leu	1.30	1.02	0.59
Met	1.47	0.97	0.39
Glu	1.44	0.75	1.00
Gln	1.27	0.80	0.97
His	1.22	1.08	0.69
Lys	1.23	0.77	0.96
Val	0.91	1.49	0.47
Ile	0.97	1.45	0.51
Phe	1.07	1.32	0.58
Tyr	0.72	1.25	1.05
Trp	0.99	1.14	0.75
Thr	0.82	1.21	1.03
Gly	0.56	0.92	1.64
Ser	0.82	0.95	1.33
Asp	1.04	0.72	1.41
Asn	0.90	0.76	1.28
Pro	0.52	0.64	1.91
Arg	0.96	0.99	0.88

Source: After T. E. Creighton. *Proteins: Structures and Molecular Properties* (W. H. Freeman, 1983), p. 235.

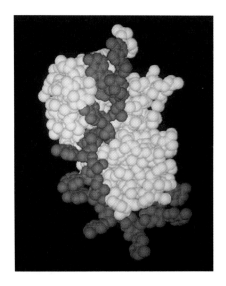

Figure 2-50
Model of ribonuclease (blue) binding an analog of an RNA substrate (red). [Courtesy of Dr. Alex McPherson.]

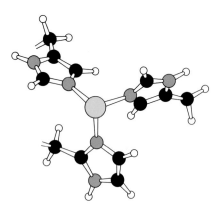

Figure 2-51
Model showing the coordination of a zinc ion (yellow) to three nitrogen atoms of histidine side chains in the catalytic site of carbonic anhydrase.

these surfaces and in these clefts enables proteins to form hydrogen bonds, electrostatic bonds, and van der Waals bonds with other molecules. Moreover, the strength of these interactions and their duration can be precisely controlled.

As was mentioned in the introduction to this chapter, nearly all reactions in biological systems are catalyzed by proteins called enzymes. We can now appreciate why proteins play the unique role of determining the pattern of chemical transformations. *The catalytic power of proteins comes from their capacity to bind substrate molecules in precise orientations and to stabilize transition states in the making and breaking of chemical bonds.* This basic principle can be made concrete by a simple example of enzymatic catalysis, the hydration of carbon dioxide by *carbonic anhydrase*.

$$H_2O + CO_2 \rightleftharpoons HCO_3^- + H^+$$

How does carbonic anhydrase accelerate this reaction by a factor of more than a million? Part of this catalytic enhancement is due to the action of a zinc ion that is coordinated to the imidazole groups of three histidine residues in the enzyme (Figure 2-51). The zinc ion is located at the bottom of a deep cleft some 15 Å from the surface of the protein. Nearby is a group of residues that recognizes and binds carbon dioxide. Water bound to the zinc ion is rapidly converted into hydroxide ion, which is precisely positioned to attack the carbon dioxide molecule bound next to it (Figure 2-52). The zinc ion helps to orient the CO_2 as well as to provide a very high local concentration of OH^-. Carbonic anhydrase, like other enzymes, is a potent catalyst because *it brings substrates into close proximity and optimizes their orientation for reaction.* Another recurring catalytic device is the *use of charged groups to polarize substrates and stabilize transition states.* Some enzymes even form covalent bonds with substrates. We shall consider enzymatic mechanisms in detail in later chapters.

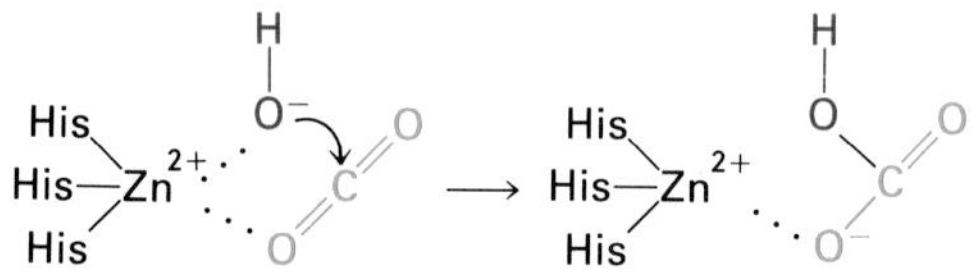

Figure 2-52
Essence of the catalytic mechanism of carbonic anhydrase. Hydroxide ion and carbon dioxide are precisely positioned for the facile formation of bicarbonate by their binding to Zn^{2+}.

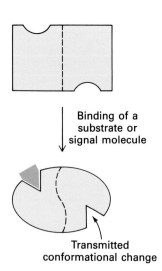

Figure 2-53
Schematic diagram of an allosteric interaction in a protein. The binding of a small molecule or macromolecule to a site in the protein leads to conformational changes that are propagated to a distant site.

Some of the most interesting and important proteins contain two or more binding sites that communicate with each other. A conformational change induced by the binding of a molecule to one site in a protein can alter other sites more than 20 Å away. Thus, proteins can be built to serve as *molecular switches* to receive, integrate, and transmit signals. Many proteins contain regulatory sites called *allosteric sites* that control their binding of other molecules and alter their catalytic rates (Figure 2-53). For example, the binding of oxygen to the heme groups of hemoglobin is altered by the binding of H^+ and CO_2 to distant sites in the protein. This dependence of oxygen binding on pH and carbon dioxide concentration makes hemoglobin a very efficient oxygen transporter (p. 156). Allosteric control mediated by conformational changes in protein molecules is central to the regulation of metabolism.

Proteins containing pairs of sites that are coupled to each other by conformational changes have the capacity to convert energy from one form to another. Suppose that a protein has a catalytic site that hydrolyzes adenosine triphosphate (ATP) to adenosine diphosphate (ADP), an energetically favored reaction (Figure 2-54). The change from a bound triphosphate to a diphosphate group induces a change at the catalytic site that is transmitted to a different binding site some distance away on the same protein. The role of this second site is to bind another protein when ADP is bound to the first site and to release it when ATP is bound. An enzyme with these properties can function as a molecular motor that converts chemical bond energy into movement, as in muscle contraction (p. 931).

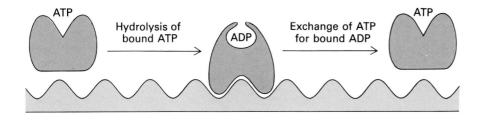

Figure 2-54
The hydrolysis of ATP at the catalytic site of an allosteric protein (red) can increase the affinity of a distant site for binding another protein (green).

SUMMARY

Proteins play key roles in virtually all biological processes. Nearly all catalysts in biological systems are proteins called enzymes. Hence, proteins determine the pattern of chemical transformations in cells. Proteins mediate a wide range of other functions, such as transport and storage, coordinated motions, mechanical support, immune protection, excitability, integration of metabolism, and the control of growth and differentiation.

The basic structural units of proteins are amino acids. All proteins in all species from bacteria to humans are constructed from the same set of twenty amino acids. The side chains of these building blocks differ in size, shape, charge, hydrogen-bonding capacity, and chemical reactivity. They can be grouped as follows: (1) aliphatic side chains—glycine, alanine, valine, leucine, isoleucine, and proline; (2) hydroxyl aliphatic side chains—serine and threonine; (3) aromatic side chains—phenylalanine, tyrosine, and tryptophan; (4) basic side chains—lysine, arginine, and histidine; (5) acidic side chains—aspartic acid and glutamic acid; (6) amide side chains—asparagine and glutamine; and (7) sulfur side chains—cysteine and methionine.

Many amino acids, usually more than a hundred, are joined by peptide bonds to form a polypeptide chain. A peptide bond links the α-carboxyl group of one amino acid and the α-amino group of the next one. Disulfide cross-links can be formed by cysteine residues. Some proteins are covalently modified and cleaved after their synthesis. Proteins have unique amino acid sequences that are genetically determined. The critical determinant of the biological function of a protein is its conformation, which is the three-dimensional arrangement of the atoms of a molecule. Three regularly repeating conformations of polypeptide chains are known: the α helix, the β pleated sheet, and the collagen helix. Segments of the α helix and the β pleated sheet are found in many proteins, as are hairpin β-turns. An important principle is that the amino acid sequence of a protein specifies its three-dimensional structure, as was first shown for ribonuclease. Reduced, unfolded ribonuclease spontaneously forms the correct disulfide pairings and regains full enzymatic activity when oxidized by air after removal of mercaptoethanol and urea. Proteins fold by the association of short polypeptide segments that transiently adopt α-helical or β-sheet forms. The strong tendency of hydrophobic residues to flee from water drives the folding of soluble proteins. Proteins are stabilized by many reinforcing hydrogen bonds and van der Waals interactions as well as by hydrophobic interactions.

Proteins are a unique class of macromolecules in being able to specifically recognize and interact with highly diverse molecules. The repertoire of twenty kinds of side chains enables proteins to fold into distinctive structures and form complementary surfaces and clefts. The catalytic power of enzymes comes from their capacity to bind substrates in precise orientations and to stabilize transition states in the making and breaking of chemical bonds. Conformational changes transmitted between distant sites in protein molecules are at the heart of the capacity of proteins to transduce energy and information.

SELECTED READINGS

Where to start

Doolittle, R. F., 1985. Proteins. *Sci. Amer.* 253(4):88–99. [A lucid overview that emphasizes molecular evolution. Reprinted in *The Molecules of Life*, a collection of readings from *Scientific American*, W. H. Freeman, 1985.]

Goldberg, M. E., 1985. The second translation of the genetic message: protein folding and assembly. *Trends Biochem. Sci.* 10:388–391.

Karplus, M., and McCammon, J. A., 1986. The dynamics of proteins. *Sci. Amer.* 254(4):42–51. [Available as *Sci. Amer.* Offprint 1569.]

Books on protein chemistry

Creighton, T. E., 1983. *Proteins: Structures and Molecular Principles.* W. H. Freeman.

Cantor, C. R., and Schimmel, P. R., 1980. *Biophysical Chemistry.* W. H. Freeman. [Chapters 2 and 5 in Part I and Chapters 20 and 21 in Part III give an excellent account of the principles of protein conformation.]

Fletterick, R. J., Schroer, T., and Matela, R. J., 1985. *Molecular Structure: Macromolecules in Three-Dimensions.* Blackwell Scientific Publications. [A fine introduction to molecular models.]

Schultz, G. E., and Schirmer, R. H., 1979. *Principles of Protein Structure.* Springer-Verlag.

Neurath, H., and Hill, R. L., (eds.), 1976. *The Proteins* (3rd ed.). Academic Press. [A multivolume treatise that contains many fine articles.]

Covalent modification of proteins

Wold, F., 1981. In vivo chemical modification of proteins (post-translational modification). *Ann. Rev. Biochem.* 50:783–814.

Glazer, A. N., DeLange, R. J., and Sigman, D. S., 1975. *Chemical Modification of Proteins.* North-Holland.

Conformation of proteins

Chothia, C., 1984. Principles that determine the structure of proteins. *Ann. Rev. Biochem.* 53:537–572.

Richardson, J. S., 1981. The anatomy and taxonomy of protein structure. *Adv. Protein Chem.* 34:167–339. [A lucid and beautifully illustrated account of three-dimensional architecture, with emphasis on supersecondary structural motifs.]

Folding of proteins

Harrison, S. C., and Durbin, R., 1985. Is there a single pathway for the folding of a polypeptide chain? *Proc. Nat. Acad. Sci.* 82:4028–4030. [The jigsaw-puzzle analogy for protein folding is presented in this incisive article.]

Kim, P. S., and Baldwin, R. L., 1982. Specific intermediates in the folding reactions of small proteins and the mechanism of protein folding. *Ann. Rev. Biochem.* 51:459–490.

Creighton, T. E., 1985. Energetics of protein structure and folding. *Biopolymers* 24:167–182.

Anfinsen, C. B., 1973. Principles that govern the folding of protein chains. *Science* 181:223–230.

Freedman, R. B., Brockway, B. E., and Lambert, N., 1984. Protein disulphide-isomerase and the formation of native disulphide bonds. *Biochem. Soc. Trans.* 12:929–932.

Prediction of protein structure

Blout, E. R., de Lozé, C., Bloom, S. M., and Fasman, G. D., 1960. The dependence of the conformations of synthetic polypeptides on amino acid composition. *J. Amer. Chem. Soc.* 82:3787–3789.

Chou, P. Y., and Fasman, G. D., 1974. Prediction of protein conformation. *Biochemistry* 13:222–244.

Kabsch, W., and Sander, C., 1983. How good are predictions of protein secondary structure? *FEBS Letters* 155:179–182.

PROBLEMS

1. Tropomyosin, a 70-kd muscle protein, is a two-stranded α-helical coiled coil. What is the length of the molecule?

2. Poly-L-leucine in an organic solvent such as dioxane is α-helical, whereas poly-L-isoleucine is not. Why do these amino acids with the same number and kinds of atoms have different helix-forming tendencies?

3. A mutation that changes an alanine residue in the interior of a protein to a valine is found to lead to a loss of activity. However, activity is regained when a second mutation at a different position changes an isoleucine residue to a glycine. How might this second mutation lead to a restoration of activity?

4. An enzyme that catalyzes disulfide-sulfhydryl exchange reactions has been isolated. Inactive scrambled ribonuclease is rapidly converted into enzymatically active ribonuclease by this enzyme. In contrast, insulin is rapidly inactivated by this enzyme. What does this important observation imply about the relation between the amino acid sequence of insulin and its three-dimensional structure?

5. A protease is an enzyme that catalyzes the hydrolysis of peptide bonds of target proteins. How might a protease bind a target protein so that its main chain becomes fully extended in the vicinity of the vulnerable peptide bond?

APPENDIX
Acid-Base Concepts

Ionization of Water

Water dissociates into hydronium (H_3O^+) and hydroxyl (OH^-) ions. For simplicity, we refer to the hydronium ion as a hydrogen ion (H^+) and write the equilibrium as

$$H_2O \rightleftharpoons H^+ + OH^-$$

The equilibrium constant K_{eq} of this dissociation is given by

$$K_{eq} = \frac{[H^+][OH^-]}{[H_2O]} \quad (1)$$

in which the terms in brackets denote molar concentrations. Because the concentration of water (55.5 M) is changed little by ionization, expression 1 can be simplified to give

$$K_w = [H^+][OH^-] \quad (2)$$

in which K_w is the ion product of water. At 25°C, K_w is 1.0×10^{-14}.

Note that the concentrations of H^+ and OH^- are reciprocally related. If the concentration of H^+ is high, then the concentration of OH^- must be low, and vice versa. For example, if $[H^+] = 10^{-2}$ M, then $[OH^-] = 10^{-12}$ M.

Definition of Acid and Base

An acid is a proton donor. A base is a proton acceptor.

$$\text{Acid} \rightleftharpoons H^+ + \text{base}$$

$$CH_3\text{—}COOH \rightleftharpoons H^+ + CH_3\text{—}COO^-$$
Acetic acid   Acetate

$$NH_4^+ \rightleftharpoons H^+ + NH_3$$
Ammonium ion   Ammonia

The species formed by the ionization of an acid is its conjugate base. Conversely, protonation of a base yields its conjugate acid. Acetic acid and acetate ion are a conjugate acid-base pair.

Definition of pH and pK

The pH of a solution is a measure of its concentration of H^+. The pH is defined as

$$\text{pH} = \log_{10}\frac{1}{[H^+]} = -\log_{10}[H^+] \quad (3)$$

The ionization equilibrium of a weak acid is given by

$$HA \rightleftharpoons H^+ + A^-$$

The apparent equilibrium constant K for this ionization is

$$K = \frac{[H^+][A^-]}{[HA]} \quad (4)$$

The pK of an acid is defined as

$$\text{p}K = -\log K = \log\frac{1}{K} \quad (5)$$

Inspection of equation 4 shows that the pK of an acid is the pH at which it is half dissociated.

Henderson-Hasselbalch Equation

What is the relationship between pH and the ratio of acid to base? A useful expression can be derived from equation 4. Rearrangement of that equation gives

$$\frac{1}{[H^+]} = \frac{1}{K}\frac{[A^-]}{[HA]} \quad (6)$$

Taking the logarithm of both sides of equation 6 gives

$$\log \frac{1}{[H^+]} = \log \frac{1}{K} + \log \frac{[A^-]}{[HA]} \quad (7)$$

Substituting pH for log 1/[H$^+$] and pK for log 1/K in equation 7 yields

$$pH = pK + \log \frac{[A^-]}{[HA]} \quad (8)$$

which is commonly known as the Henderson-Hasselbalch equation.

The pH of a solution can be calculated from equation 8 if the molar proportion of A$^-$ to HA and the pK of HA are known. Consider a solution of 0.1 M acetic acid and 0.2 M acetate ion. The pK of acetic acid is 4.8. Hence, the pH of the solution is given by

$$pH = 4.8 + \log \frac{0.2}{0.1} = 4.8 + \log 2$$
$$= 4.8 + 0.3 = 5.1$$

Conversely, the pK of an acid can be calculated if the molar proportion of A$^-$ to HA and the pH of the solution are known.

Buffering Power

An acid-base conjugate pair (such as acetic acid and acetate ion) has an important property: it resists changes in the pH of a solution. In other words, it acts as a *buffer*. Consider the addition of OH$^-$ to a solution of acetic acid (HA):

$$HA + OH^- \rightleftharpoons A^- + H_2O$$

A plot of the dependence of the pH of this solution on the amount of OH$^-$ added is called a *titration curve* (Figure 2-55). Note that there is an inflection point in the curve at pH 4.8, which is the pK of acetic acid.

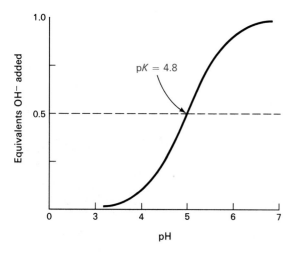

Figure 2-55
Titration curve of acetic acid.

In the vicinity of this pH, a relatively large amount of OH$^-$ produces little change in pH. In general, a weak acid is most effective in buffering against pH changes in the vicinity of its pK value.

pK Values of Amino Acids

An amino acid such as glycine contains two ionizable groups: an α-carboxyl group and a protonated α-amino group. As base is added, these two groups are titrated (Figure 2-56). The pK of the α-COOH group is 2.3, whereas that of the α-NH$_3^+$ group is 9.6. The pK values of these groups in other amino acids are similar. Some amino acids, such as aspartic acid, also contain an ionizable side chain. The pK values of ionizable side chains in amino acids range from 3.9 (aspartic acid) to 12.5 (arginine).

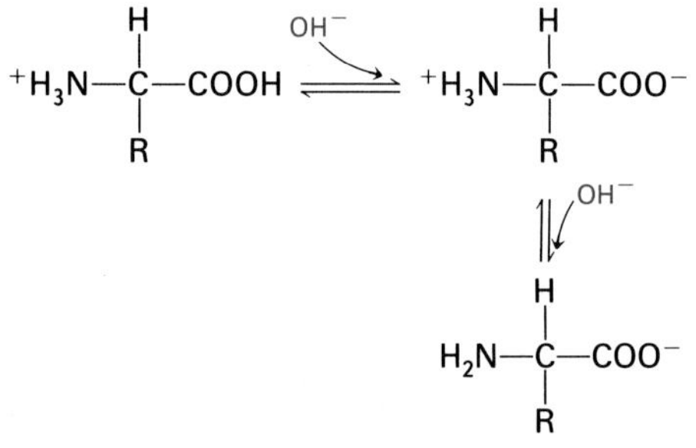

Figure 2-56
Titration of the α-carboxyl and α-amino groups of an amino acid.

Table 2-4
pK values of some amino acids

Amino acid	pK values (25°C)		
	α-COOH group	α-NH$_3^+$ group	Side chain
Alanine	2.3	9.9	
Glycine	2.4	9.8	
Phenylalanine	1.8	9.1	
Serine	2.1	9.2	
Valine	2.3	9.6	
Aspartic acid	2.0	10.0	3.9
Glutamic acid	2.2	9.7	4.3
Histidine	1.8	9.2	6.0
Cysteine	1.8	10.8	8.3
Tyrosine	2.2	9.1	10.9
Lysine	2.2	9.2	10.8
Arginine	1.8	9.0	12.5

Source: After J. T. Edsall and J. Wyman. *Biophysical Chemistry* (Academic Press, 1958), ch. 8.

CHAPTER 3

Exploring Proteins

In the preceding chapter, we saw that proteins play crucial roles in nearly all biological processes—in catalysis, transport, coordinated motion, excitability, and the control of growth and differentiation. This remarkable range of functions arises from the folding of proteins into many distinctive three-dimensional structures that bind highly diverse molecules. One of the major goals of biochemistry is to determine how amino acid sequences specify the conformations of proteins. We also want to learn how proteins bind specific substrates and other molecules, mediate catalysis, and transduce energy and information. An indispensable step in these studies is the purification of the protein of interest.

The three key approaches to analyzing and purifying proteins are electrophoresis, ultracentrifugation, and chromatography. Given a pure protein, it is possible to elucidate its amino acid sequence. The strategy is to divide and conquer, to obtain specific fragments that can readily be sequenced. Automated peptide sequencing and the application of recombinant DNA methods are providing a wealth of amino acid sequence data that are opening new vistas. For seeing beyond primary structure to conformation, for elucidating the precise positions of atoms in proteins, x-ray crystallography is the most powerful technique. The physiological context of a protein also needs to be known to fully understand how it functions. Antibodies are choice probes for locating proteins in vivo and measuring their quantities. The chapter closes with the synthesis of peptides, which makes feasible the synthesis of new drugs and antigens for inducing the formation of specific antibodies.

The exploration of proteins by this array of physical and chemical techniques has greatly enriched our understanding of the molecular basis of life and makes it possible to tackle some of the most challenging questions of biology in molecular terms.

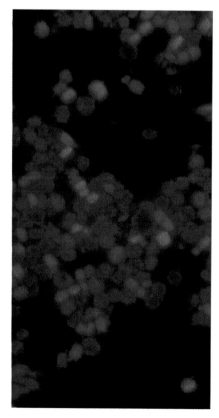

Micrograph of crystals of a light-harvesting protein of a photosynthetic bacterium. [Courtesy of Dr. Alexander Glazer.]

PROTEINS CAN BE SEPARATED BY GEL ELECTROPHORESIS AND DISPLAYED

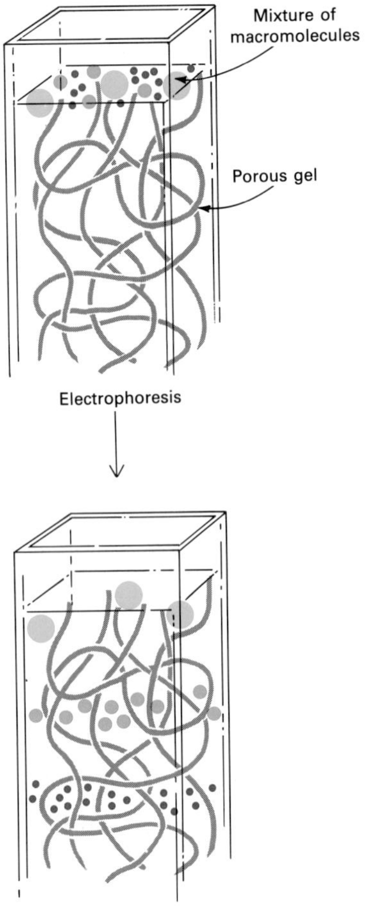

A molecule with a net charge will move in an electric field. This phenomenon, termed *electrophoresis*, offers a powerful means of separating proteins and other macromolecules, such as DNA and RNA. The velocity of migration (v) of a protein (or any molecule) in an electric field depends on the electric field strength (E), the net charge on the protein (z), and the frictional coefficient (f).

$$v = \frac{Ez}{f} \qquad (1)$$

The electric force Ez driving the charged molecule toward the oppositely charged electrode is opposed by the viscous drag fv arising from friction between the moving molecule and the medium. The frictional coefficient f depends on both the mass and shape of the migrating molecule and the viscosity of the medium.

Electrophoretic separations are nearly always carried out in gels rather than in free solution for two reasons. First, gels suppress convective currents produced by small temperature gradients, a requirement for effective separation. Second, gels serve as molecular sieves that enhance separation (Figure 3-1). Molecules that are small compared with the pores in the gel readily move through the gel, whereas molecules much larger than the pores are almost immobile. Intermediate-size molecules move through the gel with various degrees of facility. Polyacrylamide gels are choice supporting media for electrophoresis because they are chemically inert and are readily formed by the polymerization of acrylamide. Moreover, their pore sizes can be controlled by choosing various concentrations of acrylamide and methylenebisacrylamide (a cross-linking reagent) at the time of polymerization (Figure 3-2).

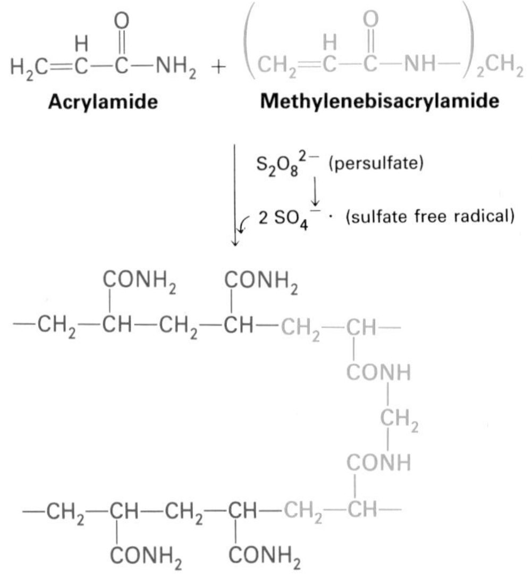

Figure 3-1
Sieving action of a porous polyacrylamide gel.

Figure 3-2
Formation of a polyacrylamide gel. The pore size can be controlled by adjusting the concentration of activated monomer (red) and crosslinker (green).

Proteins can be separated largely on the basis of mass by electrophoresis in a polyacrylamide gel under denaturing conditions. The mixture of proteins is first dissolved in a solution of sodium dodecyl sulfate (SDS), an anionic detergent that disrupts nearly all noncovalent interactions in native proteins. Mercaptoethanol or dithiothreitol is also added

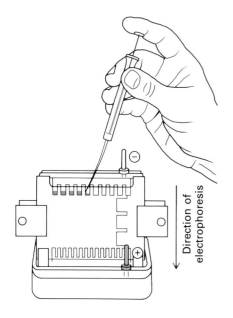

Loading of samples for electrophoresis. Typically, several samples are electrophoresed on one flat polyacrylamide gel. A microliter syringe is used to place solutions of proteins in the wells of the slab. A cover is then placed over the gel chamber and 200 volts are applied. The negatively charged SDS-protein complexes migrate in the direction of the anode, at the bottom of the gel.

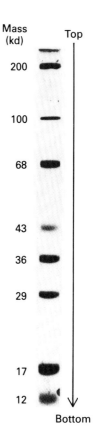

Figure 3-3
Proteins electrophoresed on an SDS-polyacrylamide gel can be visualized by staining with Coomassie blue.

to reduce disulfide bonds. Anions of SDS bind to main chains at a ratio of about one SDS for every two amino acid residues, which gives a complex of SDS with a denatured protein a large net negative charge that is roughly proportional to the mass of the protein. The negative charge acquired on binding SDS is usually much greater than the charge on the native protein; this native charge is thus rendered insignificant. The SDS complexes with the denatured proteins are then electrophoresed on a polyacrylamide gel, typically in the form of a thin verticle slab. The direction of electrophoresis is from top to bottom. Finally, the proteins in the gel can be visualized by staining them with silver or a dye such as Coomassie blue, which reveals a series of bands (Figure 3-3). Radioactive labels can be detected by placing a sheet of x-ray film over the gel, a procedure called autoradiography. *Small proteins move rapidly through the gel, whereas large ones stay at the top, near the point of application of the mixture.* The mobility of most polypeptide chains under these conditions is linearly proportional to the logarithm of their mass (Figure 3-4). This empirical relationship is not obeyed by some proteins; for example, some carbohydrate-rich proteins and membrane proteins migrate anomalously.

SDS-polyacrylamide gel electrophoresis is rapid, sensitive, and capable of a high degree of resolution. Electrophoresis and staining take about a day. As little as 0.1 μg (~2 pmol) of a protein gives a distinct band when stained with Coomassie blue, and even less (~0.02 μg) can be detected with a silver stain. Proteins that differ in mass by about 2% (e.g., 40 and 41 kd, arising from a difference of about ten residues) can usually be distinguished.

Proteins can also be separated electrophoretically on the basis of their relative contents of acidic and basic residues. The *isoelectric point* (pI) of a protein is the pH at which its net charge is zero. At this pH, its electrophoretic mobility is zero because z in equation 1 is equal to zero. For example, the pI of cytochrome c, a highly basic electron-transport protein, is 10.6, whereas that of serum albumin, an acidic protein in blood, is 4.8. Suppose that a mixture of proteins is electrophoresed in a pH gradient in a gel in the absence of SDS. Each protein will move until it reaches a position in the gel at which the pH is equal to the pI of the

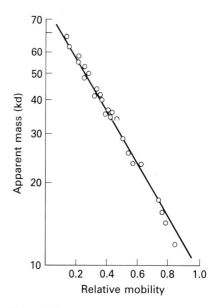

Figure 3-4
The electrophoretic mobility of many proteins in SDS-polyacrylamide gels is proportional to the logarithm of their mass. [After K. Weber and M. Osborn, *The Proteins*, 3rd ed. (Academic Press, 1975), vol. 1, p. 179.]

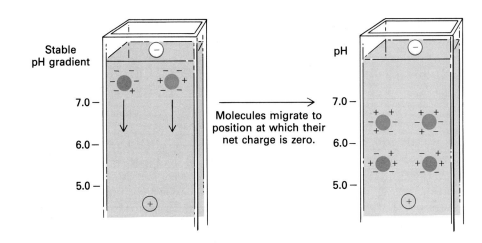

Figure 3-5
Isoelectric focusing separates proteins according to the pH at which their net charge is zero (isoelectric point). A stable pH gradient is produced in the gel before electrophoresing the mixture of proteins.

protein (Figure 3-5). This method of separating proteins according to their isoelectric point is called *isoelectric focusing*. The pH gradient in the gel is formed first by electrophoresing a mixture of *polyampholytes* (small multi-charged polymers) having many pI values. Isoelectric focusing can readily resolve proteins that differ in pI by as little as 0.01, which means that proteins differing by one net charge can be separated.

Isoelectric focusing can be combined with SDS-polyacrylamide gel electrophoresis to obtain very high-resolution separations. A sample is first subjected to isoelectric focusing. This gel is then placed horizontally on top of an SDS-polyacrylamide gel and electrophoresed vertically to yield a two-dimensional pattern of spots. In such a gel, proteins have been separated in the horizontal direction on the basis of isoelectric point, and in the vertical direction on the basis of mass. It is remarkable that more than a thousand different proteins in the bacterium *E. coli* can be resolved in a single experiment by two-dimensional electrophoresis (Figure 3-6).

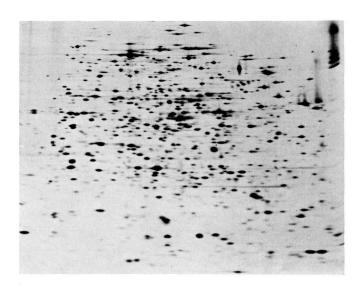

Figure 3-6
Two-dimensional electrophoresis of the proteins from *E. coli*. More than a thousand different proteins from this bacterium have been resolved. These proteins were separated according to their isoelectric pH in the horizontal direction and their apparent mass in the vertical direction. [Courtesy of Dr. Patrick H. O'Farrell.]

PROTEINS CAN BE PURIFIED ACCORDING TO SIZE, CHARGE, AND BINDING AFFINITY

Proteins can readily be visualized and differentiated by the electrophoretic methods described above. These gel techniques can also be used to obtain small quantities (micrograms) of purified polypeptides. However, they do not provide large amounts of purified proteins in their native state. Substantial quantities of purified proteins, of the order of many milligrams, are needed to fully elucidate their three-dimensional structure and their mechanism of action. Several thousand proteins have been purified in active form on the basis of such characteristics as *size, solubility, charge,* and *specific binding affinity.*

Before purifying a protein, various separation methods are tried and their efficiency is evaluated by assaying for a distinctive property of the protein of interest. The assay for an enzyme, for example, is typically a test for its specific catalytic activity. The purification procedure that is chosen is monitored at each step by SDS-polyacrylamide gel electrophoresis; if the procedure is successful, the desired protein becomes increasingly prominent in stained gels. The total amount of protein is measured, too, so that the degree of purification obtained in a particular step can be determined.

Proteins can be separated from small molecules by dialysis through a semipermeable membrane, such as a cellulose membrane with pores (Figure 3-7). Molecules having dimensions significantly greater than the pore diameter are retained inside the dialysis bag, whereas smaller molecules and ions traverse the pores of such a membrane and emerge in the dialysate outside the bag. More discriminating separations on the basis of size can be achieved by the technique of *gel-filtration chromatography* (Figure 3-8). The sample is applied to the top of a column consisting of porous beads made of an insoluble but highly hydrated polymer

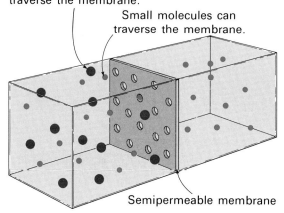

Figure 3-7
Separation of molecules on the basis of size by dialysis.

Figure 3-8
Separation of molecules on the basis of size by gel-filtration chromatography.

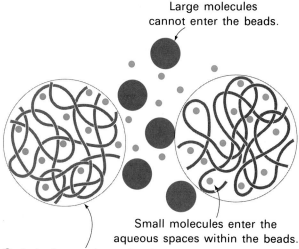

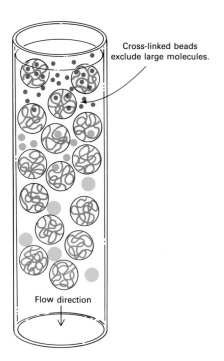

Figure 3-9
Separation of three proteins differing in size by gel-filtration chromatography.

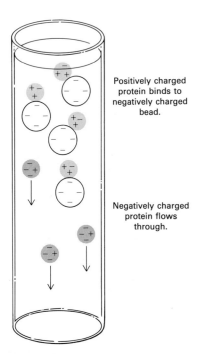

Figure 3-10
Ion-exchange chromatography separates proteins mainly according to their net charge.

such as dextran or agarose (which are carbohydrates) or polyacrylamide. Sephadex, Sepharose, and Bio-gel are commonly used commercial preparations of these beads, which are typically 100 μm (0.1 mm) in diameter. Small molecules can enter these beads, but large ones cannot. The result is that small molecules are distributed both in the aqueous solution inside the beads and between them, whereas large molecules are located only in the solution between the beads. *Large molecules flow more rapidly through this column and emerge first because a smaller volume is accessible to them* (Figure 3-9). It should be noted that the order of emergence of molecules from a column of porous beads is the reverse of the order in gel electrophoresis, in which a *continuous* polymer framework *impedes* the movement of large molecules (Figure 3-1). Much larger quantities of protein can be separated by gel-filtration chromatography than by gel electrophoresis, but a price is paid in the lower resolution of gel-filtration chromatography.

The solubility of most proteins is lowered at high salt concentrations. This effect, called *salting-out*, is very useful, though not well understood. The dependence of solubility on salt concentration differs from one protein to another. Hence, salting-out can be used to fractionate proteins. For example, 0.8 M ammonium sulfate precipitates fibrinogen, a blood-clotting protein, whereas 2.4 M is needed to precipitate serum albumin. Salting-out is also useful for concentrating dilute solutions of proteins.

Proteins can be separated on the basis of their net charge by *ion-exchange chromatography*. If a protein has a net positive charge at pH 7, it will usually bind to a column of beads containing carboxylate groups, whereas a negatively charged protein will not (Figure 3-10). A positively charged protein bound to such a column can then be eluted (released) by increasing the concentration of sodium chloride or another salt in the eluting buffer. Sodium ions compete with positively charged groups on the protein for binding to the column. Proteins that have a low density of net positive charge will tend to emerge first, followed by those having a higher charge density. Factors other than net charge, such as affinity for the supporting matrix, can also influence the behavior of proteins on ion-exchange columns. Negatively charged proteins (anionic proteins) can be separated by chromatography on positively charged diethylaminoethyl-cellulose (DEAE-cellulose) columns. Conversely, positively charged proteins (cationic proteins) can be separated on negatively charged carboxymethyl-cellulose (CM-cellulose) columns.

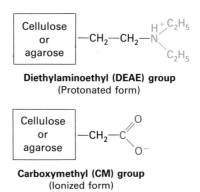

Affinity chromatography is another powerful and generally applicable means of purifying proteins. This technique takes advantage of the high affinity of many proteins for specific chemical groups. For example, the plant protein concanavalin A can be purified by passing a crude extract through a column of beads containing covalently attached glucose residues. Concanavalin A binds to such a column because it has affinity for glucose, whereas most other proteins do not. The bound concanavalin A can then be released from the column by adding a concentrated solution of glucose. The glucose in solution displaces the column-attached glucose residues from binding sites on concanavalin A (Figure 3-11). In general, affinity chromatography can be effectively used to isolate a protein that recognizes group X by (1) covalently attaching X or a derivative of it to a column, (2) adding a mixture of proteins to this column, which is then washed with buffer to remove unbound proteins, and (3) eluting the desired protein by adding a high concentration of a soluble form of X.

ULTRACENTRIFUGATION IS VALUABLE FOR SEPARATING BIOMOLECULES AND DETERMINING MOLECULAR WEIGHTS

Centrifugation is a powerful and generally applicable method for separating and analyzing cells, organelles, and biological macromolecules. A particle moving in a circle of radius r at an angular velocity ω is subject to a centrifugal (outward) field equal to $\omega^2 r$. The *centrifugal force* F_c on this particle is equal to the product of its effective mass m' and the centrifugal field.

$$F_c = m'\omega^2 r = m(1 - \bar{v}\rho)\omega^2 r \qquad (2)$$

The effective mass m' is less than the mass m because the displaced fluid exerts an opposing force. This buoyancy factor is equal to $(1 - \bar{v}\rho)$, where $\bar{v}$ is the partial specific volume of the particle and ρ is the density of the solution. A particle moves in this field at a constant velocity v when F_c is equal to the viscous drag vf, where f is the frictional coefficient of the particle. Hence, the migration velocity (sedimentation velocity) of the particle is

$$v = \frac{F_c}{f} = \frac{m(1 - \bar{v}\rho)\omega^2 r}{f} \qquad (3)$$

Note that this expression for movement in a centrifugal field is analogous to equation 1 for movement in an electric field.

Equation 3 shows that the sedimentation velocity is directly proportional to the strength of the centrifugal field. Hence, it is possible to define a measure of sedimentation that depends on the properties of the particle and solution but is independent of how fast the sample is spun. The *sedimentation coefficient s*, defined as the velocity divided by the centrifugal field, is equal to

$$s = \frac{v}{\omega^2 r} = \frac{m(1 - \bar{v}\rho)}{f} \qquad (4)$$

Sedimentation coefficients are usually expressed in *Svedberg units*. A Svedberg (S) is equal to 10^{-13} seconds. For example, suppose that a 150-kd antibody protein is spun in an ultracentrifuge at a radius of 8 cm at 75,000 revolutions per minute (rpm). The centrifugal field under these conditions is 4.9×10^8 cm s^{-2}, which is about 500,000 times the strength of the earth's gravitational field (g). If the velocity of the protein in this field is 3.4×10^{-4} cm/s, then its sedimentation coefficient is 7S.

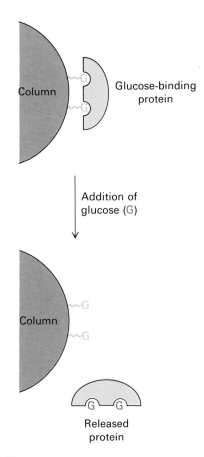

Figure 3-11
Affinity chromatography of concanavalin A (shown in yellow) on a column containing covalently attached glucose residues (G).

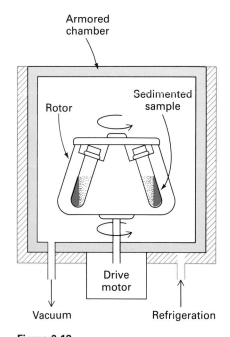

Figure 3-12
Ultracentrifuge rotor containing two sample tubes.

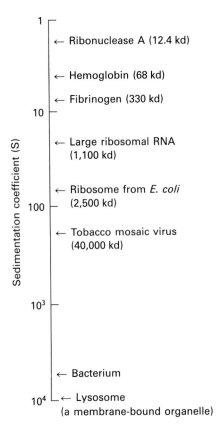

Figure 3-13
Span of S values of biomolecules and cells.

Ninhydrin

Several important conclusions can be drawn from equation 3:

1. The sedimentation velocity of a particle is proportional to its mass. A 200-kd protein moves at twice the velocity of a 100-kd protein of the same shape and density.

2. A dense particle moves more rapidly than a less dense one because the opposing buoyant force is smaller for a dense particle.

3. Shape, too, is important because it affects the viscous drag. The frictional coefficient of a compact particle is smaller than that of an extended particle of the same mass. A parachutist with a defective unopened parachute falls much more quickly than one with a functioning opened parachute.

4. The sedimentation velocity depends also on the density of the solution (ρ). Particles sink when $\bar{v}\rho < 1$, float when $\bar{v}\rho > 1$, and do not move when $\bar{v}\rho = 1$.

Let us see how centrifugation can be used to separate proteins with different sedimentation coefficients (Figure 3-14). The first step in *zonal centrifugation* (also called *band centrifugation*) is to form a density gradient in a centrifuge tube by mixing different proportions of a low-density solution (such as 5% sucrose) and a high-density one (such as 20% sucrose). The role of the density gradient here is to prevent convective flow. A small volume of the solution containing the mixture of proteins is then layered on top of this density gradient. When the rotor is spun, proteins move through the gradient and separate according to their sedimentation coefficients. Centrifugation is stopped before the fastest protein reaches the bottom of the tube. The separated bands of protein can be harvested by making a hole in the bottom of the tube and collecting drops. The drops can be assayed for protein content and catalytic activity or another functional property. This sedimentation-velocity technique readily separates proteins differing in sedimentation coefficient by a factor of two or more.

The mass (molecular weight) of a protein can be directly determined by *sedimentation equilibrium*, in which a sample is centrifuged at relatively low speed so that sedimentation is counterbalanced by diffusion. Under these conditions, a smooth gradient of protein concentration develops. The dependence of concentration on distance from the rotation axis reveals the mass of the particle. The mass m is given by

$$m = \frac{2kT}{(1 - \bar{v}\rho)\omega^2} \log_e c_2/(c_1(r_2^2 - r_1^2)) \qquad (5)$$

where c_1 and c_2 are the concentrations at distances r_1 and r_2 from the rotation axis, k is Boltzmann's constant, and T is the absolute temperature. *This sedimentation-equilibrium technique for determining mass is rigorous and can be applied under nondenaturing conditions in which the native structure of multimeric proteins is preserved.* In contrast, SDS-polyacrylamide gel electrophoresis (p. 45) provides an *estimate* of the mass of dissociated polypeptide chains under *denaturing* conditions.

AMINO ACID SEQUENCES CAN BE DETERMINED BY AUTOMATED EDMAN DEGRADATION

We turn now from the purification of proteins and analysis of their hydrodynamic properties to the elucidation of their amino acid se-

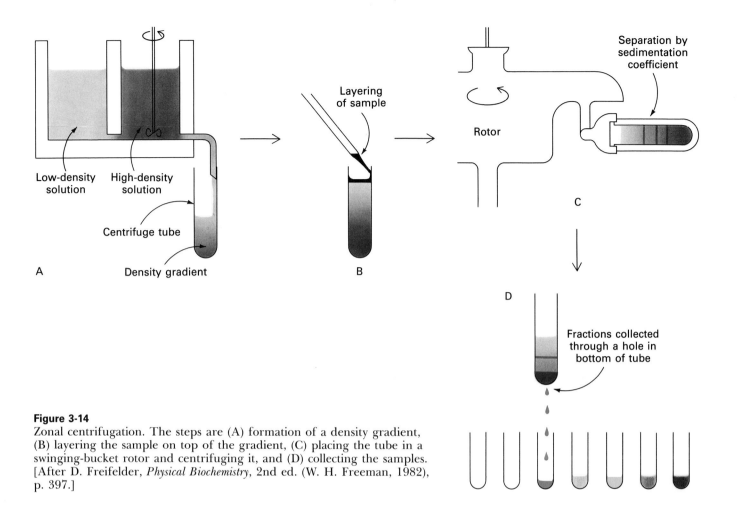

Figure 3-14
Zonal centrifugation. The steps are (A) formation of a density gradient, (B) layering the sample on top of the gradient, (C) placing the tube in a swinging-bucket rotor and centrifuging it, and (D) collecting the samples. [After D. Freifelder, *Physical Biochemistry*, 2nd ed. (W. H. Freeman, 1982), p. 397.]

quences. Let us consider how the sequence of a short peptide, such as

Ala-Gly-Asp-Phe-Arg-Gly

could be established. First, the *amino acid composition* of the peptide is determined. The peptide is hydrolyzed into its constituent amino acids by heating it in 6 N HCl at 110°C for 24 hours. Stanford Moore and William Stein showed that amino acids in hydrolysates can be separated by ion-exchange chromatography on columns of sulfonated polystyrene and quantitated by reacting them with *ninhydrin*. α-Amino acids treated this way give an intense blue color, whereas imino acids, such as proline, give a yellow color. The concentration of amino acid in a solution is proportional to the optical absorbance of the solution after heating it with ninhydrin. This technique can detect a microgram (10 nmol) of an amino acid, which is about the amount present in a thumbprint. As little as a nanogram (10 pmol) of an amino acid can be detected by means of *fluorescamine*, which reacts with the α-amino group to form a highly fluorescent product (Figure 3-15). The identity of the amino acid is revealed by its elution volume, which is the volume of buffer used to remove the amino acid from the column (Figure 3-16). A comparison of the chromatographic patterns of our sample hydrolysate with that of a standard mixture of amino acids would show that the amino acid composition of the peptide is

(Ala, Arg, Asp, Gly$_2$, Phe)

Figure 3-15
Reaction of fluorescamine with the α-amino group of an amino acid to form a fluorescent derivative.

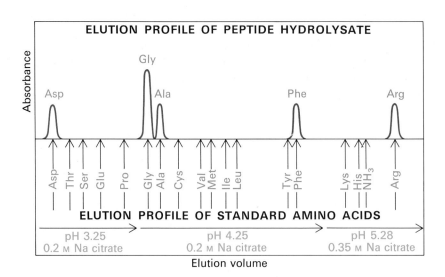

Figure 3-16
Different amino acids in a peptide hydrolysate can be separated by ion-exchange chromatography on a sulfonated polystyrene resin (such as Dowex-50). Buffers of increasing pH are used to elute the amino acids from the column. Aspartate, which has an acidic side chain, is first to emerge, whereas arginine, which has a basic side chain, is the last.

The parentheses denote that this is the amino acid composition of the peptide, not its sequence.

The amino-terminal residue of a protein or peptide can be identified by labeling it with a compound that forms a stable covalent link. *Fluorodinitrobenzene* (FDNB) was first used for this purpose by Frederick Sanger. *Dabsyl chloride* is now commonly used because it forms intensely colored derivatives that can be detected with high sensitivity. It reacts with an uncharged α-NH_2 group to form a sulfonamide derivative that is stable under conditions that hydrolyze peptide bonds (Figure 3-17). Hydrolysis of our sample dabsyl-peptide in 6 N HCl would yield a dabsyl-amino acid, which could be identified as dabsyl-alanine by its

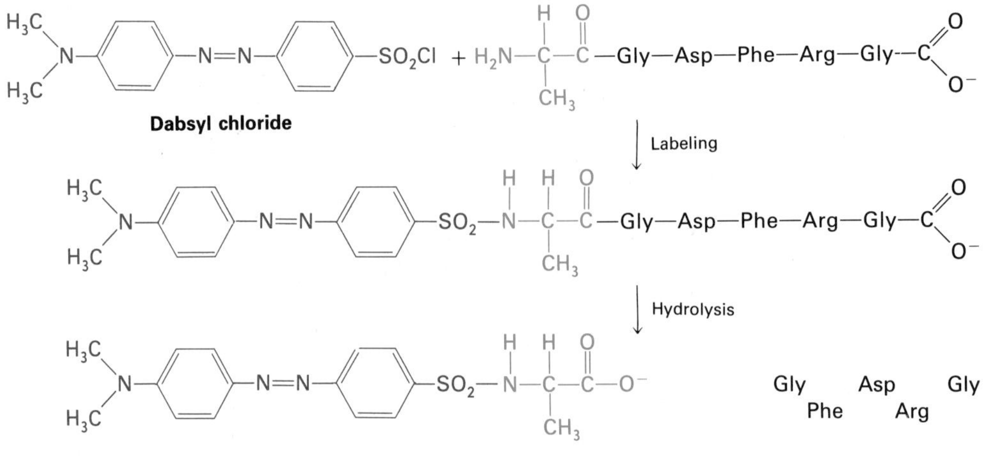

Figure 3-17
Determination of the amino-terminal residue of a peptide. Dabsyl chloride is used to label the peptide, which is then hydrolyzed. The dabsyl-amino acid (dabsyl-alanine in this example) is identified by its chromatographic characteristics.

chromatographic properties. *Dansyl chloride*, another much-used labeling reagent, forms fluorescent sulfonamide derivatives.

Although the dabsyl method for determining the amino-terminal residue is sensitive and powerful, it cannot be used repeatedly on the same peptide because the peptide is totally degraded in the acid-hydrolysis step. Pehr Edman devised a method for labeling the amino-terminal residue and cleaving it from the peptide without disrupting the peptide bonds between the other amino acid residues. The *Edman degradation* sequentially removes one residue at a time from the amino end of a peptide (Figure 3-18). *Phenyl isothiocyanate* reacts with the un-

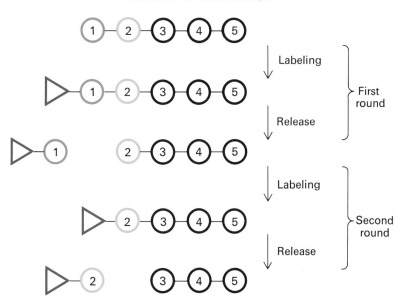

Figure 3-18
The Edman degradation. The labeled amino-terminal residue (PTH-alanine in the first round) can be released without hydrolyzing the rest of the peptide. Hence, the amino-terminal residue of the shortened peptide (Gly-Asp-Phe-Arg-Gly) can be determined in the second round. Three more rounds of the Edman degradation reveal the complete sequence of the original peptide.

charged terminal amino group of the peptide to form a phenylthiocarbamoyl derivative. Then, under mildly acidic conditions, a cyclic derivative of the terminal amino acid is liberated, which leaves an intact peptide shortened by one amino acid. The cyclic compound is a phenylthiohydantoin (PTH) amino acid, which can be identified by chromatographic procedures. Furthermore, the amino acid composition of the shortened peptide:

(Arg, Asp, Gly$_2$, Phe)

can be compared with that of the original peptide:

(Ala, Arg, Asp, Gly$_2$, Phe)

The difference between these analyses is one alanine residue, which shows that alanine is the amino-terminal residue of the original peptide. The Edman procedure can then be repeated on the shortened peptide. The amino acid analysis after the second round of degradation is

(Arg, Asp, Gly, Phe)

showing that the second residue from the amino end is glycine. This conclusion can be confirmed by chromatographic identification of PTH-glycine obtained in the second round of the Edman degradation. Three more rounds of the Edman degradation will reveal the complete sequence of the original peptide.

Analyses of protein structures have been markedly accelerated by the development of *sequenators*, which are automated instruments for the determination of amino acid sequence. In a liquid-phase sequenator, a thin film of protein in a spinning cylindrical cup is subjected to the Edman degradation. The reagents and extracting solvents are passed over the immobilized film of protein, and the released PTH-amino acid is identified by high-performance liquid chromatography (HPLC; Figure 3-19). One cycle of the Edman degradation is carried out in less

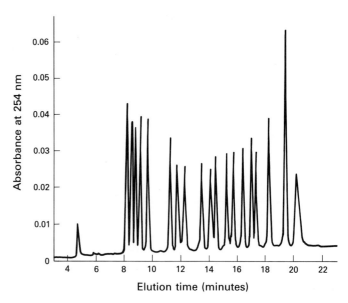

Figure 3-19
PTH-amino acids can be rapidly separated by high-pressure liquid chromatography (HPLC). In this HPLC profile, a mixture of PTH-amino acids is clearly resolved into its components. An individual amino acid can be identified by comparing its profile with this one.

than two hours. By repeated degradations, the amino acid sequence of some fifty residues in a protein can be determined. A recently devised gas-phase sequenator can analyze picomole quantities of peptides and proteins. This high sensitivity makes it feasible to analyze the sequence of a protein sample eluted from a single band of an SDS-polyacrylamide gel.

PROTEINS CAN BE SPECIFICALLY CLEAVED INTO SMALL PEPTIDES TO FACILITATE ANALYSIS

Peptides much longer than about fifty residues cannot be reliably sequenced by the Edman method because not quite all peptides in the reaction mixture release the amino acid derivative at each step. If the efficiency of release of each round were 98%, the proportion of "correct" amino acid released after sixty rounds would be only 0.3 (0.98^{60})—a hopelessly impure mix. This obstacle can be circumvented by specifically cleaving a protein into peptides not much longer than fifty residues. In essence, the strategy is to divide and conquer.

Specific cleavage can be achieved by chemical or enzymatic methods. For example, Bernhard Witkop and Erhard Gross discovered that *cyanogen bromide* (CNBr) splits polypeptide chains only on the carboxyl side of methionine residues (Figure 3-20). A protein that has ten methionines will usually yield eleven peptides on cleavage with CNBr. Highly specific cleavage is also obtained with trypsin, a proteolytic enzyme from pancreatic juice. Trypsin cleaves polypeptide chains on the carboxyl side of arginine and lysine residues (Figure 3-21). A protein

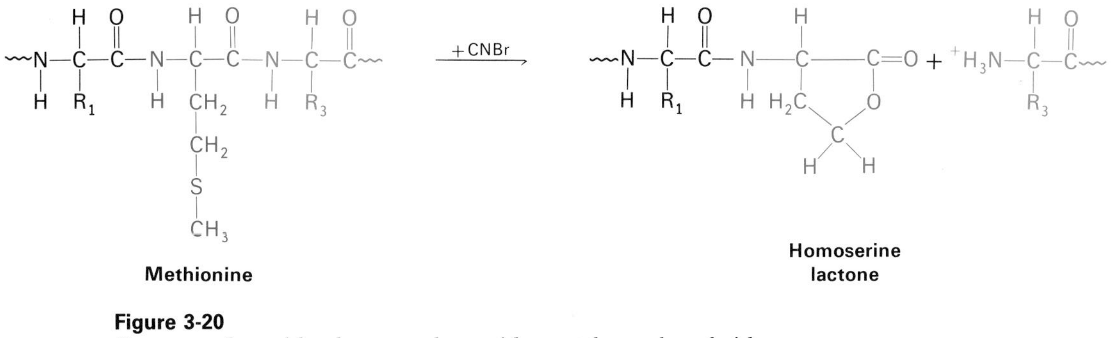

Figure 3-20
Cyanogen bromide cleaves polypeptides on the carboxyl side of methionine residues.

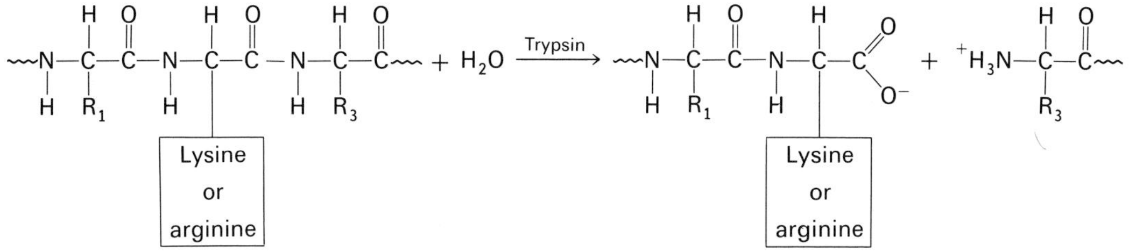

Figure 3-21
Trypsin hydrolyzes polypeptide on the carboxyl side of arginine and lysine residues.

that contains nine lysines and seven arginines will usually yield seventeen peptides on digestion with trypsin. Each of these tryptic peptides, except for the carboxyl-terminal peptide of the protein, will end with either arginine or lysine. Several other ways of specifically cleaving polypeptide chains are given in Table 3-1.

Table 3-1
Specific cleavage of polypeptides

Reagent	Cleavage site
Chemical cleavage	
Cyanogen bromide	Carboxyl side of methionine residues
O-Iodosobenzoate	Carboxyl side of tryptophan residues
Hydroxylamine	Asparagine–glycine bonds
2-Nitro-5-thiocyanobenzoate	Amino side of cysteine residues
Enzymatic cleavage	
Trypsin	Carboxyl side of lysine and arginine residues
Clostripain	Carboxyl side of arginine residues
Staphylococcal protease	Carboxyl side of aspartate and glutamate residues (glutamate only under certain conditions)

The peptides obtained by specific chemical or enzymatic cleavage are separated by chromatography. The sequence of each purified peptide is then determined by the Edman method. At this point, the amino acid sequences of segments of the protein are known, but the order of these segments is not yet defined. The necessary additional information is obtained from *overlap peptides* (Figure 3-22). An enzyme different from trypsin is used to split the polypeptide chain at other linkages. For example, chymotrypsin cleaves preferentially on the carboxyl side of aromatic and some other bulky nonpolar residues. Because these chymotryptic peptides overlap two or more tryptic peptides, they can be used to establish the order of the peptides. The entire amino acid sequence of the polypeptide chain is then known.

Figure 3-22
The peptide obtained by chymotrypic digestion overlaps two tryptic peptides, which thus establishes their order.

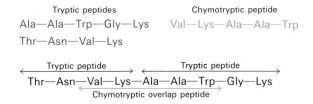

These methods apply to a protein consisting of a single polypeptide chain devoid of disulfide bonds. Additional steps are necessary if a protein has disulfide bonds or more than one chain. For a protein made up of two or more polypeptide chains held together by noncovalent bonds, denaturing agents, such as urea or guanidine hydrochloride, are used to dissociate the chains. The dissociated chains must be separated before sequence determination can begin. Polypeptide chains linked by disulfide bonds are first separated by reduction with β-mercaptoethanol or dithiothreitol. To prevent the cysteine residues

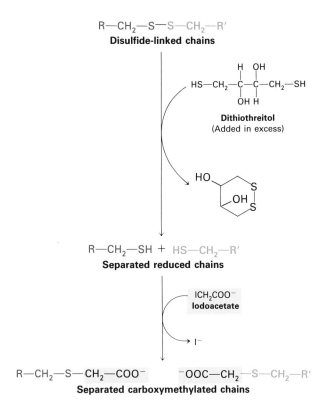

Figure 3-23
Polypeptides linked by disulfide bonds can be separated by reduction with dithiothreitol followed by alkylation.

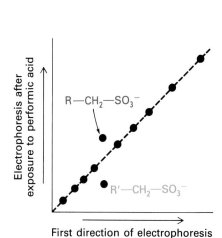

Figure 3-24
Detection of peptides joined by disulfides by diagonal electrophoresis. The mixture of peptides is electrophoresed in the horizontal direction before treatment with performic acid, and then in the vertical direction.

from recombining, they are then alkylated with iodoacetate to form stable *S*-carboxymethyl derivatives (Figure 3-23).

The positions of disulfide bonds can be determined by a *diagonal electrophoresis* technique (Figure 3-24). First, the protein is specifically cleaved into peptides under conditions in which the disulfides stay intact. The mixture of peptides is electrophoresed, and the resulting sheet is exposed to vapors of performic acid, which cleaves disulfides and converts them into cysteic acid residues. Peptides originally linked by disulfides are now independent and also more acidic because of the formation of an SO_3^- group. This mixture is electrophoresed in the perpendicular direction under the same conditions as in the first electrophoresis. Peptides that were devoid of disulfides will have the same mobility as before, and consequently all will be located on a single diagonal line. In contrast, the newly formed peptides containing cysteic acid will usually migrate differently from their parent disulfide-linked peptide and hence will lie off the diagonal.

RECOMBINANT DNA TECHNOLOGY HAS REVOLUTIONIZED PROTEIN SEQUENCING

Hundreds of proteins have been sequenced by Edman degradation of peptides derived from specific cleavages. Protein sequence determination is a demanding and time-consuming process. The elucidation of the sequence of large proteins, those with more than 1000 residues, usually requires heroic effort. Fortunately, a complementary experimental approach based on recombinant DNA technology has become available. As will be discussed in Chapter 6, long stretches of DNA can

DNA sequence	Amino acid sequence
G G G	Gly
T T C	Phe
T T G	Leu
G G A	Gly
G C A	Ala
G C A	Ala
A G G	Gly
A A G	Ser
C A C	Thr
T A T	Met
G G G	Gly
G C A	Ala

Figure 3-25
The complete nucleotide sequence of the AIDS (acquired immunodeficiency disease syndrome) virus was determined within a year after the isolation of the virus. A portion of the DNA sequence specified by the RNA genome of the virus is shown here with the corresponding amino acid sequence (deduced from a knowledge of the genetic code).

be cloned and sequenced. The sequence of the four kinds of bases in DNA—adenine (A), thymine (T), guanine (G), and cytosine (C)—directly reveals the amino acid sequence of the protein encoded by the gene or the corresponding messenger RNA molecule (Figure 3-25). The amino acid sequence deduced by reading the DNA sequence is that of the *nascent* protein, the direct product of the translational machinery in ribosomes that links amino acids in a sequence specified by a messenger RNA template.

As discussed previously, many proteins are modified after synthesis (p. 24). Some have their ends trimmed, and others arise by cleavage of a larger initial polypeptide chain. Cysteine residues in some proteins are oxidized to form disulfide links, which can be either within a chain or between polypeptide chains. Specific side chains of some proteins are altered. For example, proteins targeted to membranes usually contain carbohydrate units attached to specific asparagine side chains. Amino acid sequences derived from DNA sequences are rich in information, but they do not disclose such posttranslational modifications. Chemical analyses of proteins themselves are needed to delineate the nature of these changes, which are critical for the biological activities of most proteins. *Thus, DNA sequencing and protein chemical analyses are complementary approaches toward elucidating the structural basis of protein function.* Recombinant DNA technology is producing a wealth of amino acid sequence information at a remarkable rate. The sequences of more than 10^6 residues in about 3000 proteins are now known. At the present pace, the number of known sequences is expected to double in less than two years.

AMINO ACID SEQUENCES PROVIDE MANY KINDS OF INSIGHTS

Amino acid sequences can be highly informative in a variety of ways:

1. *The sequence of a protein of interest can be compared with all other known ones to ascertain whether significant similarities exist. Does this protein belong to one of the established families?* For example, myoglobin and hemoglobin belong to the globin family. Chymotrypsin and trypsin are members of the serine protease family, a clan of proteolytic enzymes that have a common catalytic mechanism based on a reactive serine residue. A search for kinship between a newly sequenced protein and several thousand previously sequenced ones takes about twenty minutes on a personal computer. Quite unexpected results sometimes emerge from such comparisons. For example, a viral protein that produces cancer in susceptible hosts was found to be nearly identical to a normal cellular growth factor (p. 877). This startling finding advanced the understanding of both oncogenic viruses (cancer-producing viruses) and the normal cell cycle. Comparison of amino acid sequences has also revealed that many larger proteins of higher organisms are built of domains that have come together by the fusion of gene segments. Proteins with new properties have arisen from novel combinations of these modules.

2. *Comparison of sequences of the same protein in different species yields a wealth of information about evolutionary pathways.* Genealogical relations between species can be inferred from sequence differences between their proteins, and the time of divergence of two evolutionary lines can be estimated because of the clocklike nature of random mutations. For

example, a comparison of serum albumins of primates indicates that human beings and African apes diverged only five million years ago, not thirty million years ago as was previously thought. These sequence analyses have opened a new perspective on the fossil record and the pathway of human evolution.

3. *Amino acid sequences can be searched for the presence of internal repeats.* Many proteins apparently have arisen by duplication of a primordial gene followed by its diversification. For example, antibody molecules are built of a series of similar domains, each consisting of about 108 residues (Figure 3-26). Each 25-kd light chain of antibodies is constructed from two of these modules, and each 50-kd heavy chain from four of them. The amino acid sequences of proteins express their evolutionary history.

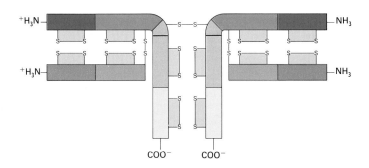

Figure 3-26
Antibody molecules consist of domains that are variations on a common theme produced by gene duplication and diversification. The pattern of disulfide bonds within the domains has been highly conserved.

4. *Amino acid sequences contain signals that determine the destination of proteins and control their processing.* Many proteins destined for export from a cell or for a membrane location contain a signal sequence, a stretch of about twenty hydrophobic residues near the amino terminus. Potential sites for the addition of carbohydrate units to asparagine residues can be identified by finding Asn-X-Ser and Asn-X-Thr in the sequence (X denotes any residue). Pairs of basic residues, such as Arg-Arg, mark potential sites of proteolytic cleavage, as in proinsulin, the precursor of insulin.

5. *Sequence data provide a basis for preparing antibodies specific for a protein of interest.* Specific antibodies can be very useful in determining the amount of a protein, ascertaining its distribution within a cell, and cloning its gene (p. 62).

6. *Amino acid sequences are also valuable for making DNA probes that are specific for the genes encoding the corresponding proteins* (p. 131). Protein sequencing is an integral part of molecular genetics, just as DNA cloning is central to the analysis of protein structure and function.

X-RAY CRYSTALLOGRAPHY REVEALS THREE-DIMENSIONAL STRUCTURE IN ATOMIC DETAIL

The understanding of protein structure and function has been greatly enriched by x-ray crystallography, a technique that can reveal the precise three-dimensional positions of most of the atoms in a protein molecule. Let us consider some basic aspects of this powerful method. First, crystals of the protein of interest are needed. Crystals can often be obtained by adding ammonium sulfate or another salt to a concentrated

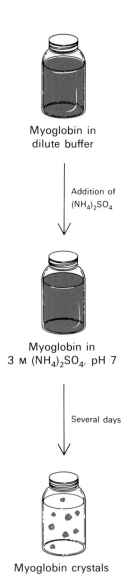

Figure 3-27
Crystallization of myoglobin.

solution of protein to reduce its solubility. For example, myoglobin crystallizes in 3 M ammonium sulfate (Figure 3-27). Slow salting-out favors the formation of highly ordered crystals instead of amorphous precipitates. Some proteins crystallize readily, whereas others do so only after much effort has been expended in finding the right conditions. Crystallization is an art; the best practitioners have great perseverance and patience, as well as a golden touch. Increasingly large and complex proteins are now being crystallized. For example, polio virus, an 8500-kd complex consisting of 240 protein subunits surrounding an RNA core, has been crystallized and its structure solved by x-ray methods.

The three components in an x-ray crystallographic analysis are a *source of x-rays*, a *protein crystal*, and a *detector* (Figure 3-28). A beam of x-rays of wavelength 1.54 Å is produced by accelerating electrons against a copper target. A narrow beam of x-rays strikes the protein crystal. Part of it goes straight through the crystal; the rest is *scattered* in various directions. The scattered (or *diffracted*) beams can be detected by x-ray film, the blackening of the emulsion being proportional to the intensity of the scattered x-ray beam, or by a solid-state electronic detector. The basic physical principles underlying the technique are:

1. *Electrons scatter x-rays.* The amplitude of the wave scattered by an atom is proportional to its number of electrons. Thus, a carbon atom scatters six times as strongly as a hydrogen atom.

2. *The scattered waves recombine.* Each atom contributes to each scattered beam. The scattered waves reinforce one another at the film or detector if they are in phase (in step) there and they cancel one another if they are out of phase.

3. *The way in which the scattered waves recombine depends only on the atomic arrangement.*

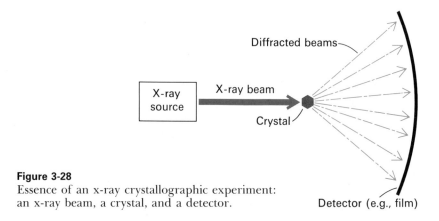

Figure 3-28
Essence of an x-ray crystallographic experiment: an x-ray beam, a crystal, and a detector.

Figure 3-29
X-ray precession photograph of a myoglobin crystal.

The protein crystal is mounted in a capillary and positioned in a precise orientation with respect to the x-ray beam and the film. Precessional motion of the crystal results in an x-ray photograph consisting of a regular array of spots called reflections. The x-ray photograph shown in Figure 3-29 is a two-dimensional section through a three-dimensional array of 25,000 spots. The intensity of each spot is measured. These *intensities* are the basic experimental data of an x-ray crystallographic analysis. The next step is to reconstruct an image of the protein from the observed intensities. In light microscopy or electron microscopy, the diffracted beams are focused by lenses to directly form an image. However, lenses for focusing x-rays do not exist. Instead, the image is formed by applying a mathematical relation called a Fourier transform. For each spot, this operation yields a wave of electron density, whose amplitude is proportional to the square root of the observed intensity of the spot. Each wave also has a *phase*—that is, the timing of its crests and troughs relative to those of other waves. The phase of each wave determines whether it reinforces or cancels the waves contributed by the other spots. These phases can be deduced from the well-understood diffraction patterns produced by heavy-atom reference markers such as uranium or mercury at specific sites in the protein.

The stage is then set for the calculation of an electron-density map, which gives the density of electrons at a large number of regularly spaced points in the crystal. This three-dimensional electron-density distribution is represented by a series of parallel sections stacked on top of each other. Each section is a transparent plastic sheet (or a layer in a computer image) on which the electron-density distribution is represented by contour lines (Figure 3-30), like the contour lines used in geological survey maps to depict altitude (Figure 3-31). The next step is to interpret the electron-density map. A critical factor is the *resolution* of the x-ray analysis, which is determined by the number of scattered intensities used in the Fourier synthesis. The fidelity of the image depends on the resolution of the Fourier synthesis, as shown by the optical analogy in Figure 3-32. A resolution of 6 Å reveals the course of the polypeptide chain but few other structural details. The reason is that polypeptide chains pack together so that their centers are between 5 and 10 Å apart. Maps at higher resolution are needed to delineate groups of atoms, which lie from 2.8 to 4.0 Å apart, and individual atoms, which are between 1.0 and 1.5 Å apart. The ultimate resolution of an x-ray analysis is determined by the degree of perfection of the crystal. For proteins, this limiting resolution is usually about 2 Å.

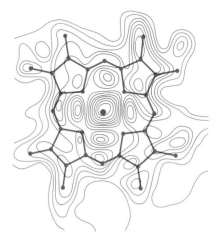

Figure 3-30
Section from the electron-density map of myoglobin showing the heme group. The peak of the center of this section corresponds to the position of the iron atom. [From J. C. Kendrew. The three-dimensional structure of a protein molecule. Copyright © 1961 by Scientific American, Inc. All rights reserved.]

Figure 3-31
Section from a U.S. Geological Survey Map of the Capitol Peak Quadrangle, Colorado.

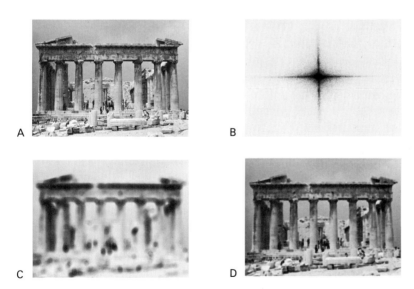

Figure 3-32
Effect of resolution on the quality of a reconstructed image, as shown by an optical analog of x-ray diffraction: (A) a photograph of the Parthenon; (B) an optical diffraction pattern of the Parthenon; (C and D) images reconstructed from the pattern in B. More data were used to obtain D than C, which accounts for the higher quality of the image in D. [Courtesy of Dr. Thomas Steitz (part A) and Dr. David De Rosier (part B).]

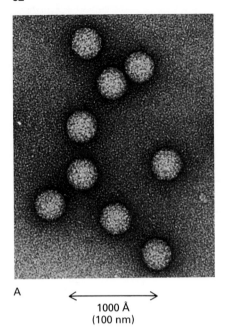

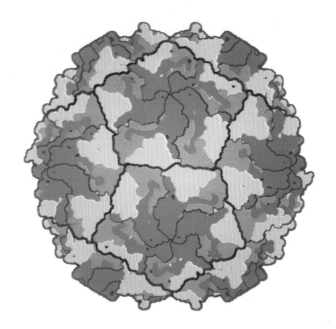

Figure 3-33
Polio virus. (A) Electron micrograph. (B) Model based on x-ray crystallographic analysis. [Part A courtesy of Dr. T. W. Jeng and Dr. Wah Chiu. Part B (computer model and photograph) courtesy of Dr. Arthur J. Olson, Research Institute of Scripps Clinic, © 1987.]

The structures of more than 200 proteins have been elucidated at atomic resolution. Knowledge of their detailed molecular architecture has provided insight into how proteins recognize and bind other molecules, how they function as enzymes, how they fold, and how they evolved. This extraordinarily rich harvest is continuing at a rapid pace and profoundly influencing the entire field of biochemistry. Moreover, x-ray crystallography is being complemented by nuclear magnetic resonance (NMR) spectroscopy, electron microscopy, and electron crystallography, in obtaining increasingly informative views of biomolecules at high resolution. We shall be looking at the three-dimensional structures of proteins and other biomolecules throughout this book and relating the architecture of these molecules to their biological function.

PROTEINS CAN BE QUANTITATED AND LOCALIZED BY HIGHLY SPECIFIC ANTIBODIES

An *antibody* is a protein synthesized by an animal in response to the presence of a foreign substance, called an *antigen* (Chapter 35). Antibodies (also called *immunoglobulins*) have specific affinity for the antigens that elicited their synthesis. Proteins, polysaccharides, and nucleic acids are effective antigens. Antibodies can also be formed to small molecules, such as synthetic peptides, provided that the small molecules are attached to a macromolecular carrier. The group recognized by an antibody is called an *antigenic determinant* (or *epitope*). Animals have a very large repertoire of antibody-producing cells, each producing antibody of a single specificity. An antigen acts by stimulating the proliferation of the small number of cells that were already forming complementary antibody. The major type of antibody in blood plasma is *immunoglobulin G*, a 150-kd protein containing two identical sites for the binding of antigen (Figure 3-34).

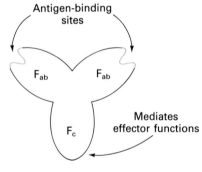

Figure 3-34
Diagram of immunoglobulin G (IgG), the major class of antibody molecules in blood plasma. IgG contains two antigen-binding F_{ab} units and an F_c unit that mediates effector functions such as the lysis of cell membranes.

Antibodies that recognize a particular protein can be obtained by injecting the protein into a rabbit twice, three weeks apart. Blood is drawn from the immunized rabbit several weeks later and centrifuged. The resulting serum, called an *antiserum*, usually contains the desired antibody. The antiserum or its immunoglobulin G fraction can be used directly. Alternatively, antibody molecules specific for the antigen can be purified by affinity chromatography. Antibodies produced in this way are *polyclonal*—that is, they are products of many different populations of antibody-producing cells and hence differ somewhat in their precise specificity and affinity for the antigen. A major advance of recent years is the discovery of a means of producing *monoclonal antibodies* of virtually any desired specificity (p. 896). Monoclonal antibodies, in contrast with polyclonal ones, are homogeneous because they are synthesized by a population of identical cells (a clone). Each such population is descended from a single *hybridoma cell* formed by fusing an antibody-producing cell with a tumor cell that has the capacity for unlimited proliferation.

Closely related proteins can be distinguished by antibodies; indeed, a difference of just one residue on the surface can be detected. Antibodies can be used as exquisitely specific analytic reagents to quantitate the amount of a protein or other antigen. In a *solid-phase immunoassay*, antibody specific for a protein of interest is attached to a polymeric support such as a sheet of polyvinylchloride (Figure 3-35). A drop of cell extract or a sample of serum or urine is laid on the sheet, which is washed after formation of the antibody-antigen complex. Antibody specific for a different site on the antigen is then added, and the sheet is again washed. This second antibody carries a radioactive or fluorescent label so that it can be detected with high sensitivity. The amount of second antibody bound to the sheet is proportional to the quantity of antigen in the sample. The sensitivity of the assay can be enhanced even further if the second antibody is attached to an enzyme such as alkaline phosphatase. This enzyme can rapidly convert an added colorless substrate into a colored product, or a nonfluorescent substrate into an intensely fluorescent product (Figure 3-36). Less than a nanogram (10^{-9} g) of a protein can readily be measured by such an *enzyme-linked immunosorbent assay* (ELISA), which is rapid and convenient. For example, pregnancy can be detected within a few days after conception by immunoassaying urine for the presence of human chorionic gonadotropin (hCG), a 37-kd protein hormone produced by the placenta.

Very small quantities of a protein of interest in a cell or in body fluid can be detected by an immunoassay technique called *Western blotting* (Figure 3-37). A sample is electrophoresed on an SDS polyacrylamide

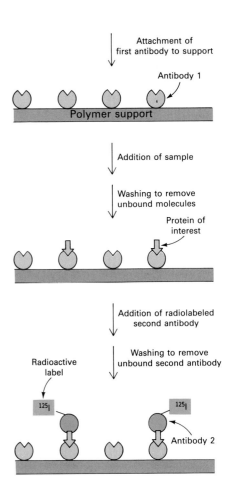

Figure 3-35
Solid-phase immunoassay. The steps are: coupling of specific antibody to a solid support, addition of the sample, washing to remove soluble compounds, and addition of a radiolabeled second antibody specific for a different site on the protein being detected.

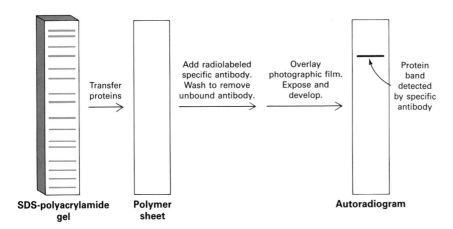

Figure 3-37
Detection of a protein on a gel by Western blotting. Proteins on an SDS-polyacrylamide gel are transferred to a polymer sheet and stained with radioactive antibody. A dark band corresponding to the protein of interest appears in the autoradiogram.

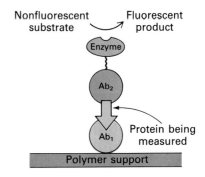

Figure 3-36
Enzyme-linked immunosorbent assay (ELISA). The steps are the same as in the immunoassay described in Figure 3-35 except that an enzyme instead of a radiolabel is attached to the second antibody. An intensely colored or fluorescent compound is formed by the catalytic action of this enzyme.

gel. The resolved proteins on the gel are transferred (by blotting) to a sheet to make them more accessible for reaction with a subsequently added antibody that is specific for the protein of interest. The antibody-antigen complex on the sheet then can be detected by rinsing the sheet with a second antibody specific for the first (e.g., goat antibody that recognizes mouse antibody). A radioactive label on the second antibody produces a dark band on x-ray film (an autoradiogram). Alternatively, an enzyme on the second antibody generates a colored product, as in the ELISA method. Western blotting makes it possible to find a protein in a complex mixture, the proverbial needle in a haystack. This technique is used advantageously in the cloning of genes (p. 133).

Antibodies are also valuable in determining the spatial distribution of antigens. Cells can be stained with fluorescent-labeled antibodies and examined by *fluorescence microscopy* to reveal the localization of a protein of interest. For example, arrays of parallel bundles are evident in cells stained with antibody specific for actin, a protein that polymerizes into filaments (Figure 3-38). Actin filaments are constituents of the cytoskel-

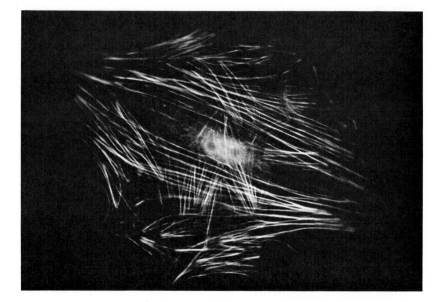

Figure 3-38
Fluorescence micrograph of actin filaments in a cell stained with an antibody specific to actin. [Courtesy of Dr. Elias Lazarides.]

eton, the internal scaffold of cells that controls their shape and movement. The finest resolution of fluorescence microscopy is about 0.2 μm (200 nm or 2000 Å) because of the wavelength of visible light. Finer spatial resolution can be achieved by electron microscopy using antibodies tagged with electron-dense markers. For example, ferritin conjugated to an antibody can readily be visualized by electron microscopy because it contains an electron-dense core of iron hydroxide. Clusters of gold can also be conjugated to antibodies to make them highly visible under the electron microscope. *Immunoelectron microscopy* can define the position of antigens to a resolution of 10 nm (100 Å) or finer (Figure 3-39).

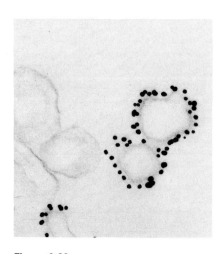

Figure 3-39
The opaque 150 Å (15 nm) diameter particles in this electron micrograph are clusters of gold atoms bound to antibody molecules. These membrane vesicles from the synapses of neurons contain a channel protein that is recognized by the specific antibody. [Courtesy of Dr. Peter Sargent.]

PEPTIDES CAN BE SYNTHESIZED BY AUTOMATED SOLID-PHASE METHODS

Synthesizing peptides of defined sequence is important for several reasons. First, *synthetic peptides can reveal the rules governing the three-dimensional conformation of proteins*. We can ask whether a particular sequence by itself folds into an α-helix, β-strand, or hairpin turn or behaves as a random coil. Second, *many hormones and other signal molecules in nature are peptides.* For example, white blood cells are attracted to bacteria by formylmethionyl peptides that come from the breakdown of bacterial proteins. Synthetic formyl-methionyl peptides have been useful in identifying the cell-surface receptor for this class of peptides. Synthetic peptides can be attached to agarose beads to prepare affinity chromatography columns for the purification of receptor proteins that specifically recognize the peptides. Third, *synthetic peptides can serve as drugs.* Vasopressin is a peptide hormone that stimulates the reabsorption of water in the distal tubules of the kidney, leading to the formation of a more concentrated urine. Patients with diabetes insipidus are deficient in vasopressin (also called antidiuretic hormone), and so they excrete large volumes of urine (more than 5 liters per day) and are continually thirsty because of this massive loss of fluid. This defect can be treated by administering 1-desamino-8-D-arginine vasopressin, a synthetic analog of the missing hormone (Figure 3-40). This synthetic peptide is degraded in vivo much more slowly than vasopressin and, additionally, does not increase the blood pressure. Fourth, *synthetic peptides can serve as antigens to stimulate the formation of specific antibodies.*

fMet peptide

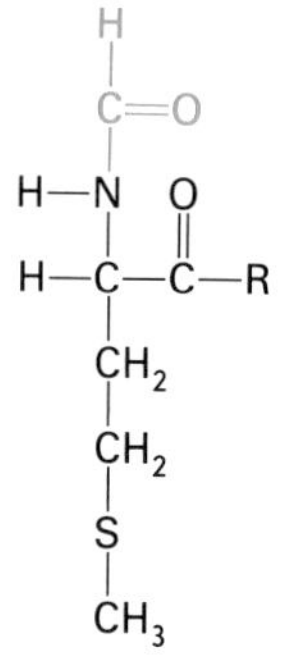

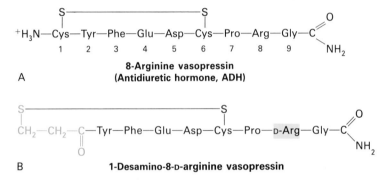

Figure 3-40
Structural formulas of (A) vasopressin (also called antidiuretic hormone), a peptide hormone that stimulates water resorption, and (B) 1-desamino-8-D-arginine vasopressin, a more stable synthetic analog.

Peptides are synthesized by linking an amino group to a carboxyl group that has been activated by reacting it with a reagent such as *dicyclohexylcarbodiimide* (DCC) (Figure 3-41). The attack of a free amino group on the activated carboxyl leads to the formation of a peptide bond and the release of dicylohexylurea. A unique product is formed only if a single amino group and a single carboxyl group are available for reaction. Hence, it is necessary to *block* (protect) all other potentially reactive groups. For example, the α-amino group of the component containing the activated carboxyl group can be blocked with a *tert*-butyloxycarbonyl (*t*-Boc) group. This *t*-Boc protecting group can be subsequently removed by exposing the peptide to dilute acid, which leaves peptide bonds intact.

Dicyclohexylcarbodiimide (DCC)

***t*-Butyloxycarbonyl amino acid (*t*-Boc amino acid)**

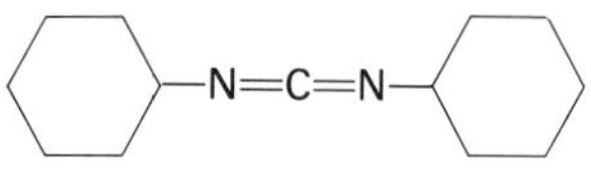

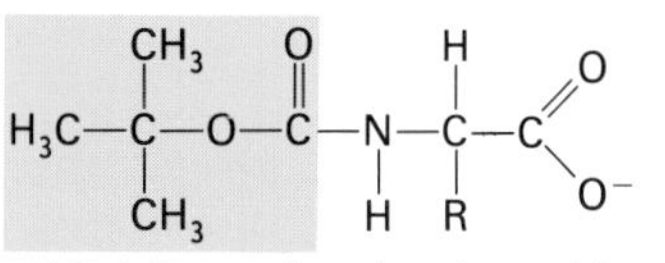

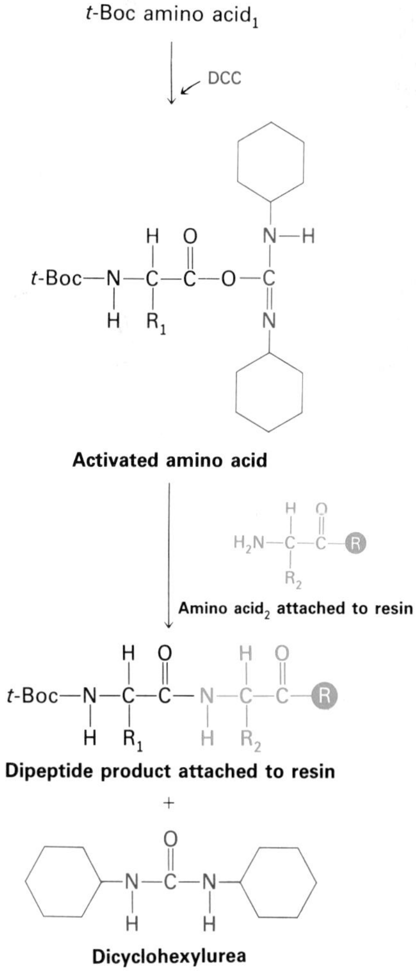

Figure 3-41
Dicyclohexylcarbodiimide is used to activate carboxyl groups for the formation of peptide bonds.

Peptides can be readily synthesized by a *solid-phase method* devised by R. Bruce Merrifield. Amino acids are added stepwise to a growing peptide chain that is linked to an insoluble matrix, such as polystyrene beads. A major advantage of this solid-phase method is that the desired product at each stage is bound to beads that can be rapidly filtered and washed and so the need to purify intermediates is obviated. All of the reactions are carried out in a single vessel, which eliminates losses due to repeated transfers of products. The carboxyl-terminal amino acid of the desired peptide sequence is first anchored to the polystyrene beads (Figure 3-42). The *t*-Boc protecting group of this amino acid is then removed. The next amino acid (in the protected *t*-Boc form) is added together with dicyclohexylcarbodiimide, the coupling agent. After formation of the peptide bond, excess reagents and dicyclohexylurea are washed away, which leaves the beads with the desired dipeptide prod-

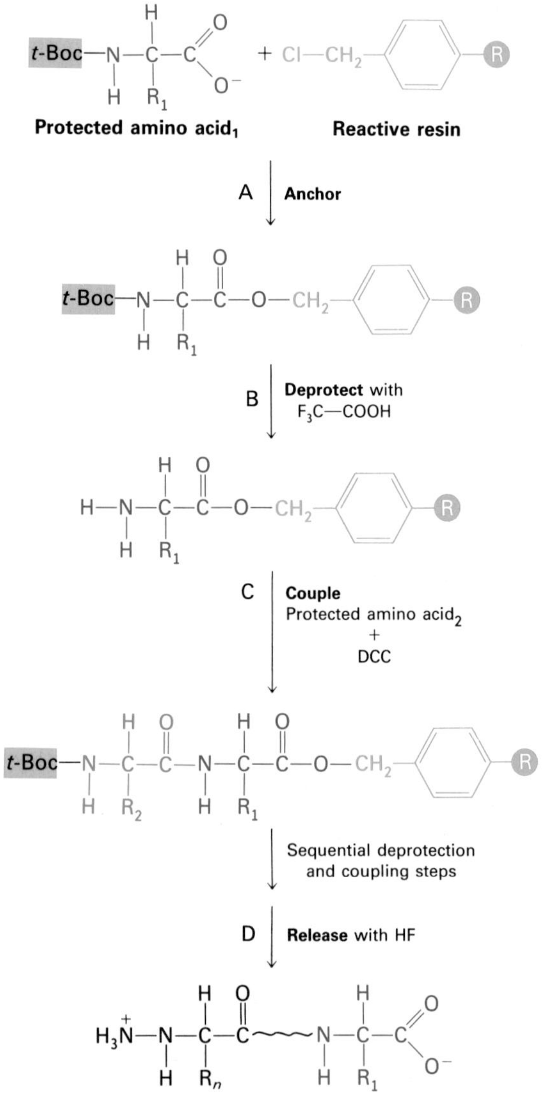

Figure 3-42
Solid-phase peptide synthesis. The steps are (A) anchoring of the C-terminal amino acid to a resin on the surface of polystyrene beads, (B) deprotection of the amino group, (C) addition of the next amino acid (in the protected form) and dicyclohexylcarbodiimide (DCC). Steps (B) and (C) are repeated for each added amino acid. The beads are washed to remove excess reagents and unwanted products after each step. Finally, (D) the completed peptide is released from the resin.

uct. Additional amino acids are linked by the same sequence of reactions. At the end of the synthesis, the peptide is released from the beads by adding HF, which cleaves the carboxyl ester anchor without disrupting peptide bonds. Protecting groups on potentially reactive side chains, such as that of lysine, are also removed at this time. This cycle of reactions can readily be automated, which makes it feasible to routinely synthesize peptides containing about 50 residues in good yield and purity. In fact, Merrifield has synthesized interferons (155 residues) that have antiviral activity and ribonuclease (124 residues) that is catalytically active.

SUMMARY

The purification of a protein is an essential step in elucidating its structure and function. Proteins can be separated from each other and from other molecules on the basis of such characteristics as size, solubility, charge, and binding affinity. SDS-polyacrylamide gel electrophoresis separates the polypeptide chains of proteins under denaturing conditions largely according to mass. Proteins can also be separated electrophoretically on the basis of net charge by isoelectric focusing in a pH gradient. Ultracentrifugation and gel-filtration chromatography resolves proteins according to size, whereas ion-exchange chromatography separates them mainly on the basis of net charge. The high affinity of many proteins for specific chemical groups is exploited in affinity chromatography, in which proteins bind to columns containing beads bearing covalently linked substrates, inhibitors, or other specifically recognized groups.

The amino acid composition of a protein can be determined by hydrolyzing it into its constituent amino acids in 6 N HCl at 110°C. The amino acids can be separated by ion-exchange chromatography and quantitated by reacting them with ninhydrin or fluorescamine. Amino acid sequences can be determined by Edman degradation, which removes one amino acid at a time from the amino end of a peptide. Phenylisothiocyanate reacts with the terminal amino group to form a phenylthiocarbamoyl derivative, which cyclizes under mildly acidic conditions to give a phenylthiohydantoin (PTH)-amino acid and a peptide shorted by one residue. This PTH-amino acid is identified by high-performance liquid chromatography (HPLC). Automated repeated Edman degradations by a sequenator can analyze sequences of about fifty residues. Longer polypeptide chains are broken into shorter ones for analysis by specifically cleaving them with a reagent such as cyanogen bromide, which splits peptide bonds on the carboxyl side of methionine residues. Enzymes such as trypsin, which cleaves on the carboxyl side of lysine and arginine residues, are also very useful in splitting proteins. Recombinant DNA techniques have revolutionized amino acid sequencing. The nucleotide sequence of DNA molecules reveals the amino acid sequence of nascent proteins encoded by them but does not disclose posttranslational modifications. Amino acid sequences are rich in information concerning the kinship of proteins, their evolutionary relations, and diseases produced by mutations. Knowledge of a sequence provides valuable clues to conformation and function.

Polypeptide chains can be synthesized by automated solid-phase methods in which the carboxyl end of the growing chain is linked to an insoluble support. The α-carboxyl group of the incoming amino acid is activated by dicyclohexylcarbodiimide and joined to the α-amino group of the growing chain. Synthetic peptides can serve as drugs and as antigens to stimulate the formation of specific antibodies. They also provide insight into relations between amino acid sequence and conformation.

Proteins can be detected and quantitated by highly specific antibodies. Enzyme-linked immunosorbent assays (ELISA) and Western blots of SDS-polyacrylamide gels are used extensively. Proteins can also be localized within cells by fluorescence microscopy and immunoelectron microscopy using labeled antibodies. X-ray crystallography has revealed the three-dimensional structures of more than a hundred proteins in atomic detail. Knowledge of molecular structure has provided insight into how proteins fold, recognize other molecules, and catalyze chemical reactions.

SELECTED READINGS

Where to start

Moore, S., and Stein, W. H., 1973. Chemical structures of pancreatic ribonuclease and deoxyribonuclease. *Science* 180:458–464.

Hunkapiller, M. W., and Hood, L. E., 1983. Protein sequence analysis: automated microsequencing. *Science* 219:650–659.

Merrifield, B., 1986. Solid phase synthesis. *Science* 232:341–347.

Books on protein chemistry

Creighton, T. E., 1983. *Proteins: Structure and Molecular Properties.* W. H. Freeman.

Cooper, T. G., 1977. *The Tools of Biochemistry.* Wiley. [A valuable guide to experimental methods in protein chemistry and in other areas of biochemistry. Principles of procedures are clearly presented.]

Hirs, C. H. W., and Timasheff, S. N., (eds.), 1983. *Enzyme Structure,* Part I. Methods in Enzymology, vol. 91. Academic Press. [An excellent collection of authoritative articles on amino acid analysis, end-group methods, chemical and enzymatic cleavage, peptide separation, sequence analysis, chemical modification, and active-site labeling. Also see volumes 47–49 in this series.]

Scopes, R., 1982. *Protein Purification: Principles and Practice.* Springer-Verlag.

Langone, J. J., and Van Vunakis, H., 1983. *Immunochemical Techniques,* Part A. Methods in Enzymology, vol. 92. Academic Press.

Physical chemistry of proteins

Cantor, C. R., and Schimmel, P. R., 1980. *Biophysical Chemistry.* W. H. Freeman. [An outstanding exposition of fundamental principles and experimental methods. Part 2 discusses many of the techniques presented in this chapter.]

Freifelder, D., 1982. *Physical Biochemistry: Applications to Biochemistry and Molecular Biology.* W. H. Freeman. [Contains a lucid discussion of ultracentrifugation.]

Amino acid sequence determination

Hunkapiller, M. W., Strickler, J. E., and Wilson, K. J., 1984. Contemporary methodology for protein structure determination. *Science* 226:304–311.

Hewick, R. M., Hunkapiller, M. W., Hood, L. E., and Dreyer, W. J., 1981. A gas-liquid solid phase peptide and protein sequenator. *J. Biol. Chem.* 256:7990–7997.

Konigsberg, W. H., and Steinman, H. M., 1977. Strategy and methods of sequence analysis. *In* Neurath, H., and Hill, R. L., (eds.), *The Proteins* (3rd ed.), vol. 3, pp. 1–178. Academic Press.

Stein, S., and Udenfriend, S., 1984. A picomole protein and peptide chemistry: some applications to the opiod peptides. *Analy. Chem.* 136:7–23.

Sequence comparisons and molecular evolution

Doolittle, R. F., 1981. Similar amino acid sequences: chance or common ancestry? *J. Mol. Biol.* 16:9–16.

Lipman, D. J., and Pearson, W. R., 1985. Rapid and sensitive protein similarity searches. *Science* 227:1435–1441. [Description of an algorithm that searches for similarities between an amino acid sequence and a large database of previously determined sequences. This program can be run on a personal computer.]

Wilson, A. C., 1985. The molecular basis of evolution. *Sci. Amer.* 253(4):164.

X-ray crystallography and NMR spectroscopy

Matthews, B. W., 1977. X-ray structure of proteins. *In* Neurath, H., and Hill, R. L., (eds.), *The Proteins* (3rd ed.), vol. 3, pp. 404–590. Academic Press.

Holmes, K. C., and Blow, D. M., 1965. *The Use of X-ray Diffraction in the Study of Protein and Nucleic Acid Structure.* Wiley-Interscience.

Glusker, J. P., and Trueblood, K. N., 1972. *Crystal Structure Analysis: A Primer.* Oxford University Press. [A lucid and concise introduction to x-ray crystallography in general.]

Kline, A. D., Braun, W., and Wuthrich, K., 1986. Studies by ^{1}H nuclear magnetic resonance and distance geometry of the solution conformation of the α-amylase inhibitor tendamistat. *J. Mol. Biol.* 189:377–382. [Determination of the three-dimensional structure of a small protein by NMR spectroscopy. This important study demonstrates the power of this approach in elucidating conformation.]

PROBLEMS

1. The following reagents are often used in protein chemistry:

CNBr	Dabsyl chloride
Urea	6 N HCl
β-Mercaptoethanol	Ninhydrin
Trypsin	Phenyl isothiocyanate
Performic acid	Chymotrypsin

   Which one is the best suited for accomplishing each of the following tasks?
   (a) Determination of the amino acid sequence of a small peptide.
   (b) Identification of the amino-terminal residue of a peptide (of which you have less than 10^{-7} g).
   (c) Reversible denaturation of a protein devoid of disulfide bonds. Which additional reagent would you need if disulfide bonds were present?
   (d) Hydrolysis of peptide bonds on the carboxyl side of aromatic residues.
   (e) Cleavage of peptide bonds on the carboxyl side of methionines.
   (f) Hydrolysis of peptide bonds on the carboxyl side of lysine and arginine residues.

2. What is the ratio of base to acid at pH 4, 5, 6, 7, and 8 for an acid with a pK of 6?

3. Anhydrous hydrazine has been used to cleave peptide bonds in proteins. What are the reaction products? How might this technique be used to identify the carboxyl-terminal amino acid?

4. The amino acid sequence of human adrenocorticotropin, a polypeptide hormone, is

   Ser-Tyr-Ser-Met-Glu-His-Phe-Arg-Trp-Gly-Lys-Pro-Val-Gly-Lys-Lys-Arg-Arg-Pro-Val-Lys-Val-Tyr-Pro-Asp-Ala-Gly-Glu-Asp-Gln-Ser-Ala-Glu-Ala-Phe-Pro-Leu-Glu-Phe

   (a) What is the approximate net charge of this molecule at pH 7? Assume that its side chains have the pK values given in Table 2-1 (p. 21) and that the pKs of the terminal —NH_3^+ and —COOH groups are 7.8 and 3.6, respectively.
   (b) How many peptides result from the treatment of the hormone with cyanogen bromide?

5. Ethyleneimine reacts with cysteine side chains in proteins to form S-aminoethyl derivatives. The peptide bonds on the carboxyl side of these modified cysteine residues are susceptible to hydrolysis by trypsin. Why?

6. The absorbance A of a solution is defined as

   $$A = \log_{10}(I_0/I)$$

   in which I_0 is the incident light intensity and I is the transmitted light intensity. The absorbance is related to the molar absorption coefficient (extinction coefficient) ϵ (in cm^{-1} M^{-1}), concentration c (in M), and path length l (in cm) by

   $$A = \epsilon l c$$

   The absorption coefficient of myoglobin at 580 nm is 15,000 cm^{-1} M^{-1}. What is the absorbance of a 1 mg/ml solution across a 1-cm path? What percentage of the incident light is transmitted by this solution?

7. Tropomyosin, a 93-kd muscle protein, sediments more slowly than does hemoglobin (65 kd). Their sedimentation coefficients are 2.6 and 4.31 S, respectively. Which structural feature of tropomyosin accounts for its slow sedimentation?

8. The relative electrophoretic mobilities of a 30-kd protein and a 92-kd protein used as standards on an SDS-polyacrylamide gel are 0.80 and 0.41, respectively. What is the apparent mass of a protein having a mobility of 0.62 on this gel?

9. The relative electrophoretic mobility of a protein on an SDS-polyacrylamide gel decreases from 0.67 to 0.64 on addition of 1 mM dithiothreitol. What is a likely reason for this shift?

10. The gene encoding a protein with a single disulfide bond undergoes a mutation that changes a serine residue into a cysteine residue. You want to find out whether the disulfide pairing in this mutant is the same as in the original protein. Propose an experiment to directly answer this question.

11. A synthetic polypeptide consisting of L-lysine residues is a random coil at pH 7 but becomes α helical as the pH is raised above 10. Account for this pH-dependent conformational transition.

12. Predict the pH dependence of the helix-coil transition of poly-L-glutamate.

13. Glycine is a highly conserved amino acid residue in the evolution of many proteins. Why?

14. Suppose that you are investigating the mechanism of action of vasopressin. You want to isolate the receptor for this hormone. Which chromatographic procedure would you use? How would you carry out the experiment?

15. The sedimentation coefficient of a tetrameric enzyme decreases by 2% on binding substrate. What is your interpretation of this finding.

16. In electrophoresis, the electric force Ez drives a charged molecule toward the oppositely charged electrode. Which parameters correspond to E and z in centrifugation?

17. What is the effect of the shape of a protein on its position relative to the axis of centrifugation in (a) a sedimentation-velocity experiment, and (b) a sedimentation-equilibrium experiment?

18. In SDS-polyacrylamide gel electrophoresis, a 90-kd single-chain protein moves less rapidly than does a 40-kd one. In contrast, the 90-kd protein emerges first from a gel-filtration column. What is the reason for this difference?

19. A complex mixture of proteins from a cell extract is analyzed by Western blotting. A 23-kd and a 57-kd band are stained by one monoclonal antibody, and the same 23-kd but a different 69-kd band are stained by a second monoclonal antibody. Propose a structural basis for this result.

20. The amino-terminal region of myoglobin is not well defined in the electron-density map obtained from an x-ray analysis of this protein. Why?

21. Fluorescence-activated cell sorting (FACS) is a powerful technique for separating cells according to their content of particular molecules. For example, a fluorescent-labeled antibody specific for a cell-surface protein can be used to detect cells containing such a molecule. Suppose that you want to isolate cells that possess a receptor enabling them to detect bacterial degradation products. However, you do not yet have an antibody directed against this receptor. Which fluorescent-labeled molecule would you prepare to identify such cells?

CHAPTER 4

DNA and RNA: Molecules of Heredity

DNA is a very long, threadlike macromolecule made up of a large number of deoxyribonucleotides, each composed of a base, a sugar, and a phosphate group. *The bases of DNA molecules carry genetic information, whereas their sugar and phosphate groups perform a structural role.* This chapter presents the key experiments that revealed that DNA is the genetic material, then describes the DNA double helix. When this structure was discovered, the complementary nature of its two chains immediately suggested that each is a template for the other in DNA replication. DNA polymerases are the enzymes that replicate DNA by taking instructions from DNA templates. These exquisitely specific enzymes replicate DNA with an error frequency of less than one in a million nucleotides. The genes of all cells and many viruses are made of DNA. Some viruses, however, use RNA (ribonucleic acid) as their genetic material. This chapter concludes with examples of the genetic role of RNA in plant viruses and animal tumor viruses.

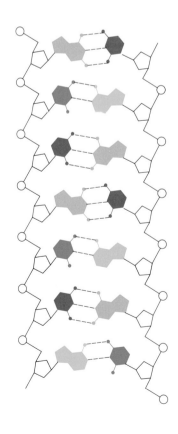

Figure 4-1
Schematic diagram of the structure of DNA. The sugar-phosphate backbone is shown in black, and the purine and pyrimidine bases are shown in color. [After A. Kornberg. The synthesis of DNA. Copyright © 1968 by Scientific American, Inc. All rights reserved.]

DNA CONSISTS OF FOUR KINDS OF BASES JOINED TO A SUGAR-PHOSPHATE BACKBONE

DNA is a polymer of deoxyribonucleotide units. A nucleotide consists of a nitrogenous base, a sugar, and one or more phosphate groups. The sugar in a deoxyribonucleotide is *deoxyribose*. The *deoxy* prefix indicates that this sugar lacks an oxygen atom that is present in ribose, the parent compound. The nitrogenous base is a derivative of *purine* or *pyrimidine*.

β-D-2-Deoxyribose   Purine   Pyrimidine

The purines in DNA are *adenine* (A) and *guanine* (G), and the pyrimidines are *thymine* (T) and *cytosine* (C).

Adenine (A)   Guanine (G)   Thymine (T)   Cytosine (C)

In a deoxyribonucleotide, the C-1 carbon atom of deoxyribose is bonded to N-1 of a pyrimidine or N-9 of a purine. The configuration of this *N*-glycosidic linkage is β (the base lies above the plane of the sugar ring). A *nucleoside* consists of a purine or pyrimidine base bonded to a sugar. The four nucleoside units in DNA are called *deoxyadenosine*, *deoxyguanosine*, *deoxythymidine*, and *deoxycytidine*. A *nucleotide* is a phosphate ester of a nucleoside. The most common site of esterification in naturally occurring nucleotides is the hydroxyl group attached to C-5 of the sugar. Such a compound is called a *nucleoside 5-phosphate* or a *5′-nucleotide*. For example, *deoxyadenosine 5′-triphosphate* (dATP) is an activated precursor in the synthesis of DNA. A primed number denotes an atom of the sugar, whereas an unprimed number denotes an atom of the purine or pyrimidine ring. The prefix *d* in dATP indicates that the sugar is deoxyribose to distinguish this compound from ATP, in which the sugar is ribose.

Deoxyadenosine
(A nucleoside)

Deoxyadenosine 5′-triphosphate
(dATP)
(A nucleotide)

The *backbone* of DNA, which is invariant throughout the molecule, consists of deoxyriboses linked by phosphate groups. Specifically, the 3'-hydroxyl of the sugar moiety of one deoxyribonucleotide is joined to the 5'-hydroxyl of the adjacent sugar by a phosphodiester bridge. The *variable part* of DNA is its *sequence of four kinds of bases (A, G, C, and T)*. The corresponding nucleotide units are called *deoxyadenylate, deoxyguanylate, deoxycytidylate,* and *deoxythymidylate.* The structure of a DNA chain is shown in Figure 4-2.

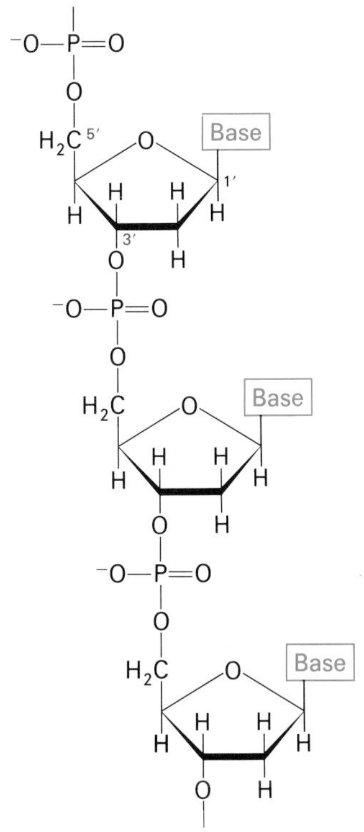

The structure of a DNA chain can be concisely represented in the following way. The symbols for the four principal deoxyribonucleosides are

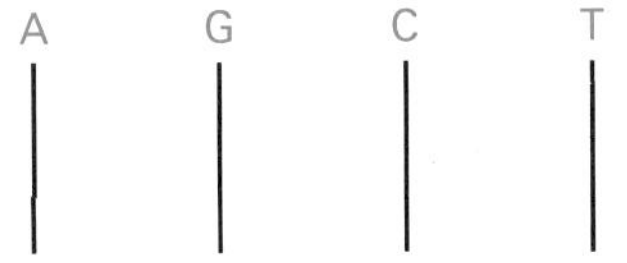

$$A \quad G \quad C \quad T$$

The bold line refers to the sugar, whereas A, G, C, and T represent the bases. The Ⓟ within the diagonal line in the diagram below denotes a phosphodiester bond. This diagonal line joins the end of one bold line and the middle of another. These junctions refer to the 5'-OH and 3'-OH, respectively. In this example, the symbol Ⓟ indicates that deoxyadenylate is linked to deoxycytidine by a phosphodiester bridge. Specifically, the 3'-OH of deoxyadenylate is joined through a phosphoryl group to the 5'-OH of deoxycytidine.

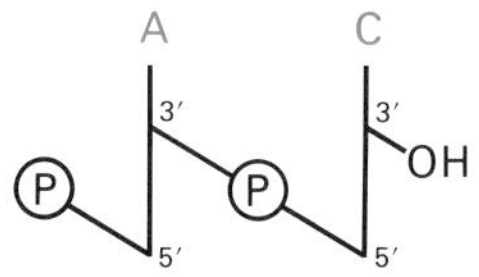

Now suppose that deoxyguanylate becomes linked to the deoxycytidine unit of this dinucleotide. The resulting trinucleotide can be represented by

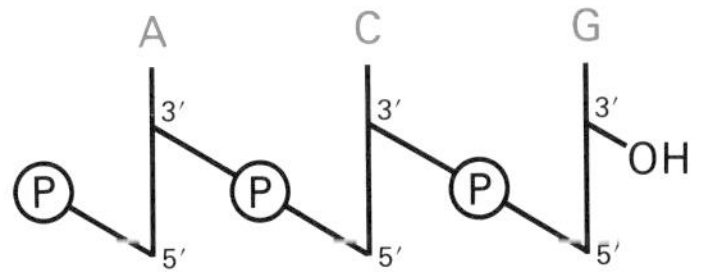

An even more abbreviated notation for this trinucleotide is pApCpG or ACG.

A DNA chain has polarity. One end of the chain has a 5'-OH group and the other a 3'-OH group that is not linked to another nucleotide. By convention, the symbol ACG means that the unlinked 5'-OH group is on deoxyadenosine, whereas the unlinked 3'-OH group is on deoxyguanosine. Thus, *the base sequence is written in the 5' → 3' direction.* Recall that the amino acid sequence of a protein is written in the amino → carboxyl direction. Note that ACG and GCA refer to different compounds, just as Glu-Phe-Ala differs from Ala-Phe-Glu.

Figure 4-2
Structure of part of a DNA chain.

TRANSFORMATION OF PNEUMOCOCCI BY DNA REVEALED THAT GENES ARE MADE OF DNA

The pneumococcus bacterium played an important part in the discovery of the genetic role of DNA. A pneumococcus is normally sur-

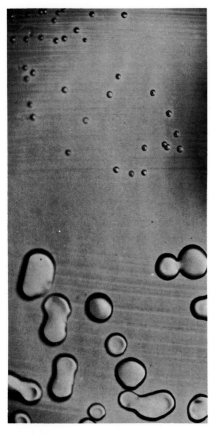

Figure 4-3
Transformation of nonpathogenic R pneumococci (small colonies) to pathogenic S pneumococci (large glistening colonies) by DNA from heat-killed S pneumococci.
[From O. T. Avery, C. M. MacLeod, and M. McCarty. *J. Exp. Med.* 79 (1944):158.]

rounded by a slimy, glistening polysaccharide capsule. This outer layer is essential for the pathogenicity of the bacterium, which causes pneumonia in humans and other susceptible mammals. Mutants devoid of a polysaccharide coat are not pathogenic. The normal bacterium is referred to as the S form (because it forms smooth colonies in a culture dish), whereas mutants without capsules are called R forms (because they form rough colonies). R mutants lack an enzyme needed for the synthesis of capsular polysaccharide.

In 1928, Fred Griffith discovered that a nonpathogenic R mutant could be *transformed* into the pathogenic S form in the following way. He injected mice with a mixture of live R and heat-killed S pneumococci. The striking finding was that this mixture was lethal to the mice, whereas either live R or heat-killed S pneumococci alone were not. The blood of the dead mice contained live S pneumococci. Thus, the heat-killed S pneumococci had somehow transformed live R pneumococci into live S pneumococci. This change was permanent: the transformed pneumococci yielded pathogenic progeny of the S form. It was then found that this R → S transformation can occur in vitro (Figure 4-3). Some of the cells in a growing culture of the R form were transformed into the S form by the addition of a *cell-free extract* of heat-killed S pneumococci. This finding set the stage for the elucidation of the chemical nature of the "transforming principle."

The cell-free extract of heat-killed S pneumococci was fractionated and the transforming activity of its components assayed. In 1944, Oswald Avery, Colin MacLeod, and Maclyn McCarty published their discovery that *"a nucleic acid of the deoxyribose type is the fundamental unit of the transforming principle of Pneumococcus Type III."* The experimental basis for their conclusion was: (1) the purified, highly active transforming principle gave an elemental chemical analysis that agreed closely with that calculated for DNA; (2) the optical, ultracentrifugal, diffusive, and electrophoretic properties of the purified material were like those of DNA; (3) there was no loss of transforming activity upon extraction of protein or lipid; (4) the polypeptide-cleaving enzymes trypsin and chymotrypsin did not affect transforming activity; (5) ribonuclease (known to digest ribonucleic acid) had no effect on the transforming principle; and (6) in contrast, transforming activity was lost following the addition of deoxyribonuclease.

This work is a landmark in the development of biochemistry. Until 1944, it was generally assumed that chromosomal proteins carry genetic information and that DNA plays a secondary role. This prevailing view was decisively shattered by the rigorously documented finding that *purified DNA has genetic specificity*. Avery gave a vivid description of this research and of its implications in a letter that he wrote in 1943 to his brother, a medical microbiologist at another university (Figure 4-4).

Further support for the genetic role of DNA came from the studies of a virus that infects the bacterium *Escherichia coli*. The T2 bacteriophage consists of a core of DNA surrounded by a protein coat. In 1951, Roger Herriott suggested that "the virus may act like a little hypodermic needle full of transforming principles; the virus as such never enters the cell; only the tail contacts the host and perhaps enzymatically cuts a small hole through the outer membrane and then the nucleic acid of the virus head flows into the cell." In 1952, this idea was tested by Alfred Hershey and Martha Chase in the following way. Phage DNA was labeled with the radioisotope ^{32}P, whereas the protein coat was labeled with ^{35}S. These labels are highly specific because DNA does not contain sulfur and the protein coat is devoid of phosphorus. A sample

For the past two years, first with MacLeod and now with Dr. McCarty, I have been trying to find out what is the chemical nature of the substance in the bacterial extract which induces this specific change. The crude extract of Type III is full of capsular polysaccharide, C (somatic) carbohydrate, nucleoproteins, free nucleic acids of both the yeast and thymus type, lipids, and other cell constituents. Try to find in the complex mixtures the active principle! Try to isolate and chemically identify the particular substance that will by itself, when brought into contact with the R cell derived from Type II, cause it to elaborate Type III capsular polysaccharide and to acquire all the aristocratic distinctions of the same specific type of cells as that from which the extract was prepared! Some job, full of headaches and heartbreaks. But at last perhaps we have it.

. . . if we prove to be right—and of course that is a big if—then it means that both the chemical nature of the inducing stimulus is known and the chemical structure of the substance produced is also known, the former being thymus nucleic acid, the latter Type III polysaccharide, and both are thereafter reduplicated in the daughter cells and after innumerable transfers without further addition of the inducing agent and the same active and specific transforming substance can be recovered far in excess of the amount originally used to induce the reaction. Sounds like a virus—may be a gene. But with mechanisms I am not now concerned. One step at a time and the first step is what is the chemical nature of the transforming principle? Some one else can work out the rest. Of course the problem bristles with implications. It touches the biochemistry of the thymus type of nucleic acids which are known to constitute the major part of chromosomes but have been thought to be alike regardless of origin and species. It touches genetics, enzyme chemistry, cell metabolism and carbohydrate synthesis. But today it takes a lot of well documented evidence to convince anyone that the sodium salt of deoxyribose nucleic acid, protein free, could possibly be endowed with such biologically active and specific properties and that is the evidence we are now trying to get. It is lots of fun to blow bubbles but it is wiser to prick them yourself before someone else tries to.

Figure 4-4
Part of a letter from Oswald Avery to his brother Roy, written in May 1943. [From R. D. Hotchkiss. In *Phage and the Origins of Molecular Biology*, J. Cairns, G. S. Stent, and J. D. Watson, eds. (Cold Spring Harbor Laboratory, 1966), pp. 185–186.]

of an *E. coli* culture was infected with labeled phage, which became attached to the bacteria during a short incubation period. The suspension was spun for a few minutes in a Waring Blendor at 10,000 rpm. This treatment subjected the phage-infected cells to very strong shearing forces, which severed the connections between the viruses and bacteria. The resulting suspension was centrifuged at a speed sufficient to throw the bacteria to the bottom of the tube. Thus, the pellet contained the infected bacteria, whereas the supernatant contained smaller particles. These fractions were analyzed for ^{32}P and ^{35}S to determine the location of the phage DNA and the protein coat. The results of these experiments were:

1. Most of the phage DNA was found in the bacteria.

2. Most of the phage protein was found in the supernatant.

3. The blendor treatment had almost no effect on the competence of the infected bacteria to produce progeny virus.

Additional experiments showed that less than 1% of the ^{35}S was transferred from the parental phage to the progeny phage. In contrast, 30% of the parental ^{32}P appeared in the progeny. These simple, incisive

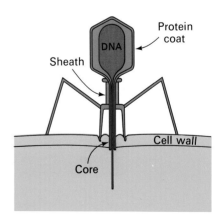

Figure 4-5
Diagram of a T2 bacteriophage injecting its DNA into a bacterial cell. [After W. B. Wood and R. S. Edgar. Building a bacterial virus. Copyright © 1967 by Scientific American, Inc. All rights reserved.]

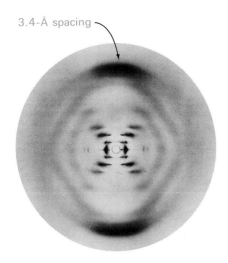

Figure 4-6
X-ray diffraction photograph of a hydrated DNA fiber (form B). The central cross is diagnostic of a helical structure. The strong arcs on the meridian arise from the stack of base pairs, which are 3.4 Å apart. [Courtesy of Dr. Maurice Wilkins.]

experiments led to the conclusion that *"a physical separation of the phage T2 into genetic and non-genetic parts is possible. . . . The sulfur-containing protein of resting phage particles is confined to a protective coat that is responsible for the adsorption of bacteria, and functions as an instrument for the injection of the phage DNA into the cell. This protein probably has no function in the growth of intracellular phage. The DNA has some function. Further chemical inferences should not be drawn from the experiments presented."*

The cautious tone of this conclusion did not detract from its impact. The genetic role of DNA soon became a generally accepted fact. The experiments of Hershey and Chase strongly reinforced what Avery, MacLeod, and McCarty had found eight years earlier in a different system. Additional support came from studies of the DNA content of single cells, which showed that in a given species *the DNA content is the same for all cells that have a diploid set of chromosomes. Haploid cells were found to have half as much DNA.*

THE WATSON-CRICK DNA DOUBLE HELIX

In 1953, James Watson and Francis Crick deduced the three-dimensional structure of DNA and immediately inferred its mechanism of replication. This brilliant accomplishment ranks as one of the most significant in the history of biology because it led the way to an understanding of gene function in molecular terms. Watson and Crick analyzed x-ray diffraction photographs of DNA fibers taken by Rosalind Franklin and Maurice Wilkins and derived a structural model that has proved to be essentially correct. The important features of their model of DNA are:

1. Two helical polynucleotide chains are coiled around a common axis. The chains run in opposite directions (Figure 4-7).

2. The purine and pyrimidine bases are on the inside of the helix, whereas the phosphate and deoxyribose units are on the outside (Figure 4-8). The planes of the bases are perpendicular to the helix axis. The planes of the sugars are nearly at right angles to those of the bases.

3. The diameter of the helix is 20 Å. Adjacent bases are separated by 3.4 Å along the helix axis and related by a rotation of 36 degrees. Hence, the helical structure repeats after ten residues on each chain; that is, at intervals of 34 Å.

4. The two chains are held together by hydrogen bonds between pairs of bases. Adenine is always paired with thymine. Guanine is always paired with cytosine (Figures 4-9 and 4-10).

5. The sequence of bases along a polynucleotide chain is not restricted in any way. *The precise sequence of bases carries the genetic information.*

The most important aspect of the DNA double helix is the specificity of the pairing of bases. Watson and Crick deduced that adenine must pair with thymine, and guanine with cytosine, because of steric and hydrogen-bonding factors. The steric restriction is imposed by the regular helical nature of the sugar-phosphate backbone of each polynucleotide chain. The glycosidic bonds that are attached to a bonded pair of bases are always 10.85

Chapter 4
DNA AND RNA:
MOLECULES OF HEREDITY

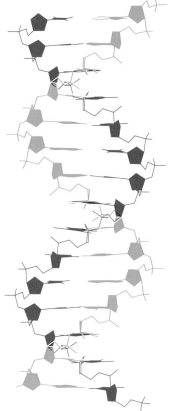

Figure 4-7
Skeletal model of double-helical DNA. The structure repeats at intervals of 34 Å, which corresponds to ten residues on each chain.

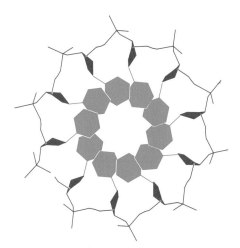

Figure 4-8
Diagram of one of the strands of a DNA double helix, viewed down the helix axis. The bases (all pyrimidines here) are inside, whereas the sugar-phosphate backbone is outside. The tenfold symmetry is evident. The bases are shown in blue and the sugars in red.

Figure 4-9
Model of an adenine-thymine base pair.

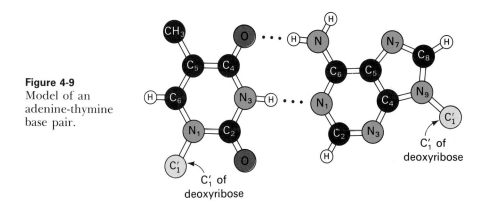

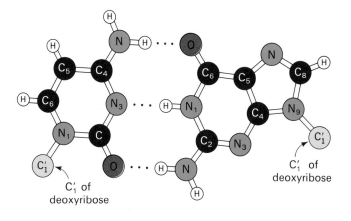

Figure 4-10
Model of a guanine-cytosine base pair.

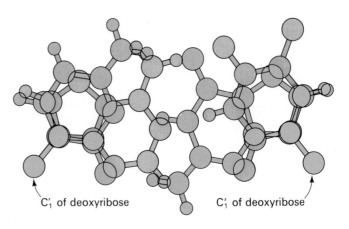

Figure 4-11
Superposition of an A-T base pair (shown in yellow) on a G-C base pair (shown in blue). Note that the positions of the glycosidic bonds and of the C-1' atom of deoxyribose are identical for the two base pairs.

Å apart (Figure 4-11). A purine-pyrimidine base pair fits perfectly in this space. In contrast, there is insufficient room for two purines. There is more than enough space for two pyrimidines, but they would be too far apart to form hydrogen bonds. Hence, one member of a base pair in a DNA helix must always be a purine and the other a pyrimidine because of steric factors. The base pairing is further restricted by hydrogen-bonding requirements. The hydrogen atoms in the purine and pyrimidine bases have well-defined positions. Adenine cannot pair with cytosine because there would be two hydrogens near one of the bonding positions and none at the other. Likewise, guanine cannot pair with thymine. In contrast, adenine forms two hydrogen bonds with thymine, whereas guanine forms three with cytosine (Figures 4-9 and 4-10). The orientations and distances of these hydrogen bonds are optimal for achieving strong attraction between the bases.

This base-pairing scheme was strongly supported by the results of earlier studies of the base compositions of DNAs from different species. In 1950, Erwin Chargaff found that the *ratios of adenine to thymine and of guanine to cytosine were nearly 1.0 in all species studied.* The meaning of these equivalences was not evident until the Watson-Crick model was proposed. Only then could it be seen that they reflect an essential facet of DNA structure and function—the specificity of base pairing.

The double-helical structure of DNA shown in Figure 4-12 is *the B form of DNA (B-DNA).* As will be discussed in a later chapter (p. 650), DNA can assume different helical forms, such as A-DNA and Z-DNA. Under physiologic conditions, DNA is almost entirely in the Watson-Crick B form.

THE COMPLEMENTARY CHAINS ACT AS TEMPLATES FOR EACH OTHER IN DNA REPLICATION

The double-helical model immediately suggested a mechanism for the replication of DNA. Watson and Crick (1953*b*) published their hypothesis a month after they had presented their structural model in a beautifully simple and lucid paper:

> . . . If the actual order of the bases on one of the pairs of chains were given, one could write down the exact order of the bases on the other one, because of the specific pairing. Thus one chain is, as it were, the

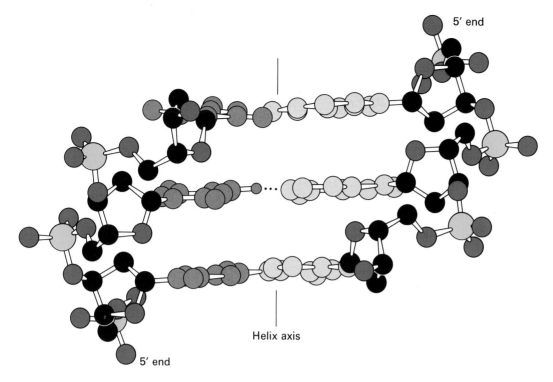

Figure 4-12
Model of a double-helical DNA molecule showing three base pairs (gray). Note that the two strands run in opposite directions.

complement of the other, and it is this feature which suggests how the deoxyribonucleic acid molecule might duplicate itself.

Previous discussions of self-duplication have usually involved the concept of a template, or mould. Either the template was supposed to copy itself directly or it was to produce a "negative," which in its turn was to act as a template and produce the original "positive" once again. In no case has it been explained in detail how it would do this in terms of atoms and molecules.

Now our model for deoxyribonucleic acid is, in effect, a *pair* of templates, each of which is complementary to the other. We imagine that prior to duplication the hydrogen bonds are broken, and the two chains unwind and separate. Each chain then acts as a template for the formation onto itself of a new companion chain, so that eventually we shall have *two* pairs of chains, where we only had one before. Moreover, the sequence of the pairs of bases will have been duplicated exactly.

DNA REPLICATION IS SEMICONSERVATIVE

Watson and Crick proposed that one of the strands of each daughter DNA molecule is newly synthesized, whereas the other is passed on unchanged from the parent DNA molecule. This distribution of parental atoms is called *semiconservative*. A critical test of this hypothesis was carried out by Matthew Meselson and Franklin Stahl. The parent DNA was labeled with ^{15}N, the heavy isotope of nitrogen, to make it denser than ordinary DNA. This was accomplished by growing *E. coli* for many generations in a medium that contained $^{15}NH_4Cl$ as the sole nitrogen source. The bacteria were abruptly transferred to a medium that contained ^{14}N, the ordinary isotope of nitrogen. The question asked was: What is the distribution of ^{14}N and ^{15}N in the DNA molecules after successive rounds of replication?

The distribution of ^{14}N and ^{15}N was revealed by the newly developed technique of *density-gradient equilibrium sedimentation*. A small amount of DNA was dissolved in a concentrated solution of cesium chloride having a density close to that of the DNA ($\sim$1.7 g cm^{-3}). This solution was centrifuged until it was nearly at equilibrium. The opposing processes of sedimentation and diffusion created a gradient in the concentration of cesium chloride across the centrifuge cell. The result was a stable density gradient, ranging from 1.66 to 1.76 g cm^{-3}. The DNA molecules in this density gradient were driven by centrifugal force into the region where the solution's density was equal to their own. High-molecular-weight DNA yielded a narrow band that was detected by its absorption of ultraviolet light. A mixture of ^{14}N DNA and ^{15}N DNA molecules gave clearly separate bands because they differ in density by about 1% (Figure 4-13).

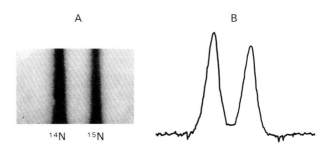

Figure 4-13
Resolution of ^{14}N DNA and ^{15}N DNA by density-gradient centrifugation: (A) ultraviolet absorption photograph of a centrifuge cell; (B) densitometric tracing of the absorption photograph. [From M. Meselson and F. W. Stahl. *Proc. Nat. Acad. Sci.* 44(1958):671.]

DNA was extracted from the bacteria at various times after they were transferred from a ^{15}N to a ^{14}N medium. Analysis of these samples by the density-gradient technique showed that there was a single band of DNA after one generation (Figure 4-14). The density of this band was precisely halfway between those of ^{14}N DNA and ^{15}N DNA. *The absence of ^{15}N DNA indicated that parental DNA was not preserved as an intact unit on replication.* The absence of ^{14}N DNA indicated that all of the daughter DNA molecules derived some of their atoms from the parent DNA. This proportion had to be one-half, because the density of the hybrid DNA band was halfway between those of ^{14}N DNA and ^{15}N DNA.

After two generations, there were equal amounts of two bands of DNA. One was hybrid DNA, the other was ^{14}N DNA. Meselson and Stahl concluded from these incisive experiments *"that the nitrogen of a DNA molecule is divided equally between two physically continuous subunits; that, following duplication, each daughter molecule receives one of these; and that the subunits are conserved through many duplications."* Their results agreed perfectly with the Watson-Crick model for DNA replication (Figure 4-15).

THE DOUBLE HELIX CAN BE REVERSIBLY MELTED

The two strands of a DNA helix readily come apart when the hydrogen bonds between its paired bases are disrupted. This can be accomplished

Chapter 4
DNA AND RNA:
MOLECULES OF HEREDITY

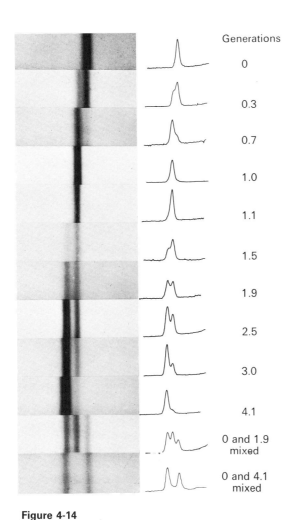

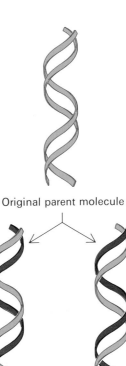

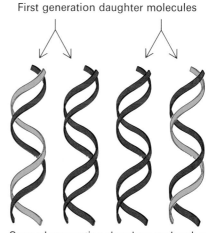

Figure 4-14
Detection of semiconservative replication in *E. coli* by density-gradient centrifugation. The position of a band of DNA depends on its content of ^{14}N and ^{15}N. After 1.0 generation, all of the DNA molecules are hybrids containing equal amounts of ^{14}N and ^{15}N. No parental DNA (^{15}N) is left after 1.0 generation. [From M. Meselson and F. W. Stahl. *Proc. Nat. Acad. Sci.* 44(1958):671.]

Figure 4-15
Schematic diagram of semiconservative replication. Parental DNA is shown in green and newly synthesized DNA in red. [After M. Meselson and F. W. Stahl. *Proc. Nat. Acad. Sci.* 44(1958):671.]

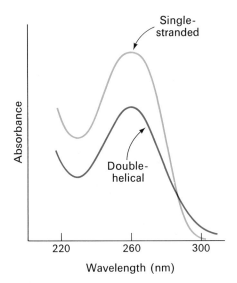

Figure 4-16
The absorbance of a DNA solution at wavelength 260 nm increases when the double helix is melted into single strands.

by heating a solution of DNA or by adding acid or alkali to ionize its bases. The unwinding of the double helix is called *melting* because it occurs abruptly at a certain temperature. The *melting temperature* (T_m) is defined as the temperature at which half of the helical structure is lost. The abruptness of the transition indicates that the DNA double helix is a *highly cooperative structure*, held together by many reinforcing bonds; it is stabilized by the stacking of bases as well as by base pairing. The melting of DNA is readily monitored by measuring its absorbance of light at wavelength 260 nm. The unstacking of the base pairs results in increased absorbance, an effect called *hyperchromism* (Figure 4-16).

The melting temperature of a DNA molecule depends markedly on its base composition. DNA molecules rich in GC base pairs have a higher T_m than those having an abundance of AT base pairs (Figure 4-17). In fact, the T_m of DNA from many species varies linearly with GC

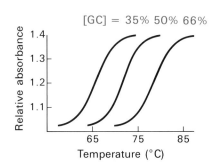

Figure 4-17
DNA melting curves. The absorbance relative to that at 25°C is plotted against temperature. (The wavelength of the incident light was 260 nm.) The T_m is 69°C for *E. coli* DNA (50% GC pairs) and 76°C for *P. aeruginosa* DNA (68% GC pairs).

content, rising from 77°C to 100°C as the fraction of GC pairs increases from 20% to 78%. GC base pairs are more stable than AT pairs because their bases are held together by three hydrogen bonds rather than by two. In addition, adjacent GC base pairs interact more strongly with one another than do adjacent AT base pairs. Hence, *the AT-rich regions of DNA are the first to melt* (Figure 4-18). The double helix is melted in vivo by the action of specific proteins (p. 672).

Separated complementary strands of DNA spontaneously reassociate to form a double helix when the temperature is lowered below T_m. This renaturation process is sometimes called *annealing*. The facility with which double helices can be melted and then reassociated is crucial for the biological functions of DNA.

DNA MOLECULES ARE VERY LONG

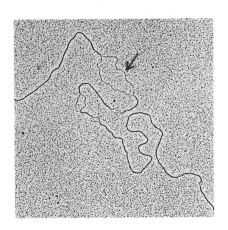

Figure 4-18
Electron micrograph of a DNA molecule partly unwound by alkali. Single-stranded regions appear as loops that stain less intensely than double-stranded segments. These unwound regions are rich in AT base pairs. One of them is marked by an arrow. [From R. B. Inman and M. Schnos. *J. Mol. Biol.* 49(1970):93.]

A striking characteristic of naturally occurring DNA molecules is their length. DNA molecules must be very long to encode the large number of proteins present in even the simplest cells. The *E. coli* chromosome, for example, is a single molecule of double-helical DNA consisting of four million base pairs. The mass of this DNA molecule is 2.6×10^6 kd. It has a *highly asymmetric shape* when taken out of the cell. The length of *E. coli* DNA is 14×10^6 Å, but its diameter is only 20 Å. The 1.4-mm length of this DNA molecule corresponds to a macroscopic dimension, whereas its width of 20 Å is on the atomic scale. Bruno Zimm found that the largest chromosome of *Drosophila melanogaster* contains a single DNA molecule of 6.2×10^7 base pairs, which has a length of 2.1 cm.

Such highly asymmetric molecules are very susceptible to cleavage by shearing forces. Unless special precautions are taken in their handling, they easily break into segments whose masses are a thousandth of that of the original molecule.

DNA molecules from many bacteria and viruses have been directly visualized by electron microscopy (Figure 4-19). The dimensions of some of these DNA molecules are given in Table 4-1. It should be noted that even the smallest DNA molecules are highly elongate. The DNA from polyoma virus, for example, consists of 5100 base pairs and has a length of 1.7 μm (17,000 Å). Hemoglobin, which is roughly spherical, has a diameter of 65 Å; collagen, one of the longest proteins, has a length of 3000 Å. These comparisons emphasize the remarkable length and asymmetry of DNA molecules.

Table 4-1
Sizes of DNA molecules

Organism	Base pairs (in thousands, or kb)	Length (μm)
Viruses		
Polyoma or SV40	5.1	1.7
λ phage	48.6	17
T2 phage	166	56
Vaccinia	190	65
Bacteria		
Mycoplasma	760	260
E. coli	4,000	1,360
Eucaryotes		
Yeast	13,500	4,600
Drosophila	165,000	56,000
Human	2,900,000	990,000

Source: After A. Kornberg, *DNA Replication* (W. H. Freeman and Company, 1980), p. 20.

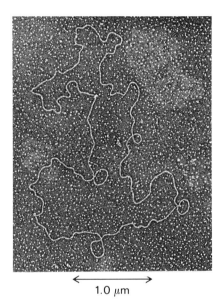

Figure 4-19
Electron micrograph of a DNA molecule from λ bacteriophage (RF II form). [Courtesy of Dr. Thomas Broker.]

SOME DNA MOLECULES ARE CIRCULAR AND SUPERCOILED

Electron microscopy has shown that intact DNA molecules from many sources are circular (Figure 4-19). The finding that *E. coli* has a circular chromosome was anticipated by genetic studies that revealed that *the gene-linkage map of this bacterium is circular*. The term "circular" refers to the continuity of the DNA chain, not to its geometrical form. DNA molecules in vivo necessarily have a very compact shape. Note that the length of the *E. coli* chromosome is about a thousand times as long as the greatest diameter of the bacterium.

Not all DNA molecules are circular. DNA from the T7 bacteriophage, for example, is *linear*. The DNA molecules of some viruses, such as the λ bacteriophage, *interconvert between linear and circular forms*. The linear form is present inside the virus particle, whereas the circular form is present in the host cell (see p. 860).

A new property appears in the conversion of a linear DNA duplex into a closed circular molecule. The axis of the double helix can itself be twisted to form a *superhelix*. A circular DNA without any superhelical

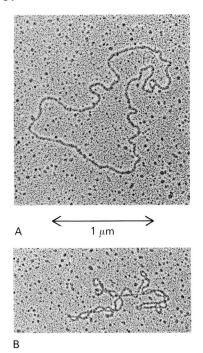

Figure 4-20
Electron micrographs of DNA from mitochondria: (A) relaxed circular form; (B) supercoiled circular form. [Courtesy of Dr. David Clayton.]

turns is known as a *relaxed molecule.* Supercoiling is biologically important for two reasons. First, *a supercoiled DNA has a more compact shape than its relaxed counterpart* (Figure 4-20). Supercoiling is critical for the packaging of DNA in the cell. Second, supercoiling affects the capacity of the double helix to unwind, and thereby affects its interactions with other molecules. These topological features of DNA will be discussed further in a later chapter (p. 660).

DNA IS REPLICATED BY POLYMERASES THAT TAKE INSTRUCTIONS FROM TEMPLATES

We turn now to the molecular mechanism of DNA replication. In 1958, Arthur Kornberg and his colleagues isolated an enzyme from *E. coli* that catalyzes the synthesis of DNA. They named the enzyme DNA polymerase; it is now called *DNA polymerase I* because other DNA polymerases have since been found. DNA replication is mediated by the intricate and coordinated interplay of more than twenty protein We focus here on DNA polymerase I to illustrate some new principles.

DNA polymerase I is a 103-kd single polypeptide chain. *It catalyzes the step-by-step addition of deoxyribonucleotide units to a DNA chain:*

$$(DNA)_{n \text{ residues}} + dNTP \rightleftharpoons (DNA)_{n+1} + PP_i$$

(The abbreviation dNTP denotes any deoxyribonucleoside triphosphate, and PP_i denotes the pyrophosphate group.) DNA polymerase I requires the following components to synthesize a chain of DNA (Figure 4-21):

1. All four of the activated precursors—the *deoxyribonucleoside 5'-triphosphates dATP, dGTP, dTTP, and dCTP*—must be present. Mg^{2+} ion is also required.

2. DNA polymerase I adds deoxyribonucleotides to the 3'-hydroxyl terminus of a preexisting DNA chain. In other words, a *primer chain* with a free 3'-OH group is required.

3. A *DNA template* is essential. The template can be single- or double-stranded DNA. Double-stranded DNA is an effective template only if its sugar-phosphate backbone is broken at one or more sites.

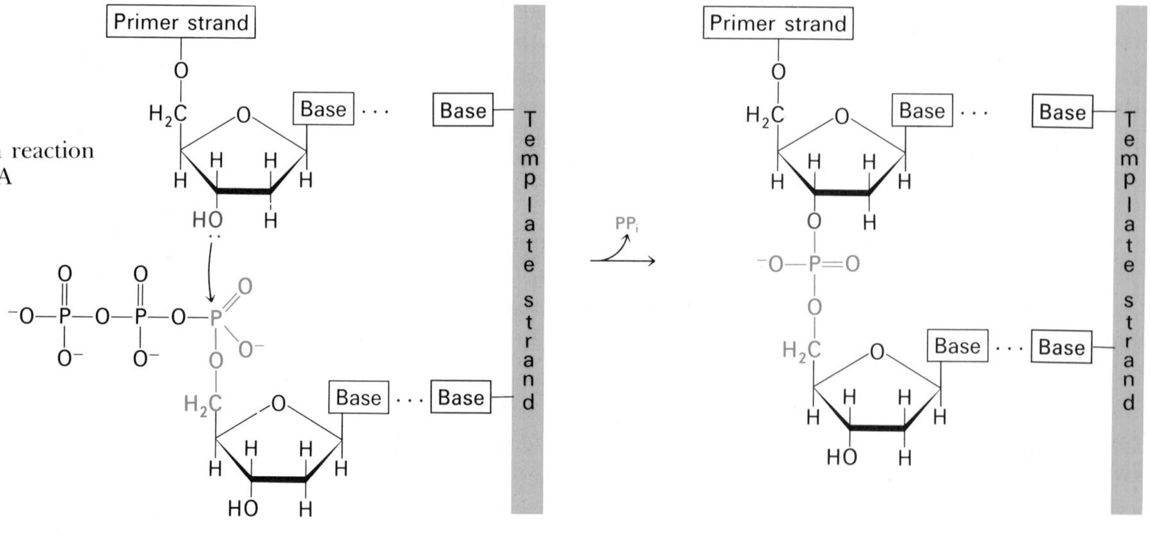

Figure 4-21
Chain-elongation reaction catalyzed by DNA polymerases.

The chain-elongation reaction catalyzed by DNA polymerase is *a nucleophilic attack of the 3'-OH terminus of the primer on the innermost phosphorus atom of a deoxyribonucleoside triphosphate*. A phosphodiester bridge is formed and pyrophosphate is concomitantly released. The subsequent hydrolysis of pyrophosphate by inorganic pyrophosphatase, a ubiquitous enzyme, drives the polymerization forward. *Elongation of the DNA chain proceeds in the 5' → 3' direction* (Figure 4-22).

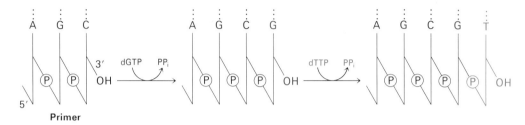

Figure 4-22
DNA polymerases catalyze elongation of DNA chains in the 5' → 3' direction.

DNA polymerase catalyzes the formation of a phosphodiester bond only if the base on the incoming nucleotide is complementary to the base on the template strand. The probability of making a covalent link is very low unless the incoming base forms a Watson-Crick type of base pair with the base on the template strand. Thus, DNA polymerase is a *template-directed enzyme*. The enzyme takes instructions from the template and synthesizes a product with a base sequence complementary to that of the template. Indeed, DNA polymerase I was the first template-directed enzyme to be discovered. Another striking feature of DNA polymerase I is that it corrects mistakes in DNA by removing mismatched nucleotides. These properties of DNA polymerase I contribute to the remarkably high fidelity of DNA replication, which has an error rate of less than 10^{-8} per base pair.

SOME VIRUSES HAVE SINGLE-STRANDED DNA DURING PART OF THEIR LIFE CYCLE

Not all DNA is double stranded. Robert Sinsheimer discovered that the DNA in φX174, *a small virus that infects* E. coli, *is single stranded*. Several experimental results led to this unexpected conclusion. First, the base ratios of φX174 DNA do not conform to the rule that [A] = [T] and [G] = [C]. Second, a solution of φX174 DNA is much less viscous than a solution of the same concentration of *E. coli* DNA. The hydrodynamic properties of φX174 DNA are like those of a randomly coiled polymer. In contrast, the DNA double helix behaves hydrodynamically as a quite rigid rod. Third, the amino groups of the bases of φX174 DNA react readily with formaldehyde, whereas the bases in double helical DNA are virtually inaccessible to this reagent.

The finding of this single-stranded DNA raised doubts concerning the universality of the semiconservative replicative scheme proposed by Watson and Crick. However, it was soon shown that φX174 DNA is single stranded for only a part of the life cycle of the virus. Sinsheimer found that infected *E. coli* cells contain a *double-stranded form of φX174*

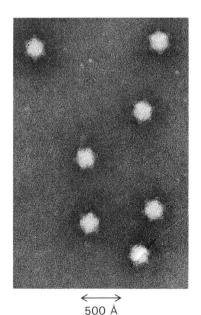

Figure 4-23
Electron micrograph of φX174 virus particles. [Courtesy of Dr. Robley Williams.]

DNA. This double-helical DNA is called the *replicative form* because it serves as the template for the synthesis of the DNA of the progeny virus. The generality of the Watson-Crick scheme for replication was reinforced by the finding of this double-stranded viral DNA intermediate.

THE GENES OF SOME VIRUSES ARE MADE OF RNA

Genes in all procaryotic and eucaryotic organisms are made of DNA. In viruses, genes are made of either DNA or RNA (ribonucleic acid). RNA, like DNA, is a long, unbranched polymer consisting of nucleotides joined by $3' \rightarrow 5'$ phosphodiester bonds (Figure 4-24). The covalent structure of RNA differs from that of DNA in two respects. As indicated by its name, the sugar units in RNA are *riboses* rather than deoxyriboses. Ribose contains a 2'-hydroxyl group not present in deoxyribose. The other difference is that one of the four major bases in RNA is *uracil* (U) instead of thymine (T). Uracil, like thymine, can form a base pair with adenine but it lacks the methyl group present in thymine. RNA molecules can be single-stranded or double-stranded. RNA cannot form a double helix of the B-DNA type because of steric interference by the 2'-hydroxyl groups of its ribose units. However, RNA can adopt a modified double helical form in which the base pairs are tilted about 20 degrees away from the perpendicular to the helix axis, a structure like A-DNA (p. 652).

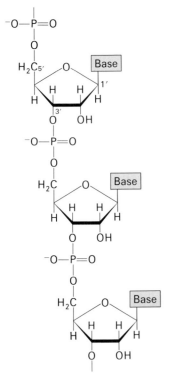

Figure 4-24
Structure of part of an RNA chain.

Tobacco mosaic virus, which infects the leaves of tobacco plants, is one of the best-characterized RNA viruses. It consists of a single strand of RNA (6390 nucleotides) surrounded by a protein coat of 2130 identical subunits (see Chapter 34 for a discussion of its structure and assembly). The protein can be separated from the RNA by treatment of the virus with phenol. *The isolated viral RNA is infective, whereas the viral protein is not.* Synthetic hybrid virus particles provide additional evidence that the genetic specificity of the virus resides exclusively in its RNA. A variety of strains of tobacco mosaic virus are known. A synthetic hybrid virus was prepared from the RNA of strain 1 and the protein of strain 2. Another was prepared from the RNA of strain 2 and the protein of strain 1. After infection, *the progeny virus always consisted of RNA and protein corresponding to the RNA in the infecting hybrid virus* (Figure 4-25).

Tobacco mosaic virus replicates in an infected plant cell by first synthesizing a (−) RNA strand that is complementary to the (+) RNA strand in the virus particle. The (−) RNA strand then serves as the template for the synthesis of a large number of (+) RNA strands that become packaged in new virus particles that are released by the cell. These syntheses are catalyzed by RNA polymerases that take instructions from RNA templates (*RNA-directed RNA polymerases*).

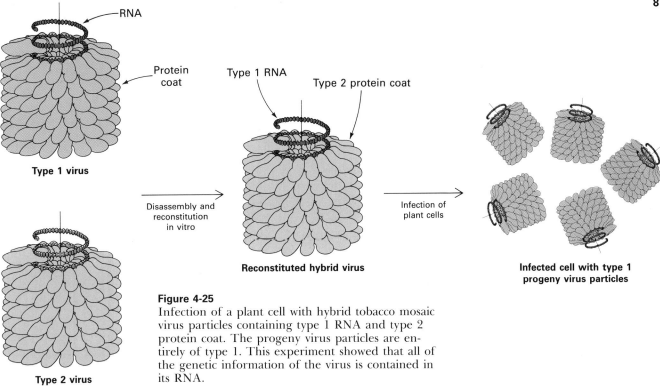

Figure 4-25
Infection of a plant cell with hybrid tobacco mosaic virus particles containing type 1 RNA and type 2 protein coat. The progeny virus particles are entirely of type 1. This experiment showed that all of the genetic information of the virus is contained in its RNA.

RNA TUMOR VIRUSES REPLICATE THROUGH DOUBLE-HELICAL DNA INTERMEDIATES

A number of RNA viruses produce malignant tumors after being injected into susceptible animal hosts. *Rous sarcoma virus* is one of the best-studied members of this group of *RNA tumor viruses*, which contain a single strand of RNA (p. 875). A striking feature of RNA tumor viruses is that they replicate through *DNA* intermediates (Figure 4-26). The RNA of the virus particle, called the (+) strand, is delivered into the host cell. This (+) RNA is the template for the synthesis of a complementary (−) DNA strand by a *reverse transcriptase*, an enzyme that is brought into the cell by the virus particle for this special purpose. Reverse transcriptase is an *RNA-directed DNA polymerase*. In this case, genetic information flows from RNA to DNA, the *reverse* of the normal

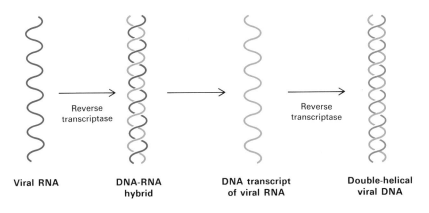

Figure 4-26
RNA tumor viruses replicate through double-helical DNA intermediates. DNA complementary to viral RNA is synthesized by reverse transcriptase, an enzyme brought into the cell by the infecting virus particle.

direction of information transfer (hence the name of the enzyme catalyzing this unusual step). The (−) DNA then serves as a template for the synthesis of (+) DNA. The resulting double-helical DNA version of the viral genome becomes incorporated into the chromosomal DNA of the host and is replicated along with the normal cellular DNA in the course of cell division. At some later time, the integrated viral genome is expressed to form viral (+) RNA and viral proteins, which assemble into new virus particles. RNA tumor viruses are also called *retroviruses* because their genetic information flows from RNA to DNA.

SUMMARY

DNA is the molecule of heredity in all procaryotic and eucaryotic organisms. In viruses, the genetic material is either DNA or RNA. All cellular DNA consists of two very long, helical polynucleotide chains coiled around a common axis. The two strands of the double helix run in opposite directions. The sugar-phosphate backbone of each strand is on the outside of the double helix, whereas the purine and pyrimidine bases are inside. The two chains are held together by hydrogen bonds between pairs of bases. Adenine (A) is always paired with thymine (T), and guanine (G) is always paired with cytosine (C). Hence, one strand of a double helix is the complement of the other. Genetic information is encoded in the precise sequence of bases along a strand. Most DNA molecules are circular. The axis of the double helix of a circular DNA can itself be coiled to form a superhelix. Supercoiled DNA is more compact than is relaxed DNA.

In the replication of DNA, the two strands of a double helix unwind and separate as new chains are synthesized. Each parent strand acts as a template for the formation of a new complementary strand. Thus, the replication of DNA is semiconservative—each daughter molecule receives one strand from the parent DNA molecule. The replication of DNA is a complex process carried out by many proteins, including several DNA polymerases. The activated precursors in the synthesis of DNA are the four deoxyribonucleoside 5′-triphosphates. The new strand is synthesized in the 5′ → 3′ direction by a nucleophilic attack by the 3′-hydroxyl terminus of the primer strand on the innermost phosphorus atom of the incoming deoxyribonucleoside triphosphate. Most important, DNA polymerases catalyze the formation of a phosphodiester bond only if the base on the incoming nucleotide is complementary to the base on the template strand. In other words, DNA polymerases are template-directed enzymes.

Some viruses have single-stranded DNA during a part of their life cycle. The DNA in ϕX174, a small virus that infects *E. coli*, is single stranded. In the infected host, however, a complementary strand is made, to form a double-helical replicative intermediate. The genes of some viruses, such as tobacco mosaic virus, are made of single-stranded RNA. An RNA-directed RNA polymerase mediates the replication of this viral RNA. RNA tumor viruses, such as Rous sarcoma virus, replicate through double-helical DNA intermediates. The RNA of these tumor viruses is transcribed into DNA by reverse transcriptase, an RNA-directed DNA polymerase. These RNA tumor viruses are also known as retroviruses because their genetic information flows from RNA to DNA, the reverse of the normal flow.

SELECTED READINGS

Where to start

Felsenfeld, G., 1985. DNA. *Sci. Amer.* 253(4):58–67. [Also printed in *The Molecules of Life*, an excellent series of articles. W. H. Freeman, 1985.]

Darnell, J. E., Jr., 1985. RNA. *Sci. Amer.* 253(4):26–36. [Also printed in *The Molecules of Life*.]

Dickerson, R. E., 1983. The DNA helix and how it is read. *Sci. Amer.* 249(6):94–111. [Offprint 1545].

Discovery of the major concepts

Avery, O. T., MacLeod, C. M., and McCarty, M., 1944. Studies on the chemical nature of the substance inducing transformation of pneumococcal types. Induction of transformation by a deoxyribonucleic acid fraction isolated from Pneumococcus Type III. *J. Exp. Med.* 79:137–158.

Hershey, A. D., and Chase, M., 1952. Independent functions of viral protein and nucleic acid in growth of bacteriophage. *J. Gen. Physiol.* 36:39–56.

Watson, J. D., and Crick, F. H. C., 1953a. Molecular structure of nucleic acid. A structure for deoxyribose nucleic acid. *Nature* 171:737–738.

Watson, J. D., and Crick, F. H. C., 1953b. Genetic implications of the structure of deoxyribonucleic acid. *Nature* 171:964–967.

Kornberg, A., 1960. Biologic synthesis of deoxyribonucleic acid. *Science* 131:1503–1508.

Meselson, M., and Stahl, F. W., 1958. The replication of DNA in *Escherichia coli*. *Proc. Nat. Acad. Sci.* 44:671–682.

Taylor, J. H., (ed.), 1965. *Selected Papers on Molecular Genetics*. Academic Press. [Contains the classic papers listed above.]

DNA structure

Saenger, W., 1984. *Principles of Nucleic Acid Structure*. Springer-Verlag. [An outstanding advanced account of the three-dimensional structure of nucleotides, DNA, and RNA. Contains many excellent illustrations.]

Dickerson, R. E., Drew, H. R., Conner, B. N., Wing, R. M., Fratini, A. V., and Kopka, M. L., 1982. The anatomy of A-, B-, and Z-DNA. *Science* 216:475–485.

Cantor, C. R., and Schimmel, P. R., 1980. *Biophysical Chemistry* (3 vols). W. H. Freeman. [Chapters 3 and 6 (in Part I) and Chapters 22, 23, and 24 (in Part III) give an excellent account of the conformation of nucleic acids.]

DNA replication

Kornberg, A., 1980. *DNA Replication*. W. H. Freeman. [An outstanding and highly readable book. Also see the *1982 Supplement to DNA Replication*, an update.]

Reminiscences and historical accounts

Watson, J. D., 1968. *The Double Helix*. Atheneum. [A lively, personal account of the discovery of the structure of DNA and its biological implications.]

McCarty, M., 1985. *The Transforming Principle: Discovering that Genes Are Made of DNA*. Norton. [A warm and lucid account of one of the major discoveries of this century by a sensitive participant.]

Cairns, J., Stent, G. S., and Watson, J. D., (eds.), 1966. *Phage and the Origins of Molecular Biology*. Cold Spring Harbor Laboratory. [A fascinating collection of reminiscences by some of the architects of molecular biology.]

Olby, R., 1974. *The Path to the Double Helix*. University of Washington Press.

Portugal, F. H., and Cohen, J. S., 1977. *A Century of DNA: A History of the Discovery of the Structure and Function of the Genetic Substance*. MIT Press.

Judson, H., 1979. *The Eighth Day of Creation*. Simon and Schuster.

PROBLEMS

1. Write the complementary sequence (in the standard $5' \to 3'$ notation) for:
   (a) GATCAA.
   (b) TCGAAC.
   (c) ACGCGT.
   (d) TACCAT.

2. The composition (in mole-fraction units) of one of the strands of a double-helical DNA is [A] = 0.30 and [G] = 0.24.
   (a) What can you say about [T] and [C] for the same strand?
   (b) What can you say about [A], [G], [T], and [C] of the complementary strand?

3. The DNA of a deletion mutant of λ bacteriophage has a length of 15 μm instead of 17 μm. How many base pairs are missing from this mutant?

4. What result would Meselson and Stahl have obtained if the replication of DNA were conservative (i.e., the parental double helix stayed together)? Give the expected distribution of DNA molecules after 1.0 and 2.0 generations for conservative replication.

5. Griffith used heat-killed S pneumococci to transform R mutants. Studies years later showed that double-stranded DNA is needed for efficient transformation and that high temperatures melt the DNA double

helix. Why were Griffith's experiments nevertheless successful?

6. Strains of *Bacillus subtilis* that can be transformed by foreign DNA are termed *competent*. Others, termed *non-competent*, are insusceptible to transformation. How might these strains differ from one another?

7. Bacteriophage M13 infects *E. coli* differently from the way bacteriophage T2 does. The M13 protein coat is removed in the inner membrane of the bacterial cell, where it is sequestered and subsequently used for the envelopment of progeny DNA. Why would M13 have been much less suitable than T2 was for the experiments carried out by Hershey and Chase?

8. Suppose that you want to radioactively label DNA but not RNA in dividing and growing bacterial cells. Which radioactive molecule would you add to the culture medium?

9. Suppose that you want to prepare DNA in which the backbone phosphorus atoms are uniformly labeled with ^{32}P. Which precursors should be added to a solution containing DNA polymerase I and primed template DNA? Specify the position of radioactive atoms in these precursors.

10. A solution contains DNA polymerase I and the Mg^{2+} salts of dATP, dGTP, dCTP, and dTTP. The DNA molecules listed below are added to aliquots of this solution. Which of them would lead to DNA synthesis?
    (a) A single-stranded closed circle containing 1000 nucleotide units.
    (b) A double-stranded closed circle containing 1000 nucleotide pairs.
    (c) A single-stranded closed circle of 1000 nucleotides base paired to a linear strand of 500 nucleotides with a free 3′-OH terminus.
    (d) A double-stranded linear molecule of 1000 nucleotide pairs with a free 3′-OH at each end.

11. Suppose that you want to assay reverse transcriptase activity. If polyriboadenylate is the template in the assay, what should you use as the primer? Which radioactive nucleotide should you use to follow chain elongation?

12. Reverse transcriptase has ribonuclease activity as well as polymerase activity. What is the role of its ribonuclease activity?

13. You have purified a virus that infects turnip leaves. Treatment of a sample with phenol removes viral proteins. Application of the residual material to scraped leaves results in the formation of progeny virus particles. You infer that the infectious substance is a nucleic acid.
    (a) Propose a simple and highly sensitive means of determining whether the infectious nucleic acid is DNA or RNA.
    (b) Is it likely that the virus particle carries an enzyme essential for its replication?

14. Spontaneous deamination of cytosine bases in DNA occurs at low but measurable frequency. Cytosine is converted into uracil by loss of its amino group. After this conversion, which base pair occupies this position in each of the daughter strands resulting from one round of replication? two rounds of replication?

CHAPTER 5

Flow of Genetic Information

We turn now from the storage and transmission of genetic information to its expression. Genes specify the kinds of proteins that are made by cells. However, DNA is not the direct template for protein synthesis. Rather, the templates for protein synthesis are RNA (ribonucleic acid) molecules. This chapter begins with an account of the discovery that a class of RNA molecules called *messenger RNAs* (mRNAs) are the information-carrying intermediates in protein synthesis. Other RNA molecules, such as transfer RNA (tRNA) and ribosomal RNA (rRNA) are part of the protein-synthesizing machinery. All forms of cellular RNA are synthesized by RNA polymerases that take instructions from DNA templates. This process of *transcription* is followed by *translation*, the synthesis of proteins according to instructions given by mRNA templates. Thus, the flow of genetic information in normal cells is

$$\text{DNA} \xrightarrow{\text{transcription}} \text{RNA} \xrightarrow{\text{translation}} \text{protein}$$

This brings us to the genetic code, the relation between the sequence of bases in DNA (or its mRNA transcript) and the sequence of amino acids in a protein. The code, which is nearly the same in all organisms, is beautiful in its simplicity. A sequence of three bases, called a codon, specifies an amino acid. Codons in mRNA are read sequentially by tRNA molecules, which serve as adaptors in protein synthesis. Protein synthesis takes place on ribosomes, which are complex assemblies of rRNAs and more than fifty kinds of proteins. Newly synthesized proteins contain signals that enable them to be targeted to specific destinations. The last theme considered in this chapter is the interrupted character of most eucaryotic genes, which are mosaics of introns and exons. Both are transcribed, but introns are cut out of newly synthesized RNA molecules, leaving mature RNA molecules with continuous exons. The existence of introns and exons has profound implications for evolution.

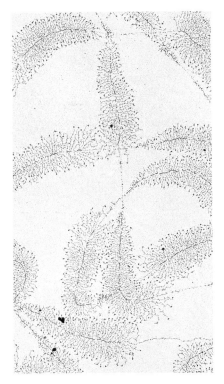

Figure 5-1
Transcription of DNA to form precursors of ribosomal RNA. This electron micrograph shows a tandem array of nascent RNA molecules emerging from a thread of DNA. [From O. L. Miller, Jr., and B. R. Beatty. Portrait of a gene. *J. Cell Physiol.* 74(supp. 1, 1969):225.]

SEVERAL KINDS OF RNA PLAY KEY ROLES IN GENE EXPRESSION

RNA is a long, unbranched macromolecule consisting of nucleotides joined by $3' \rightarrow 5'$ phosphodiester bonds (Figure 4-24). As the name indicates, the sugar unit in RNA is ribose. The four major bases in RNA are adenine (A), uracil (U), guanine (G), and cytosine (C). Adenine can pair with uracil, and guanine with cytosine. The number of nucleotides in RNA ranges from as few as seventy-five to many thousands. *RNA molecules are usually single stranded*, except in some viruses. Consequently, an RNA molecule need not have complementary base ratios. In fact, the proportion of adenine differs from that of uracil, and the proportion of guanine differs from that of cytosine, in most RNA molecules. However, *RNA molecules do contain regions of double-helical structure that are produced by the formation of hairpin loops* (Figure 5-2). In these regions, A pairs with U, and G pairs with C. The base pairing in RNA hairpins is frequently imperfect. G can also form a base pair with U, but it is less strong than the GC base pair (p. 744). Some of the apposing bases may not be complementary at all, and one or more bases along a single strand may be looped out to facilitate the pairing of the others. The proportion of helical regions in different kinds of RNA varies over a wide range; a value of 50% is typical.

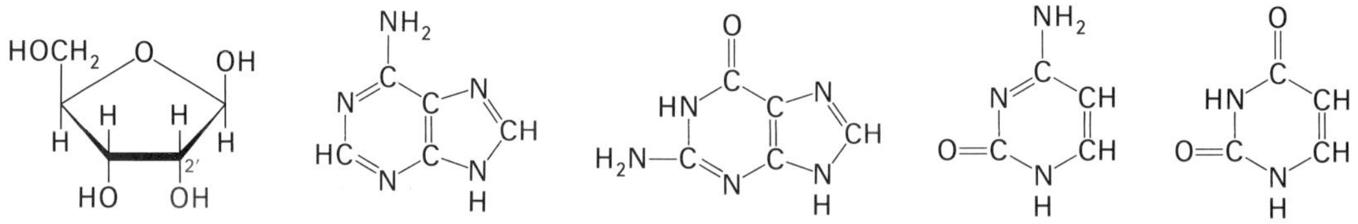

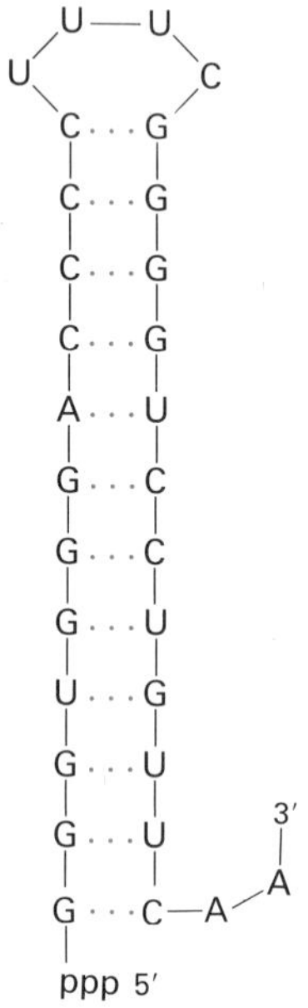

Cells contain several kinds of RNA (Table 5-1). *Messenger RNA* (mRNA) is the template for protein synthesis. An mRNA molecule is produced for each gene or group of genes that is to be expressed. Consequently, mRNA is a very heterogeneous class of molecules. In *E. coli*, the average length of an mRNA molecule is about 1.2 kb. *Transfer*

Figure 5-2
RNA can fold back on itself to form double-helical regions.

Table 5-1
RNA molecules in *E. coli*

Type	Relative amount (%)	Sedimentation coefficient (S)	Mass (kd)	Number of nucleotides
Ribosomal RNA (rRNA)	80	23	1.2×10^3	3700
		16	0.55×10^3	1700
		5	3.6×10^1	120
Transfer RNA (tRNA)	15	4	2.5×10^1	75
Messenger RNA (mRNA)	5	Heterogeneous		

RNA (tRNA) carries amino acids in an activated form to the ribosome for peptide-bond formation, in a sequence determined by the mRNA template. There is at least one kind of tRNA for each of the twenty amino acids. Transfer RNA consists of about seventy-five nucleotides (having a mass of about 25 kd), which makes it the smallest of the RNA molecules. *Ribosomal RNA* (rRNA) is the major component of ribosomes, but its precise role in protein synthesis is not yet known. The finding of catalytic RNA (p. 113) makes this question even more intriguing. In *E. coli*, there are three kinds of rRNA, called 23S, 16S, and 5S RNA because of their sedimentation behavior. One molecule of each of these species of rRNA is present in each ribosome. Ribosomal RNA is the most abundant of the three types of RNA. Transfer RNA comes next, followed by messenger RNA, which constitutes only 5% of the total RNA. Eucaryotic cells contain additional small RNA molecules. *Small nuclear RNA* (snRNA) molecules, for example, participate in the splicing of RNA exons. A small RNA molecule in the cytosol plays a role in the targeting of newly synthesized proteins.

FORMULATION OF THE CONCEPT OF MESSENGER RNA

The concept of mRNA was formulated by Francois Jacob and Jacques Monod in a classic paper published in 1961. Because proteins are synthesized in the cytoplasm rather than in the nucleus of eucaryotic cells, it was evident that there must be a chemical intermediate, which they called the structural messenger, specified by the genes. What is the nature of this intermediate? An important clue came from their studies of the control of protein synthesis in *E. coli* (Chapter 32). Certain enzymes in *E. coli*, such as those that participate in the uptake and utilization of lactose, are inducible—that is, the amount of these enzymes increases more than a thousandfold if an inducer (such as isopropyl-thiogalactoside) is present. The kinetics of induction were very revealing. The addition of an inducer elicited maximal synthesis of the lactose enzymes within a few minutes. Furthermore, the removal of the inducer resulted in the cessation of the synthesis of these enzymes in an equally short time. These experimental findings were incompatible with the presence of stable templates for the formation of these enzymes. Hence, Jacob and Monod surmised that *the messenger must be a very short-lived intermediate.* They then proposed that the messenger should have the following properties:

1. The messenger should be a polynucleotide.

2. The base composition of the messenger should reflect the base composition of the DNA that specifies it.

3. The messenger should be very heterogeneous in size because genes (or groups of genes) vary in length. They correctly assumed that three nucleotides code for one amino acid and calculated that the molecular weight of a messenger should be at least a half million.

4. The messenger should be transiently associated with ribosomes, the sites of protein synthesis.

5. The messenger should be synthesized and degraded very rapidly.

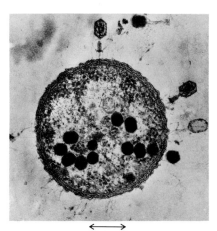

Figure 5-3
Electron micrograph of an *E. coli* cell infected by T2 viruses. [Courtesy of Dr. Lee Simon.]

It was apparent to Jacob and Monod that none of the known RNA fractions at that time met these criteria. Ribosomal RNA, then generally assumed to be the template for protein synthesis, was too homogeneous in size. Also, its base composition was similar in species that had very different DNA base ratios. Transfer RNA also seemed an unlikely candidate for the same reasons. In addition, it was too small. However, there were suggestions in the literature of a third class of RNA that appeared to meet the above criteria for the messenger. In *E. coli* infected with T2 bacteriophage, there was a new RNA fraction of appropriate size that had a very short half-life. Most interesting, the base composition of this new RNA fraction was like that of the viral DNA rather than like that of *E. coli* DNA.

EXPERIMENTAL EVIDENCE FOR MESSENGER RNA, THE INFORMATIONAL INTERMEDIATE IN PROTEIN SYNTHESIS

The hypothesis of a short-lived messenger RNA as the information-carrying intermediate in protein synthesis was tested shortly after the concept was formulated. Sydney Brenner, Francois Jacob, and Matthew Meselson carried out experiments on *E. coli* infected with T2 bacteriophage. Nearly all of the proteins made by the cell after infection are genetically determined by the phage. The synthesis of these proteins is not accompanied by an overall synthesis of RNA. However, a minor RNA fraction with a short half-life appears soon after infection. In fact, this RNA fraction has a nucleotide composition like that of the phage DNA. The fraction seemed optimal for a test of the messenger hypothesis because it appeared with a sudden switch in the kinds of proteins synthesized by the cell, and because neither rRNA nor tRNA are synthesized after infection.

How did this switch in the kinds of proteins made after infection occur? One possibility was that the phage DNA specifies a new set of ribosomes. In this model, genes control the synthesis of specialized ribosomes, and each ribosome can make only one kind of protein. An alternative model, the one proposed by Jacob and Monod, was that ribosomes are nonspecialized structures that receive genetic information from the gene in the form of an unstable messenger RNA. The experiments of Brenner, Jacob, and Meselson were designed to determine whether new ribosomes are synthesized after infection or whether new RNA joins preexisting ribosomes.

Bacteria were grown in a medium containing heavy isotopes (^{15}N and ^{13}C), infected with phage, and then immediately transferred to a medium containing light isotopes (^{14}N and ^{12}C). Ribosomes synthesized before and after infection could be separated by density-gradient centrifugation because their densities differed ("heavy" and "light," respectively). Furthermore, new RNA was labeled by the radioisotope ^{32}P- or ^{14}C-uracil and new protein by ^{35}S. These experiments showed that

1. *Ribosomes were not synthesized after infection*, as evidenced by the absence of "light" ribosomes.

2. RNA *was* synthesized after infection. Most of the radioactively labeled RNA emerged in the "heavy" ribosome peak. Thus, *most of the new RNA was associated with preexisting ribosomes*. Additional experiments showed that this new RNA turns over rapidly during the growth of phage.

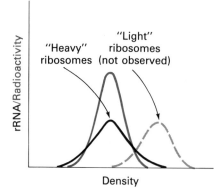

Figure 5-4
Density-gradient centrifugation of ribosomes of normal and T2 phage-infected bacteria. ^{32}P-labeled RNA synthesized *after* infection banded with ribosomes formed prior to infection ("heavy" ribosomes). Ribosomes were not synthesized after infection.

3. The radioisotope ^{35}S appeared transiently in the "heavy" ribosome peak, which showed that *new proteins were synthesized in preexisting ribosomes*.

These experiments led to the conclusion that *ribosomes are nonspecialized structures which synthesize, at a given time, the protein dictated by the messenger they happen to contain*. Studies of uninfected bacterial cells also showed that messenger RNA is the information-carrying link between gene and protein. In a very short time, the concept of messenger RNA became a central facet of molecular biology.

HYBRIDIZATION STUDIES SHOWED THAT MESSENGER RNA IS COMPLEMENTARY TO ITS DNA TEMPLATE

In 1961, Sol Spiegelman developed a new technique called *hybridization* to answer the following question: Is the RNA synthesized after infection with T2 phage complementary to T2 DNA? It was known from the work of Julius Marmur and Paul Doty that heating double-helical DNA above its melting temperature results in the formation of single strands. If the mixture is cooled slowly, these strands reassociate to form a double-helical structure with biological activity. Marmur and Doty also found that double-helical molecules are formed only from strands derived from the same or from closely related organisms. This observation suggested to Spiegelman that a double-stranded DNA-RNA hybrid might be formed from a mixture of single-stranded DNA and RNA if their base sequences were complementary (Figure 5-5). The experimental design was this:

1. A sample of T2 mRNA (the RNA synthesized after infection of *E. coli* with T2 phage) was prepared with a ^{32}P label. T2 DNA labeled with ^{3}H was prepared separately.

2. A mixture of the T2 mRNA and T2 DNA was heated to 100°C, which melted the double-helical DNA into single strands. This solution of single-stranded RNA and DNA was slowly cooled to room temperature.

3. The cooled mixture was analyzed by density-gradient centrifugation. The samples were centrifuged for several days in swinging-bucket rotors. The plastic sample tubes were then punctured at the bottom and drops were collected for analysis.

Three bands were found (Figure 5-6). The densest of these was single-stranded RNA. A second band contained double-helical DNA. A third band consisting of double-stranded DNA-RNA hybrid molecules was present near the DNA band. Thus, T2 mRNA formed a hybrid with T2 DNA. In contrast, T2 mRNA did not hybridize with DNA derived from a variety of bacteria and unrelated viruses, even if their base ratios were like those of T2 DNA. Subsequent experiments showed that the mRNA fraction of uninfected cells hybridized with the DNA derived from that particular organism but not from unrelated ones. These incisive experiments revealed that *the base sequence of mRNA is complementary to that of its DNA template. Furthermore, a powerful tool was developed for tracing the flow of genetic information in cells and for determining whether two nucleic acid molecules are similar.*

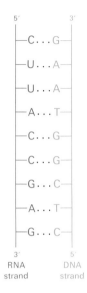

Figure 5-5
An RNA-DNA hybrid can be formed if the RNA and DNA have complementary sequences.

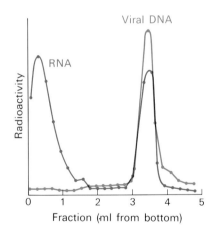

Figure 5-6
The RNA produced after *E. coli* has been infected with T2 phage is complementary to the viral DNA. In this hybridization experiment, RNA was labeled with ^{32}P, whereas T2 DNA was labeled with ^{3}H. The distribution of radioactivity in a cesium chloride density gradient shows that much of the RNA synthesized after infection bands with the T2 DNA. [After S. Spiegelman. Hybrid nucleic acids. Copyright © 1964 by Scientific American, Inc. All rights reserved.]

RIBOSOMAL RNA AND TRANSFER RNA ARE ALSO SYNTHESIZED ON DNA TEMPLATES

This hybridization technique was then used to determine whether rRNA and tRNA are also synthesized on DNA templates. The formation of RNA-DNA hybrids was detected by a filter assay rather than by density-gradient centrifugation because it is simpler, more sensitive, and much more rapid. Single-stranded RNA passes through a nitrocellulose filter, whereas double-helical DNA and RNA-DNA hybrids are retained by this filter. Various kinds of *E. coli* RNA labeled with ^{32}P were added to unlabeled *E. coli* DNA. These mixtures were heated, slowly cooled, and then filtered through nitrocellulose. The radioactivity retained on the filter was measured. The results were unequivocal: *RNA-DNA hybrids were formed with both rRNA (5S, 16S, and 23S) and tRNA, which shows that complementary sequences for these RNA molecules are present in the* E. coli *genome.*

ALL CELLULAR RNA IS SYNTHESIZED BY RNA POLYMERASES

The concept of mRNA stimulated the search for an enzyme that synthesizes RNA according to instructions given by a DNA template. In 1960, Jerard Hurwitz and Samuel Weiss independently discovered such an enzyme, which they named *RNA polymerase*. The enzyme from *E. coli* requires the following components for the synthesis of RNA:

1. *Template.* The preferred template is *double-stranded DNA*. Single-stranded DNA can also serve as a template. RNA, whether single or double stranded, is not an effective template, nor are RNA-DNA hybrids.

2. *Activated precursors.* All four *ribonucleoside triphosphates*—ATP, GTP, UTP, and CTP—are required.

3. *Divalent metal ion.* Mg^{2+} or Mn^{2+} are effective. Mg^{2+} meets this requirement in vivo.

RNA polymerase catalyzes the initiation and elongation of RNA chains. The reaction catalyzed by this enzyme is

$$(RNA)_{n \text{ residues}} + \text{ribonucleoside triphosphate} \rightleftharpoons (RNA)_{n+1 \text{ residues}} + PP_i$$

The synthesis of RNA is like that of DNA in several respects (Figure 5-8). First, the direction of synthesis is $5' \rightarrow 3'$. Second, the mechanism of elongation is similar: the 3'-OH group at the terminus of the growing chain makes a nucleophilic attack on the innermost phosphate of the incoming nucleoside triphosphate. Third, the synthesis is driven forward by the hydrolysis of pyrophosphate. In contrast with DNA polymerase, RNA polymerase does not require a primer. Another difference is that the DNA template is fully conserved in RNA synthesis, whereas it is semiconserved in DNA synthesis. Also, RNA polymerase lacks the nuclease capability used by DNA polymerase to excise mismatched nucleotides.

All three types of cellular RNA—mRNA, tRNA, and rRNA—are synthesized in *E. coli* by the same RNA polymerase according to instruc-

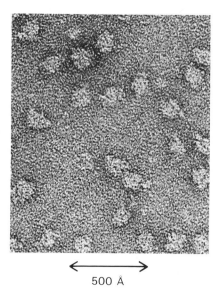

Figure 5-7
Electron micrograph of RNA polymerase from *E. coli*. [Courtesy of Dr. Robley Williams and Dr. Michael Chamberlin.]

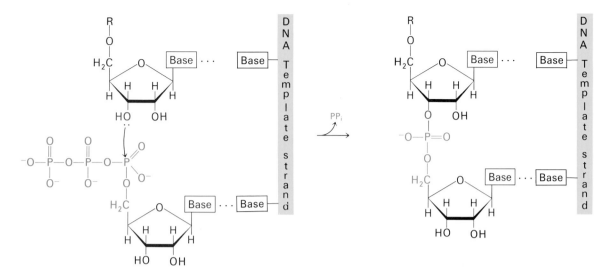

Figure 5-8
Mechanism of the chain-elongation reaction catalyzed by RNA polymerase.

tions given by a DNA template. In mammalian cells, there is a division of labor among several different kinds of RNA polymerases. We shall return to these RNA polymerases in Chapter 29. It is noteworthy that some viruses code for RNA-synthesizing enzymes that are very different from those of their host cells. For example, the RNA replicase that is encoded by Qβ, an RNA phage, is an RNA-dependent RNA polymerase because it takes instructions from an RNA template rather than from a DNA template (Chapter 34). In contrast, the cellular enzymes that synthesize RNA are *DNA-dependent RNA polymerases*.

RNA POLYMERASE TAKES INSTRUCTIONS FROM A DNA TEMPLATE

RNA polymerase, like the DNA polymerases described in the preceding chapter, takes instructions from a DNA template. The earliest evidence was the finding that the *base composition* of newly synthesized RNA is the complement of that of the DNA template strand, as exemplified by the RNA synthesized from single-stranded ϕX174 DNA as template (Table 5-2). Hybridization experiments also revealed that RNA synthesized by RNA polymerase is complementary to its DNA template. The strongest evidence for the fidelity of transcription came from base-sequence studies showing that the RNA sequence is the precise complement of the DNA template sequence (Figure 5-9).

Table 5-2
Base composition of RNA synthesized from a viral DNA template

DNA template (plus strand of ϕX174)		RNA product	
A	0.25	0.25	U
T	0.33	0.32	A
G	0.24	0.23	C
C	0.18	0.20	G

```
5'-GCGGCGACGCGCAGUUAAUCCCACAGCCGCCAGUUCCGCUGGCGGCAUUUU-3'   mRNA
3'-CGCCGCTGCGCGTCAATTAGGGTGTCGGCGGTCAAGGCGACCGCCGTAAAA-5' ⎫
5'-GCGGCGACGCGCAGTTAATCCCACAGCCGCCAGTTCCGCTGGCGGCATTTT-3' ⎭ DNA
```

Figure 5-9
The base sequence of mRNA is the complement of that of the DNA template strand. The sequence shown here is from the tryptophan operon, a segment of DNA containing the genes for five enzymes that catalyze the synthesis of tryptophan.

TRANSCRIPTION BEGINS NEAR PROMOTER SITES AND ENDS AT TERMINATOR SITES

DNA templates contain regions called *promoter sites* that specifically bind RNA polymerase and determine where transcription begins. In *bacteria*, two sequences on the 5' (upstream) side of the first nucleotide to be transcribed are important (Figure 5-10). One of them, called the *Pribnow box*, has the consensus sequence TATAAT and is centered at −10 (ten nucleotides on the 5' side of the first one transcribed, which is denoted by +1). The other, called the −35 *region*, has the consensus sequence TTGACA. The first nucleotide transcribed is usually a purine.

Consensus sequence—
The base sequences of promoter sites are not all identical. However, they do possess common features, which can be represented by an idealized consensus sequence. Each base in the consensus sequence TATAAT is found in a majority of procaryotic promoters. Nearly all promoter sequences differ from this consensus sequence at only two or fewer bases.

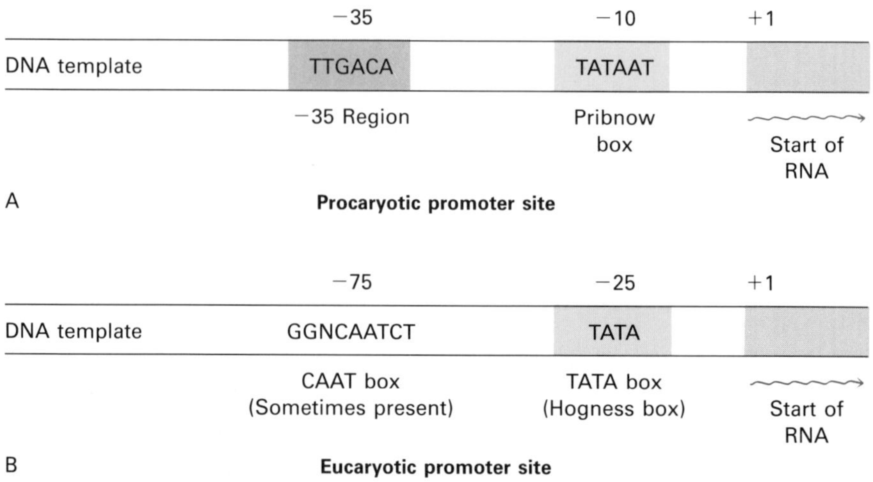

Figure 5-10
Promoter sites for transcription in (A) procaryotes and (B) eucaryotes. Consensus sequences are shown. The first nucleotide to be transcribed is numbered +1. The adjacent nucleotide on the 5' side is numbered −1.

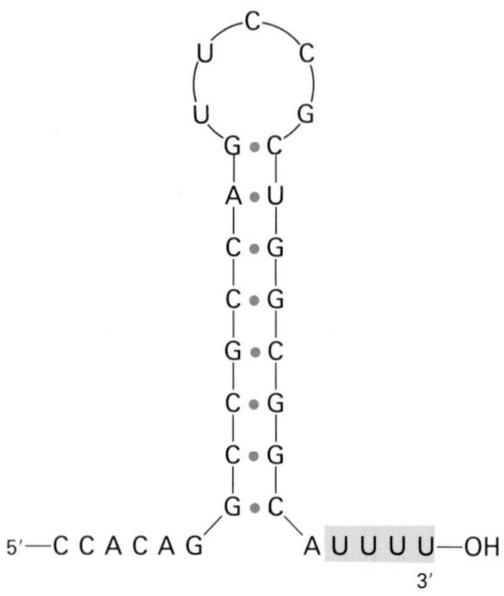

Eucaryotic genes encoding proteins have promoter sites with a TATAAA consensus sequence centered at about −25 (Figure 5-10B). This *TATA box* (also called *Hogness box*) is like the procaryotic Pribnow box except that it is farther upstream. Many eucaryotic promoters also exhibit a *CAAT* consensus sequence centered at about −75. Transcription of eucaryotic genes is further stimulated by *enhancer sequences*, which can be quite distant (up to several kilobases) from the start site, on either its 5' or its 3' side.

RNA polymerase proceeds along the DNA template and transcribes one of its strands until a terminator is reached. The termination signal in *E. coli* is a *base-paired hairpin* on the newly synthesized RNA molecule (Figure 5-11). This hairpin is formed by base-pairing of self-complementary sequences that are rich in G and C. Nascent RNA spontaneously dissociates from RNA polymerase when this hairpin is followed by a string of U residues. Alternatively, RNA synthesis can be terminated by the action of *rho*, a protein. Less is known about the termination of transcription in eucaryotes. A more detailed discussion of the initiation and termination of transcription will be given in later chapters. The important point now is that discrete start and stop signals for transcription are encoded in the DNA template.

Figure 5-11
Base sequence of the 3' end of an mRNA transcript in *E. coli*. A stable hairpin structure is followed by a sequence of U residues.

TRANSFER RNA IS THE ADAPTOR MOLECULE IN PROTEIN SYNTHESIS

We have seen that mRNA is the template for protein synthesis. How does it direct amino acids to become joined in the correct sequence? In 1958, Francis Crick wrote:

> One's first naive idea is that the RNA will take up a configuration capable of forming twenty different "cavities," one for the side chain of each of the twenty amino acids. If this were so, one might expect to be able to play the problem backwards—that is, to find the configuration of RNA by trying to form such cavities. All attempts to do this have failed, and on physical-chemical grounds the idea does not seem in the least plausible.

He observed that RNA does not have the knobby hydrophobic surfaces that could distinguish valine from leucine and isoleucine, nor does it have properly positioned charged groups to discriminate between positively and negatively charged amino acid side chains. Crick then proposed an entirely different mechanism for recognition of the mRNA template:

> ... RNA presents mainly a sequence of sites where hydrogen bonding could occur. One would expect, therefore, that whatever went onto the template in a *specific* way did so by forming hydrogen bonds. It is therefore a natural hypothesis that the amino acid is carried to the template by an adaptor molecule, and that the adaptor is the part which actually fits onto the RNA. In its simplest form one would require twenty adaptors, one for each amino acid.

This highly innovative hypothesis soon became an established fact. *The adaptor in protein synthesis is tRNA.* The structure and reactions of these remarkable adaptor molecules will be considered in detail in Chapter 30. For the moment it suffices to note that tRNA contains an *amino acid attachment site* and a *template-recognition site* (Figures 5-12 and 5-13). A tRNA molecule carries a specific amino acid in an activated form to the site of protein synthesis. The carboxyl group of this amino acid is esterified to the 3'- or 2'-hydroxyl group of the ribose unit at the 3' end of the tRNA chain. The esterified amino acid may migrate between the 2'- and 3'-hydroxyl groups during protein synthesis. The joining of an amino acid to a tRNA to form an aminoacyl-tRNA is catalyzed by a specific enzyme called an aminoacyl-tRNA synthetase (or activating enzyme). This esterification reaction is driven by ATP. There is at least one specific synthetase for each of the twenty amino acids. The template-recognition site on tRNA is a sequence of three bases called the *anticodon* (Figure 5-13). The anticodon on tRNA recognizes a complementary sequence of three bases on mRNA, called the *codon.*

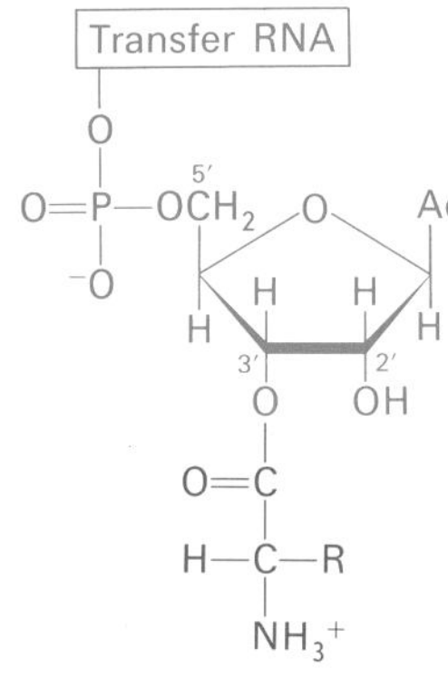

AMINO ACIDS ARE CODED BY GROUPS OF THREE BASES STARTING FROM A FIXED POINT

The *genetic code* is the relation between the sequence of bases in DNA (or its RNA transcripts) and the sequence of amino acids in proteins. Experiments by Francis Crick, Sydney Brenner, and others established the following features of the genetic code by 1961:

1. *What is the coding ratio?* A single-base code can specify only four kinds of amino acids because there are four kinds of bases in DNA. Sixteen kinds of amino acids can be specified by a two-base code

Figure 5-12
Mode of attachment of an amino acid (shown in red) to a tRNA molecule. The amino acid is esterified to the 3'-hydroxyl group of the terminal adenosine of tRNA. A tRNA having an attached amino acid is an aminoacyl-tRNA or a "charged" tRNA, whereas a tRNA without an attached amino acid is "uncharged."

Figure 5-13
Symbolic diagram of an aminoacyl-tRNA showing the amino acid attachment site and the anticodon, which is the template-recognition site.

($4 \times 4 = 16$), whereas sixty-four kinds of amino acids can be determined by a three-base code ($4 \times 4 \times 4 = 64$). Proteins are built from a basic set of twenty amino acids, and so it was evident from this simple calculation that three or more bases are probably needed to specify one amino acid. Genetic experiments then showed that *an amino acid is in fact coded by a group of three bases.* This group of bases is called a *codon.*

2. *Is the code nonoverlapping or overlapping?* In a nonoverlapping triplet code, each group of three bases in a sequence ABCDEF . . . specifies only one amino acid—ABC specifies the first, DEF the second, and so forth—whereas in a completely overlapping triplet code, ABC specifies the first amino acid, BCD the second, CDE the third, and so forth.

These alternatives were distinguished by studies of the sequence of amino acids in mutants. Suppose that the base C is mutated to C'. In a nonoverlapping code, only amino acid 1 will be changed. In a completely overlapping code, amino acids 1, 2, and 3 will all be altered by a mutation of C to C'. Amino acid sequence studies of tobacco mosaic virus mutants and abnormal hemoglobins showed that alterations usually affected only a single amino acid. Hence, it was concluded that the *genetic code is nonoverlapping.*

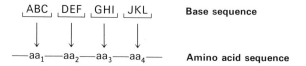

3. *How is the correct group of three bases read?* One possibility a priori is that one of the four bases (denoted as Q) serves as a "comma" between groups of three bases:

. . . QABCQDEFQGHIQJKLQ . . .

This turned out not to be the case. Rather, *the sequence of bases is read sequentially from a fixed starting point.* There are no commas.

```
          Start
           ↓
       ABC|DEF|GHI|JKL|MNO
       aa₁—aa₂—aa₃—aa₄—aa₅
```

Suppose that a mutation deletes base G:

```
          Start        G (deleted)
           ↓           ↗
       ABC|DEF|HIJ|KLM|NOP
       aa₁—aa₂—aa₃—aa₄—aa₅
       Normal   Altered
```

The first two amino acids in the resulting polypeptide chain will be normal, but the rest of the base sequence will be read incorrectly because *the reading frame has been shifted* by the deletion of G. Suppose instead that a base Z has been added between F and G:

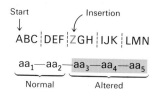

This addition also disrupts the reading frame starting at the codon for amino acid 3. In fact, genetic studies of addition and deletion mutants revealed many of the features of the genetic code.

4. As mentioned earlier, there are sixty-four possible base triplets and twenty amino acids. Is there just one triplet for each of the twenty amino acids or are some amino acids coded by more than one triplet? Genetic studies showed that most of the sixty-four triplets do code for amino acids. Subsequent biochemical studies demonstrated that sixty-one of the sixty-four triplets specify particular amino acids. Thus, *for most amino acids, there is more than one code word*. In other words, the genetic code is *degenerate*.

DECIPHERING THE GENETIC CODE: SYNTHETIC RNA CAN SERVE AS MESSENGER

What is the relation between sixty-four kinds of code words and the twenty kinds of amino acids? In principle, this question can be directly answered by comparing the amino acid sequence of a protein with the corresponding base sequence of its gene or mRNA. However, this approach was not experimentally feasible in 1961 because the base sequences of genes and mRNA molecules were entirely unknown. The breaking of the genetic code then did not seem imminent, but the situation suddenly changed. Marshall Nirenberg discovered that *the addition of polyuridylate (poly U) to a cell-free protein-synthesizing system led to the synthesis of polyphenylalanine*. Poly U evidently served as a messenger RNA. The first code word was deciphered: UUU codes for phenylalanine. This remarkable experiment pointed the way to the complete elucidation of the genetic code.

Let us consider this landmark experiment in more detail. The two essential components were a cell-free system that actively synthesizes protein and a synthetic polyribonucleotide that serves as the messenger RNA. A cell-free protein-synthesizing system was obtained from *E. coli* in the following way. Bacterial cells were gently broken open by grinding them with finely powdered alumina to yield a cell sap. Cell-wall and cell-membrane fragments were then removed by centrifugation. The resulting extract contained DNA, mRNA, tRNA, ribosomes, enzymes, and other cell constituents. Protein was synthesized by this cell-free system when it was supplemented with ATP, GTP, and amino acids. At least one of the added amino acids was radioactive so that its incorporation into protein could be detected. This mixture was incubated at 37°C for about an hour. Trichloroacetic acid was then added to terminate the reaction and precipitate the proteins, leaving the free amino acids in the supernatant. The precipitate was washed and its radioactivity was then measured to determine how much labeled amino acid was incorporated into newly synthesized protein.

A crucial feature of this system is that protein synthesis can be halted by the addition of deoxyribonuclease, which destroys the template for the synthesis of new mRNA (Figure 5-14). The mRNA present at the time of addition of deoxyribonuclease has a short life, and so protein synthesis stops within a few minutes. Nirenberg then found that protein synthesis resumed on addition of a crude fraction of mRNA. Thus, here was a *cell-free protein-synthesizing system that was responsive to the addition of mRNA*.

The other critical component in this experiment was a synthetic polyribonucleotide—namely, poly U. Poly U was synthesized by using *poly-*

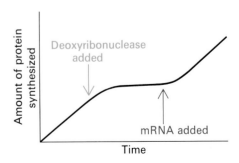

Figure 5-14
Protein synthesis in a cell-free system stops a few minutes after the addition of deoxyribonuclease and resumes following the addition of mRNA.

nucleotide phosphorylase, an enzyme discovered in 1955 by Marianne Grunberg-Manago and Severo Ochoa. This enzyme catalyzes the synthesis of polyribonucleotides from ribonucleoside diphosphates:

$$(RNA)_n + \text{ribonucleoside diphosphate} \rightleftharpoons (RNA)_{n+1} + P_i$$

The reactions catalyzed by this enzyme and by RNA polymerase are very different. In this reaction, the activated substrates are ribonucleoside diphosphates rather than triphosphates. Orthophosphate is a product instead of pyrophosphate. Hence, this reaction cannot be driven to the right by the hydrolysis of pyrophosphate. Indeed, the equilibrium in vivo is in the direction of RNA degradation, not synthesis. A critical difference is that *polynucleotide phosphorylase does not utilize a template.* The RNA synthesized by this enzyme has a composition dictated only by the ratios of the ribonucleotides present in the incubation mixture and a sequence that is essentially random. For this reason, polynucleotide phosphorylase proved to be a most valuable experimental tool in deciphering the genetic code. Thus, poly U was synthesized by incubating high concentrations of UDP in the presence of this enzyme. Copolymers of two ribonucleotides, say U and A, with a random sequence were prepared by incubating UDP and ADP with this enzyme.

A variety of synthetic ribonucleotides were added to the cell-free protein-synthesizing system, and the incorporation of ^{14}C-labeled L-phenylalanine was measured. The results were striking:

Polyribonucleotide added	^{14}C counts per minute
None	44
Poly A	50
Poly C	38
Poly U	39,800

The same experiment was carried out with a different ^{14}C-labeled amino acid in each incubation mixture. It was found that *poly A led to the synthesis of polylysine* and that *poly C led to the synthesis of polyproline.* Thus, three code words were deciphered. The code word GGG could not be deciphered in the same way because poly G does not act as a template, probably because it forms a triple-stranded helical structure. Polyribonucleotides with extensive regions of ordered structure are ineffective templates for protein synthesis.

Code word	Amino acid
UUU	Phenylalanine
AAA	Lysine
CCC	Proline

Polynucleotides consisting of two kinds of bases were then used as templates. For example, a random copolymer of U and G contains eight different triplets: UUU, UUG, UGU, GUU, UGG, GUG, GGU, and GGG. With a template containing 0.76 U and 0.24 G, phenylalanine was incorporated to the greatest extent, as expected, because the triplet UUU was the most prevalent one. Valine, leucine, and cysteine came next, followed by tryptophan and glycine. Hence, it was concluded that valine, leucine, and cysteine are specified by codons that contain 2 U and 1 G, whereas tryptophan and glycine are specified by codons that contain 1 U and 2 G. The same type of experiment was carried out with other random copolymers to determine the *compositions* of codons for all twenty amino acids.

Table 5-3
Amino acid incorporation stimulated by a random copolymer of U (0.76) and G (0.24)

Amino acid	Relative amount incorporated	Inferred codon composition
Phenylalanine	100	UUU
Valine	37	2U, 1G
Leucine	36	2U, 1G
Cysteine	35	2U, 1G
Tryptophan	14	1U, 2G
Glycine	12	1U, 2G

TRINUCLEOTIDES PROMOTE THE BINDING OF SPECIFIC TRANSFER RNA MOLECULES TO RIBOSOMES

The use of mixed copolymers as templates revealed the *composition but not the sequence of codons* corresponding to particular amino acids (except for UUU, AAA, and CCC). Valine is coded by a triplet with 2 U and 1 G, but is it UUG, UGU, or GUU? This question was answered by two entirely different experimental approaches: (1) the use of synthetic polyribonucleotides with an ordered sequence, and (2) the codon-dependent binding of specific tRNA molecules to ribosomes.

In 1964, Nirenberg discovered that *trinucleotides promote the binding of specific tRNA molecules to ribosomes in the absence of protein synthesis.* For example, the addition of pUUU (the *p* indicates that the 5' end of the trinucleotide is phosphorylated) led to the binding of phenylalanine tRNA, whereas pAAA markedly enhanced the binding of lysine tRNA, as did pCCC for proline tRNA. Dinucleotides were ineffective in stimulating the binding of tRNA to ribosomes. These studies showed that a *trinucleotide (like a triplet in mRNA) specifically binds the particular tRNA molecule for which it is the code word.* A simple and rapid binding assay was devised: tRNA molecules bound to ribosomes are retained by a cellulose nitrate filter, whereas unbound tRNA passes through the filter. The kind of tRNA retained by the filter was identified by using tRNA molecules charged with a particular ^{14}C-labeled amino acid.

All sixty-four trinucleotides were synthesized by organic-chemical or enzymatic techniques. For each trinucleotide, the binding of tRNA molecules corresponding to all twenty amino acids was assayed. For example, it was found that pUUG stimulated the binding of leucine tRNA only, pUGU stimulated the binding of cysteine tRNA only, and pGUU stimulated the binding of valine tRNA only. Hence, it was concluded that the codons UUG, UGU, and GUU correspond to leucine, cysteine, and valine, respectively. For a few codons, no tRNA was strongly bound, whereas, for a few others, more than one kind of tRNA was bound. For most codons, clear-cut binding results were obtained. In fact, about fifty codons were deciphered by this simple and elegant experimental approach.

COPOLYMERS WITH A DEFINED SEQUENCE WERE ALSO INSTRUMENTAL IN BREAKING THE CODE

At about the same time, H. Gobind Khorana succeeded in synthesizing polyribonucleotides with a defined repeating sequence. A variety of copolymers with two, three, and four kinds of bases were synthesized by a combination of organic-chemical and enzymatic techniques. Let us consider the strategy for the synthesis of poly (GUA), for example. This ordered copolymer has the sequence

GUAGUAGUAGUAGUAGUAGUA. . .

First, Khorana synthesized by organic-chemical methods two complementary deoxyribonucleotides, each with nine residues: d(TAC)$_3$ and d(GTA)$_3$. Partially overlapping duplexes formed on mixing these oligonucleotides then served as templates for the synthesis by DNA polymerase of long repeating double-helical DNA chains. The next step was to obtain long polyribonucleotide chains with a sequence complementary to one of the two DNA strands. The DNA strand to be transcribed could be selected by adding the three complementary ribonucleoside triphosphates. For example, when GTP, UTP, and ATP were added to the incubation mixture, the polyribonucleotide poly (GUA) was synthesized from the poly d(TAC) template strand. The other strand was not transcribed because CTP, one of the required substrates, was missing. Thus, *organic synthesis, followed by template-directed syntheses carried out by DNA polymerase and then RNA polymerase, yielded two long polyribonucleotides having defined repeating sequences* (Figure 5-15).

These regular copolymers were used as templates in the cell-free protein-synthesizing system. Let us examine some of the results. A copolymer consisting of an alternating sequence of two bases P and Q

PQP QPQ PQP QPQ PQP . . .

contains two codons, PQP and QPQ. The polypeptide product should therefore contain an alternating sequence of two kinds of amino acids (abbreviated as aa$_1$ and aa$_2$):

aa$_1$—aa$_2$—aa$_1$—aa$_2$—aa$_1$—

Whether aa$_1$ or aa$_2$ is amino-terminal in the polypeptide product depends on whether the reading frame starts at P or Q. When poly (UG) was the template, a polypeptide with an alternating sequence of cysteine and valine was synthesized.

UGU GUG UGU GUG UGU GUG

Cys—Val—Cys—Val—Cys—Val

This result unequivocally confirmed the triplet nature of the code and showed that either UGU or GUG codes for cysteine and that the other of these two triplets codes for valine. When this result was considered together with tRNA binding data, it was evident that UGU codes for cysteine and that GUG codes for valine. The polypeptides synthesized in response to several other alternating copolymers of two bases were

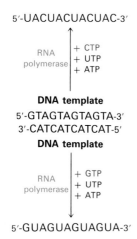

Figure 5-15
One strand or the other of this double-helical DNA template can be selected for transcription by adding just three ribonucleoside triphosphates.

Template	Product	Codon assignments	
Poly (UC)	Poly (Ser-Leu)	UCU ⟶ Ser	CUC ⟶ Leu
Poly (AG)	Poly (Arg-Gln)	AGA ⟶ Arg	GAG ⟶ Gln
Poly (AC)	Poly (Thr-His)	ACA ⟶ Thr	CAC ⟶ His

Now consider a template consisting of three bases in a repeating sequence, poly (PQR). If the reading frame starts at P, the resulting polypeptide should contain only one kind of amino acid, the one coded by the triplet PQR:

Start
↓
PQR|PQR|PQR|PQR|PQR|PQR...

$aa_1 - aa_1 - aa_1 - aa_1 - aa_1 - aa_1$

However, if the reading frame starts at Q, the polypeptide synthesized should contain a different amino acid, the one coded by the triplet QRP:

Start
↓
P|QRP|QRP|QRP|QRP|QRP|QRP...

$aa_2 - aa_2 - aa_2 - aa_2 - aa_2 - aa_2$

If the reading frame starts at R, a third kind of polypeptide should be formed, containing only the amino acid coded by the triplet RPQ:

Start
↓
PQ|RPQ|RPQ|RPQ|RPQ|RPQ|RPQ...

$aa_3 - aa_3 - aa_3 - aa_3 - aa_3 - aa_3$

Thus, the expected products are three different homopolypeptides. In fact, this was observed for most templates consisting of repeating sequences of three nucleotides. For example, poly (UUC) led to the synthesis of polyphenylalanine, polyserine, and polyleucine. This result, taken together with the outcome of other experiments, showed that UUC codes for phenylalanine, UCU codes for serine, and CUU codes for leucine. The polypeptides synthesized in response to other templates of this type are given in Table 5-4. Note that poly (GUA) and poly (GAU) each elicited the synthesis of two rather than three homopolypeptides. The reason will be evident shortly.

Khorana also synthesized several copolymers with repeating tetranucleotides, such as poly (UAUC). This template led to the synthesis of a polypeptide with the repeating sequence Tyr-Leu-Ser-Ile, irrespective of the reading frame:

UAU|CUA|UCU|AUC|UAU|CUA|UCU|AUC

Tyr—Leu—Ser—Ile—Tyr—Leu—Ser—Ile

Four codon assignments were deduced from this result.

A very different result was obtained when poly (GUAA) was the template. The only products were dipeptides and tripeptides. Why not longer chains? The reason is that one of the triplets present in this polymer—namely, UAA—codes not for an amino acid but rather for the termination of protein synthesis:

GUA|AGU|AAG|UAA|GUA|A...

Val—Ser—Lys—Stop

Only di- and tripeptides were formed also when poly (AUAG) was the template, because UAG is a second signal for chain termination:

AUA|GAU|AGA|UAG|AUA|G...

Ile—Asp—Arg—Stop

Table 5-4
Homopolypeptides synthesized using messengers containing repeating trinucleotide sequences

Messenger	Homopolypeptides synthesized
Poly (UUC)	Phe, Ser, Leu
Poly (AAG)	Lys, Glu, Arg
Poly (UUG)	Cys, Leu, Val
Poly (CCA)	Gln, Thr, Asn
Poly (GUA)	Val, Ser
Poly (UAC)	Tyr, Thr, Leu
Poly (AUC)	Ile, Ser, His
Poly (GAU)	Met, Asp

Now let us look again at Table 5-4. Two rather than three homopolypeptides were synthesized with poly (GUA) as a template because the third reading frame corresponds to the sequence

G | UAG | UAG | UAG . . .
Stop — Stop — Stop

which is a repeating sequence of termination signals. What about poly (GAU)? Again, only two homopolypeptides were synthesized with this template, because the third reading frame corresponds to

GA | UGA | UGA | UGA | U . . .
Stop — Stop — Stop

UGA is yet another signal for chain termination. In fact, *UAG, UAA, and UGA are the only three codons that do not specify an amino acid.*

Khorana's synthesis of polynucleotides with a defined sequence was an outstanding accomplishment. The use of these polymers as templates for protein synthesis, together with Nirenberg's studies of the trinucleotide-stimulated binding of tRNA to ribosomes, resulted in the complete elucidation of the genetic code by 1966, an outcome that would have been deemed a quixotic dream only six years earlier.

MAJOR FEATURES OF THE GENETIC CODE

All sixty-four codons have been deciphered (Table 5-5). Sixty-one triplets correspond to particular amino acids, whereas three code for chain termination. Because there are twenty amino acids and sixty-one triplets that code for them, it is evident that the code is highly *degenerate*. In other words, *many amino acids are designated by more than one triplet*. Only tryptophan and methionine are coded by just one triplet. The other eighteen amino acids are coded by two or more. Indeed, leucine, arginine, and serine are specified by six codons each. Under normal physiological conditions, *the code is not ambiguous:* a codon designates only one amino acid.

Codons that specify the same amino acid are called *synonyms*. For example, CAU and CAC are synonyms for histidine. Note that synonyms are not distributed haphazardly throughout the table of the genetic code (Table 5-5). An amino acid specified by two or more synonyms occupies a single box (unless there are more than four synonyms). The amino acids in a box are specified by codons that have the same first two bases but differ in the third base, as exemplified by GUU, GUC, GUA, and GUG. Thus, *most synonyms differ only in the last base of the triplet.* Inspection of the code shows that XYC and XYU always code for the same amino acid, whereas XYG and XYA usually code for the same amino acid. The structural basis for these equivalences of codons will become evident when we consider the nature of the anticodons of tRNA molecules (p. 743).

What is the biological significance of the extensive degeneracy of the genetic code? One possibility is that *degeneracy minimizes the deleterious effects of mutations*. If the code were not degenerate, then twenty codons would designate amino acids and forty-four would lead to chain termination. The probability of mutating to chain termination would therefore be much higher with a nondegenerate code than with the actual code. It is important to recognize that chain-termination mutations usu-

Rosetta stone, inscribed in hieroglyphs, demotic, and Greek. [© Archiv/Photo Researchers, Inc.]

Table 5-5
The genetic code

First position (5' end)	Second position				Third position (3' end)
	U	C	A	G	
U	Phe	Ser	Tyr	Cys	U
	Phe	Ser	Tyr	Cys	C
	Leu	Ser	Stop	Stop	A
	Leu	Ser	Stop	Trp	G
C	Leu	Pro	His	Arg	U
	Leu	Pro	His	Arg	C
	Leu	Pro	Gln	Arg	A
	Leu	Pro	Gln	Arg	G
A	Ile	Thr	Asn	Ser	U
	Ile	Thr	Asn	Ser	C
	Ile	Thr	Lys	Arg	A
	Met	Thr	Lys	Arg	G
G	Val	Ala	Asp	Gly	U
	Val	Ala	Asp	Gly	C
	Val	Ala	Glu	Gly	A
	Val	Ala	Glu	Gly	G

Note: Given the position of the bases in a codon, it is possible to find the corresponding amino acid. For example, the codon 5' AUG 3' on mRNA specifies methionine, whereas CAU specifies histidine. UAA, UAG, and UGA are termination signals. AUG is part of the initiation signal, in addition to coding for internal methionines.

ally lead to inactive proteins, whereas substitutions of one amino acid for another are usually rather harmless. *Degeneracy of the code may also be significant in permitting DNA base composition to vary over a wide range without altering the amino acid sequence of the proteins encoded by the DNA.* The [G] + [C] content of bacterial DNA ranges from less than 30% to more than 70%. DNA rich in GC has a higher melting temperature than DNA rich in AT. As might be expected, the DNA of algae residing in hot springs has a high content of GC. DNA molecules with quite different [G] + [C] contents could code for the same proteins if different synonyms were consistently used.

START AND STOP SIGNALS FOR PROTEIN SYNTHESIS

It has already been mentioned that *UAA, UAG, and UGA designate chain termination.* These codons are read not by tRNA molecules but rather by specific proteins called release factors (p. 758). The start signal for protein synthesis is more complex. Polypeptide chains in bacteria start with a modified amino acid—namely, formylmethionine (fMet). A specific tRNA, the initiator tRNA, carries fMet. This fMet-tRNA recognizes the codon AUG or, less frequently, GUG. However, AUG is also the codon for an internal methionine, and GUG is the codon for an internal valine. Hence, the signal for the first amino acid in the polypeptide chain must be more complex than for all subsequent ones. *AUG*

Formylmethionine (fMet)

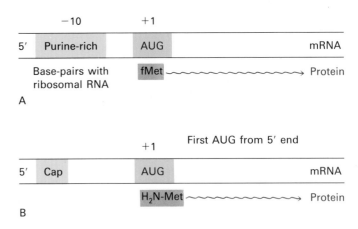

Figure 5-16
Start signals for the initiation of protein synthesis in (A) procaryotes and (B) eucaryotes. In eucaryotic mRNAs the 5' end, called a *cap*, contains modified bases (Chapter 29).

(*or GUG*) *is part of the initiation signal* (Figure 5-16). In bacteria, the initiating AUG (or GUG) is preceded several nucleotides away by a purine-rich sequence that base-pairs with a complementary sequence in a ribosomal RNA molecule (p. 753). In eucaryotes, the AUG closest to the 5' end of an mRNA is usually the start signal for protein synthesis. This particular AUG is read by an initiator tRNA charged with methionine.

THE GENETIC CODE IS NEARLY UNIVERSAL

The genetic code was deciphered by studies of trinucleotides and synthetic mRNA templates in cell-free systems derived from bacteria. Is the genetic code the same in all organisms? Analyses of spontaneous and specifically designed mutations in viruses, bacteria, and higher organisms have been highly informative. The base sequences of many wild-type and mutant genes are known, as are the amino acid sequences of their encoded proteins. In each case, the nucleotide change in the gene and the amino acid change in the protein are as predicted by the genetic code. Furthermore, mRNAs can be correctly translated by the protein-synthesizing machinery of very different species. For example, human hemoglobin mRNA is correctly translated by a wheat-germ extract. As will be discussed in the next chapter, bacteria efficiently express recombinant DNA molecules encoding human proteins such as insulin. These experimental findings strongly suggested that the genetic code is universal.

However, there was surprise when the sequence of human mitochondrial DNA became known. Human mitochondria read UGA as a codon for tryptophan rather than as a stop signal (Table 5-6). Another difference is that AGA and AGG are read as stop signals rather than as codons for arginine, and AUA is read as a codon for methionine instead of isoleucine. Mitochondria of other species, such as those of yeast, also have a genetic code that differs slightly from the standard one. Mitochondria can have a different genetic code from the rest of the cell because mitochondrial DNA encodes a distinct set of tRNAs. Do any cellular protein-synthesizing systems deviate from the standard genetic code? Recent studies have revealed that ciliated protozoa read AGA and AGG as stop signals rather than as codons for arginine. Thus, *the genetic code is nearly but not absolutely universal*. Variations clearly exist in mitochondria and in species, such as ciliates, that branched off very early in eucaryotic evolution. It is interesting to note that two of the

Table 5-6
Distinctive codons of human mitochondria

Codon	Standard code	Mitochondrial code
UGA	Stop	Trp
UGG	Trp	Trp
AUA	Ile	Met
AUG	Met	Met
AGA	Arg	Stop
AGG	Arg	Stop

codon reassignments in human mitochondria diminish the information content of the third base of the triplet (e.g., both AUA and AUG specify methionine). Most variations from the standard genetic code are in the direction of a simpler code.

Why has the code remained nearly invariant through billions of years of evolution, from bacteria to humans? A mutation that altered the reading of mRNA would change the amino acid sequence of most, if not all, of the proteins synthesized by that particular organism. Many of these changes would undoubtedly be deleterious, and so there would be strong selection against a mutation with such pervasive consequences.

THE SEQUENCES OF GENES AND THEIR ENCODED PROTEINS ARE COLINEAR

Before the genetic code was deciphered, high-resolution gene-mapping studies carried out by Seymour Benzer in the 1950s had revealed that genes are unbranched structures. This important result was in harmony with the established fact that DNA consists of a linear sequence of base pairs. Polypeptide chains also have unbranched structures. The following question was therefore asked: *Is there a linear correspondence between a gene and its polypeptide product?*

Charles Yanofsky approached this problem by using mutants of *E. coli* that produced an altered enzyme molecule. Numerous mutants of the α chain of tryptophan synthetase (an enzyme catalyzing the final step in the synthesis of tryptophan) were isolated and their positions on the genetic map for the α chain were determined by measuring frequencies of recombination. Some of these mutants were very close to each other on the genetic map, whereas others were far apart within the same gene. The next task was to determine the location of the amino acid substitution for each of these mutants. First, the amino acid sequence of the wild-type α chain was determined. Then, the position and nature of the amino acid change in each mutant was identified. *The order of the mutants on the genetic map proved to be the same as the order of the corresponding changes in the amino acid sequence of the polypeptide product.* The same result was subsequently obtained in many other systems. Thus, *genes and their polypeptide products are colinear.*

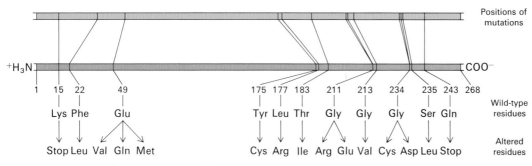

Figure 5-17
Colinearity of the gene and the amino acid sequence of the α chain of tryptophan synthetase. The locations of mutations in DNA (shown in yellow) were determined by genetic-mapping techniques. The positions of the altered amino acids in the amino acid sequence (shown in blue) are in the same order as the corresponding mutations. [After C. Yanofsky. Gene structure and protein structure. Copyright © 1967 by Scientific American, Inc. All rights reserved.]

MOST EUCARYOTIC GENES ARE MOSAICS OF INTRONS AND EXONS

In bacteria, polypeptide chains are encoded by a continuous array of triplet codons in DNA. It was assumed for many years that genes in higher organisms also are continuous. This view was unexpectedly shattered in 1977, when investigators in several laboratories discovered that several genes are *discontinuous*. For example, the gene for the β chain of hemoglobin is interrupted within its amino acid coding sequence by a long *noncoding intervening sequence* of 550 base pairs and a short one of 120 base pairs. Thus, the *β-globin gene is split into three coding sequences*.

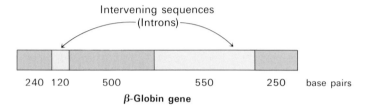

This unexpected structure was revealed by electron-microscopic studies of hybrids formed between β-globin mRNA and a segment of mouse DNA containing a β-globin gene (Figure 5-18). The DNA duplex is

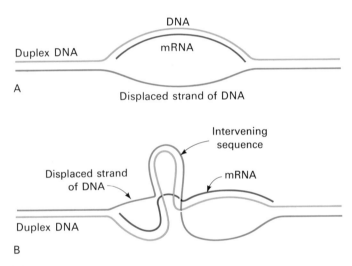

Figure 5-18
Detection of intervening sequences by electron microscopy. An mRNA molecule (shown in red) is hybridized to genomic DNA containing the corresponding gene. (A) A single loop of single-stranded DNA (shown in blue) is seen if the gene is continuous. (B) Two loops of single-stranded DNA (blue) and a loop of double-stranded DNA (blue and green) are seen if the gene contains an intervening sequence.

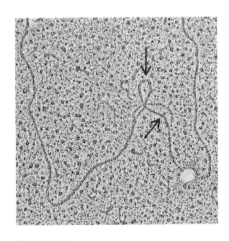

Figure 5-19
Electron micrograph of a hybrid between β-globin mRNA and a fragment of genomic DNA containing the β-globin gene. The thick loops of double-helical DNA that buckle out of the structure are the intervening sequences in DNA that are absent from mRNA (as in part B of Figure 5-18). The upper arrow points to the large intervening sequence, and the lower arrow to the small one. [Courtesy of Dr. Philip Leder.]

partly melted to allow mRNA to hybridize to the complementary strand of DNA. The single-stranded region of DNA then loops out and appears in electron micrographs as a thin line, in contrast with double-stranded DNA or DNA-RNA hybrid regions, which have a thick appearance. If the β-globin gene were continuous, a single loop would be seen. However, three loops are clearly evident in electron micrographs of these hybrids (Figure 5-19), which indicates that the gene is interrupted by at least one stretch of DNA that is absent from the mRNA. The large differences between maps of the β-globin gene and of a reverse transcript of β-globin mRNA showed that the genomic DNA contains untranslated sequences between the coding regions.

At what stage in gene expression are intervening sequences removed? Newly synthesized RNA chains isolated from nuclei are much larger than the mRNA molecules derived from them. In fact, the primary transcript of the β-globin gene contains two untranslated regions. *These intervening sequences in the 15S primary transcript are excised and the coding sequences are simultaneously linked by a precise splicing enzyme to form the mature 9S mRNA* (Figure 5-20). The coding sequences of split genes are called *exons* (for expressed regions), whereas their untranslated intervening sequences are known as *introns*. More generally, introns are sequences that are spliced out in the formation of mature RNA molecules.

Another split eucaryotic gene is the one for ovalbumin in chickens, which is made up of eight exons separated by seven long introns (Figure 5-21). Even more striking is the collagen gene, which contains more than forty exons. A common feature in the expression of these genes is that their exons are ordered in the same sequence in mRNA as in DNA. Thus, *split genes, like continuous genes, are colinear with their polypeptide products.*

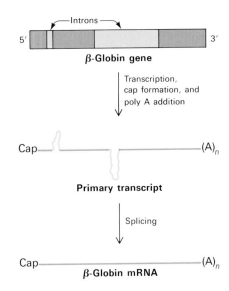

Figure 5-20
Transcription of the β-globin gene and removal of the intervening sequences in the primary RNA transcript. Cap formation and the addition of poly A are discussed in a later chapter (p. 720).

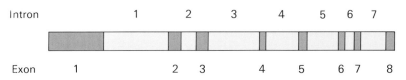

Figure 5-21
Structure of the chick ovalbumin gene. The introns (noncoding regions) are shown in yellow and the exons (translated regions) in blue.

Splicing is a complex operation that is carried out by *spliceosomes*, which are assemblies of proteins and small RNA molecules (p. 724). This enzymatic machinery recognizes signals in the nascent RNA that specify the splice sites. *Introns nearly always begin with GU and end with an AG that is preceded by a pyrimidine-rich tract (Figure 5-22). This consensus sequence is part of the signal for splicing.*

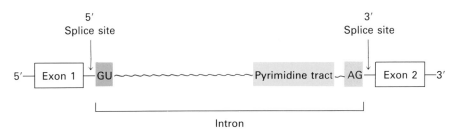

Figure 5-22
Consensus sequence for the splicing of mRNA precursors.

MANY EXONS ENCODE PROTEIN DOMAINS

Most genes of higher eucaryotes, such as birds and mammals, are split. Lower eucaryotes, such as yeast, have a much higher proportion of continuous genes. In eubacteria, such as *E. coli*, no split genes have been

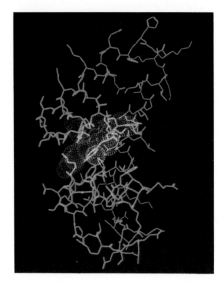

Figure 5-23
The core of myoglobin is encoded by the central exon of its gene. The polypeptide chain encoded by this exon is shown in green and the van der Waals outline of the oxygen-binding heme group in red.

found. Have introns been inserted into genes in the evolution of higher organisms? Or have introns been removed from genes to form the streamlined genomes of procaryotes and simple eucaryotes? Comparisons of the DNA sequences of genes encoding proteins that are highly conserved in evolution strongly suggest that *introns were present in ancestral genes and were lost in the evolution of organisms that have become optimized for very rapid growth, such as eubacteria and yeast.* The positions of introns in some genes are at least 10^9 years old. Furthermore, a common mechanism of splicing developed before the divergence of fungi, plants, and vertebrates, as shown by the finding that mammalian cell extracts can splice yeast RNA.

Many exons encode discrete structural and functional units of proteins (Figure 5-23). For example, the central exon of myoglobin and hemoglobin genes encodes a heme-binding region that can reversibly bind O_2 (p. 150). Other exons specify α-helical segments that anchor proteins in cell membranes. An entire domain of a protein may be encoded by a single exon. An attractive hypothesis is that *new proteins arose in evolution by the rearrangement of exons encoding discrete structural elements, binding sites, and catalytic sites.* Exon shuffling is a rapid and efficient means of generating novel genes because it preserves functional units, but allows them to interact in new ways. Introns are extensive regions in which DNA can break and recombine with no deleterious effect on encoded proteins. In contrast, the exchange of sequences between different exons usually leads to loss of function. *Gene duplication* is another means of increasing the genetic potentialities of an organism. The duplicated gene can undergo diversification while the original one continues to serve a vital function. Finally, genes can be altered by *point mutations*, which change a single nucleotide at a time and usually lead to the replacement of a single amino acid residue.

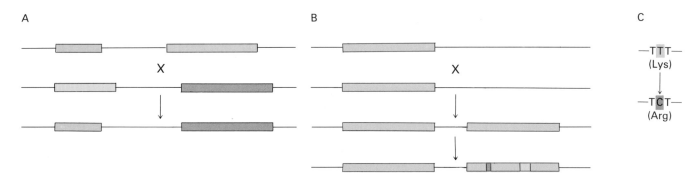

Figure 5-24
The genetic repertoire can be expanded by (A) exon shuffling, (B) gene duplication and diversification, and (C) point mutations.

Another advantage conferred by split genes is the potentiality for generating a series of related proteins by splicing a nascent RNA transcript in different ways. For example, a precursor of an antibody-producing cell forms an antibody that is anchored in the cell's plasma membrane (Figure 5-25). Stimulation of such a cell by a specific foreign antigen that is recognized by the attached antibody leads to cell differentiation and proliferation. The activated antibody-producing cells then splice their nascent RNA transcript in an alternative manner to form soluble antibody molecules that are secreted rather than retained on the cell surface. We see here a clear-cut example of a benefit conferred by the complex arrangement of introns and exons in higher organisms. *Alternative splicing is a means of forming a set of proteins that are variations of a basic motif according to a developmental program.*

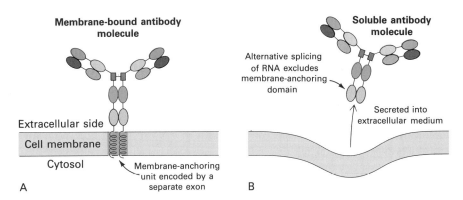

Figure 5-25
Alternative splicing generates mRNAs that are templates for different forms of a protein: (A) a membrane-bound antibody on the surface of a lymphocyte, and (B) its soluble counterpart, exported from the cell. Domains encoded by distinct exons are depicted in different colors. The membrane-bound antibody is anchored to the plasma membrane by a helical segment (highlighted in yellow) that is encoded by its own exon.

RNA PROBABLY CAME BEFORE DNA AND PROTEINS IN EVOLUTION

The flow of genetic information from DNA to RNA to protein depends on the intricate interplay of enzymes and other proteins with nucleic acids. Likewise, the replication of DNA and RNA is mediated by the interaction of polymerases and other proteins with nucleic acid templates. How did nucleic acid molecules early in the evolution of life replicate in the absence of enzymes? A likely solution to this enigma comes from the recent discovery that *RNA molecules as well as proteins can be enzymes.* Thomas Cech found that the precursor of a ribosomal RNA in *Tetrahymena* (a ciliated protozoan) undergoes self-splicing (p. 725). The intron in this precursor RNA molecule is precisely removed by a catalytic activity of the RNA itself. This liberated intron then loses a short 5′-terminal sequence to form a 395-nucleotide RNA molecule that catalyzes the transformation of *other* RNA molecules. This intron-derived RNA catalyzes the cleavage and joining of RNA chains at specific sites without itself being consumed (p. 214). Hence, it is a true enzyme. Protein catalysts, known for a century, are now joined by RNA catalysts (*ribozymes*).

This revolutionary finding enables us to envision an RNA world early in the evolution of life prior to the appearance of DNA and protein. Walter Gilbert has proposed that RNA molecules first catalyzed their own replication and developed a repertoire of enzymatic activities. In the next stage, RNA molecules began to synthesize proteins, which emerged as superior enzymes because their twenty side chains are more versatile than the four bases of RNA. Finally, DNA was formed by reverse transcription of RNA. DNA replaced RNA as the genetic material because its double-helix is a more stable and reliable store of genetic information than is single-stranded RNA. At this point, RNA was left with roles it has retained to this day as the information carrier (mRNA) and adapter (tRNA) in protein synthesis and as components of assemblies that mediate gene expression (e.g., rRNA in ribosomes). The present intricate (indeed baroque) mechanism of information transfer from gene to protein is an ancient epic, which probably began when RNA alone wrote the script, directed the action, and played all the key parts.

SUMMARY

The flow of genetic information in normal cells is from DNA to RNA to protein. The synthesis of RNA from a DNA template is called transcription, whereas the synthesis of a protein from an RNA template is termed translation. Cells contain several kinds of RNA: messenger RNA (mRNA), transfer RNA (tRNA), ribosomal RNA (rRNA), and small nuclear RNA (snRNA). Most RNA molecules are single stranded, but many contain extensive double-helical regions that arise from the folding of the chain into hairpins. The smallest RNA molecules are the tRNAs, which contain as few as 75 nucleotides, whereas the largest ones are some mRNAs, which may have more than 5000 nucleotides. All cellular RNA is synthesized by RNA polymerase according to instructions given by DNA templates. The activated intermediates are ribonucleoside triphosphates. The direction of RNA synthesis is $5' \rightarrow 3'$, like that of DNA synthesis. RNA polymerase differs from DNA polymerase in not requiring a primer. Another difference is that the DNA template is fully conserved in RNA synthesis, whereas it is semiconserved in DNA synthesis. Many RNA molecules are cleaved and chemically modified after transcription.

The base sequence of a gene is colinear with the amino acid sequence of its polypeptide product. The genetic code is the relation between the sequence of bases in DNA (or its RNA transcript) and the sequence of amino acids in proteins. Amino acids are coded by groups of three bases (called codons) starting from a fixed point. Sixty-one of the sixty-four codons specify particular amino acids, whereas the other three codons (UAA, UAG, and UGA) are signals for chain termination. Thus, for most amino acids there is more than one code word. In other words, the code is degenerate. Codons specifying the same amino acid are called synonyms. Most synonyms differ only in the last base of the triplet. The genetic code, which is nearly the same in all organisms, was deciphered after the discovery that the polyribonucleotide poly U codes for polyphenylalanine. Various synthetic polyribonucleotides then were used as mRNAs in cell-free protein-synthesizing systems. Natural mRNAs contain start and stop signals for translation, just as genes do for directing where transcription begins and ends.

Most genes in higher eucaryotes are discontinuous. Coding sequences (exons) in these split genes are separated by intervening sequences (introns), which are removed in the conversion of the primary transcript into mRNA and other functional mature RNA molecules. For example, the β-globin gene in mammals contains two introns. Nascent RNA molecules contain signals that specify splice sites. Split genes, like continuous genes, are colinear with their polypeptide products. A striking feature of many exons is that they encode functional domains in proteins. New proteins probably arose in evolution by the shuffling of exons. Introns may have been present in primordial genes, but were lost in the evolution of such fast-growing organisms as bacteria and yeast. The recent discovery that certain RNA molecules undergo self-splicing and can serve as enzymes suggests that RNA came before DNA and proteins in evolution.

SELECTED READINGS

Where to start

Miller, O. L., Jr., 1973. The visualization of genes in action. *Sci. Amer.* 228(3):34–42. [Available as *Sci. Amer.* Offprint 1267.]

Crick, F. H. C., 1966. The genetic code III. *Sci. Amer.* 215(4):55–62. [Offprint 1052. A view of the code when it was almost completely elucidated.]

Chambon, P., 1981. Split genes. *Sci. Amer.* 244(5):60–71.

Yanofsky, C., 1967. Gene structure and protein structure. *Sci. Amer.* 216(5):80–94. [Available as *Sci. Amer.* Offprint 1074. Presents the evidence for colinearity.]

Books

Darnell, J., Lodish, H., and Baltimore, D., 1986. *Molecular Cell Biology*. Scientific American Books. [Contains an excellent presentation of gene expression in eucaryotes.]

Lewin, B., 1987. *Genes* (3rd ed.). Wiley. [A lucid account of the flow of genetic information in procaryotes and eucaryotes.]

Watson, J. D., Hopkins, N. H., Roberts, J. W., Steitz, J. A., and Weiner, A. M., 1987. *Molecular Biology of the Gene* (4th ed.). Benjamin/Cummings. [Volume 1 of this outstanding work deals with general principles and volume 2 with eucaryotic systems.]

Discovery of messenger RNA

Jacob, F., and Monod, J., 1961. Genetic regulatory mechanisms in the synthesis of proteins. *J. Mol. Biol.* 3:318–356.

Brenner, S., Jacob, F., and Meselson, M., 1961. An unstable intermediate carrying information from genes to ribosomes for protein synthesis. *Nature* 190:576–581.

Hall, B. D., and Spiegelman, S., 1961. Sequence complementarity of T2-DNA and T2-specific RNA. *Proc. Nat. Acad. Sci.* 47:137–146.

Genetic code

Crick, F. H. C., Barnett, L., Brenner, S., and Watts-Tobin, R. J., 1961. General nature of the genetic code for proteins. *Nature* 192:1227–1232.

Khorana, H. G., 1968. Nucleic acid synthesis in the study of the genetic code. In *Nobel Lectures: Physiology or Medicine* (1963–1970), pp. 341–369. American Elsevier (1973).

Nirenberg, M., 1968. The genetic code. In *Nobel Lectures: Physiology or Medicine* (1963–1970), pp. 372–395. American Elsevier (1973).

Crick, F. H. C., 1958. On protein synthesis. *Symp. Soc. Exp. Biol.* 12:138–163. [A brilliant anticipatory view of the problem of protein synthesis. The adaptor hypothesis is presented in this article.]

Woese, C. R., 1967. *The Genetic Code*. Harper & Row.

Crothers, D. M., 1982. Nucleic acid aggregation geometry and the possible evolutionary origin of ribosomes and the genetic code. *J. Mol. Biol.* 162:379–391.

Colinearity of gene and protein

Yanofsky, C., Carlton, B. C., Guest, J. R., Helinski, D. R., and Henning, U., 1964. On the colinearity of gene structure and protein structure. *Proc. Nat. Acad. Sci.* 51:266–272.

Sarabhai, A. S., Stretton, O. W., Brenner, S., and Bolle, A., 1964. Colinearity of gene with polypeptide chain. *Nature* 201:13–17.

Introns, exons, and split genes

Gilbert, W., 1985. Genes-in-pieces revisited. *Science* 228:823–824.

Cochet, M., Gannon, F., Hen, R., Maroteaux, L., Perrin, F., and Chambon, P., 1979. Organization and sequence studies of the 17-piece chicken conalbumin gene. *Nature* 282:567–574.

Tilghman, S. M., Tiemeier, D. C., Seidman, J. G., Peterlin, B. M., Sullivan, M., Maijel, J. V., and Leder, P., 1978. Intervening sequence of DNA identified in the structural portion of a mouse β-globin gene. *Proc. Nat. Acad. Sci.* 75:725–729.

Craik, C. S., Rutter, W. J., and Fletterick, R., 1983. Splice junctions: association with variation in protein structure. *Science* 220:1125–1129.

Padgett, R. A., Grabowski, P. J., Konarska, M. M., Seiler, S., and Sharp, P. A., 1986. Splicing of messenger RNA precursors. *Ann. Rev. Biochem.* 55:1119.

Catalytic activity of RNA

Zaug, A. J., and Cech, T. R., 1986. The intervening sequence RNA of *Tetrahymena* is an enzyme. *Science* 231:470–475.

Altman, S., 1984. Aspects of biochemical catalysis. *Cell* 36:237–239.

Cech, T. R., and Bass, B. L., 1986. Biological catalysis by RNA. *Ann. Rev. Biochem.* 55:599.

Molecular evolution

Wilson, A. C., 1985. The molecular basis of evolution. *Sci. Amer.* 253(4):164.

Gilbert, W., 1986. The RNA world. *Nature* 319:618.

Lewin, R., 1986. RNA catalysis gives fresh perspective on the origin of life. *Science* 231:545–546.

Sharp, P. A., 1985. On the origin of RNA splicing and introns. *Cell* 42:397–400.

Cech, T. R., 1986. A model for the RNA-catalyzed replication of RNA. *Proc. Nat. Acad. Sci. USA* 83:4360–4363.

Marchionni, M., and Gilbert, W., 1986. The triosephosphate isomerase gene from maize: introns antedate the plant-animal divergence. *Cell* 46:133–141.

PROBLEMS

1. Compare DNA polymerase I and RNA polymerase from *E. coli* in regard to each of the following features:
   (a) Activated precursors.
   (b) Direction of chain elongation.
   (c) Conservation of the template.
   (d) Need for a primer.

2. Write the sequence of the mRNA molecule synthesized from a DNA template strand having the sequence

   5'-ATCGTACCGTTA-3'

3. RNA is readily hydrolyzed by alkali, whereas DNA is not. Why?

4. How does cordycepin (3'-deoxyadenosine) block the synthesis of RNA?

5. What amino acid sequence is encoded by the following base sequence of an mRNA molecule? Assume that the reading frame starts at the 5' end.

   5'-UUGCCUAGUGAUUGGAUG-3'

6. What is the sequence of the polypeptide formed on addition of poly (UUAC) to a cell-free protein-synthesizing system?

7. A protein chemist told a molecular geneticist that he had found a new mutant hemoglobin in which aspartate replaced lysine. The molecular geneticist expressed surprise and sent his friend scurrying back to the laboratory.
   (a) Why was the molecular geneticist dubious about the reported amino acid substitution?
   (b) Which amino acid substitutions would have been more palatable to the molecular geneticist?

8. The RNA transcript of a region of G4 phage DNA contains the sequence 5'-AAAUGAGGA-3'. This sequence encodes three different polypeptides. What are they?

9. Proteins generally have low contents of Met and Trp, intermediate ones of His and Cys, and high ones of Leu and Ser. What is the relation between the number of codons of an amino acid and its frequency of occurrence in proteins? What might be the selective advantage of this relation?

10. A transfer RNA with a UGU anticodon is enzymatically charged with a ^{14}C-labeled cysteine. The cysteine unit is then chemically modified to alanine (using Raney nickel, which removes the sulfur atom of cysteine). The altered aminoacyl-tRNA is added to a protein-synthesizing system containing normal components except for this tRNA. The mRNA added to this mixture contains the following sequence:

    5'-UUUUGCCAUGUUUGUGCU-3'

    What is the sequence of the corresponding radiolabeled peptide?

11. Valine is specified by four codons. How might the relative frequencies of their usage in an alga isolated from a volcanic hot spring differ from those of an alga isolated from an Antarctic bay?

12. The amino acid sequences of a yeast protein and a human protein carrying out the same function are found to be 60% identical. However, the corresponding DNA sequences are only 45% identical. Account for this differing degree of identity.

13. The genes for the green-absorbing and red-absorbing visual pigments mediating human color vision are located next to each other on the X chromosome. These genes are very similar to one another.
    (a) What is the consequence in a male of having only one visual pigment gene on the X chromosome?
    (b) A small proportion of the population has three visual pigment genes adjacent to each other on the X chromosome. Two of them are identical. How might this arise? What is its potential evolutionary significance?

CHAPTER 6

Exploring Genes: Analyzing, Constructing, and Cloning DNA

Recombinant DNA technology, a new approach to exploring the central molecules of life, came into being in the early 1970s. It has revolutionized biochemistry by providing powerful means of analyzing and altering genes and proteins. The genetic endowment of organisms can now be precisely changed in designed ways. Recombinant DNA technology is a fruit of several decades of basic research on DNA, RNA, and viruses. It depends, first, on having enzymes that can cut, join, and replicate DNA and reverse transcribe RNA. Earlier chapters have already discussed DNA polymerases and reverse transcriptase. This chapter begins with restriction enzymes, which cut very long DNA molecules into specific fragments that can be manipulated. The availability of many kinds of restriction enzymes and of DNA ligases, enzymes that join DNA strands, makes it feasible to treat DNA sequences as modules and to move them at will from one DNA molecule to another. Thus, recombinant DNA technology is based on nucleic acid enzymology.

A second foundation is the base-pairing language that mediates nucleic acid recognition. We have already seen that hybridization with complementary DNA or RNA probes is a sensitive and powerful means of detecting specific nucleotide sequences. In recombinant DNA technology, base pairing is used to construct new combinations of DNA as well as to detect particular sequences. This revolutionary technology is also critically dependent on the existence of viruses and detailed knowledge concerning their interplay with susceptible hosts. Viruses are the ultimate parasites. They efficiently deliver their own DNA (or RNA) into hosts, subverting them either to replicate the viral genome and produce viral protein or to incorporate viral DNA into the host genome. Likewise, plasmids, which are accessory chromosomes, have been indispensable in recombinant DNA technology.

Figure 6-1
Crystals of human insulin produced by bacteria harboring recombinant DNA encoding the hormone. [From R. E. Chance, E. P. Kroeff, and J. A. Hoffmann. In *Insulins, Growth Hormone, and Recombinant DNA Technology*, J. L. Gueriguian, ed. (Raven Press, 1981), p. 77.]

> *Palindrome*—
> A word, sentence, or verse that reads the same from right to left as it does from left to right.
>
> Radar
> Madam, I'm Adam
> Able was I ere I saw Elba
> Roma tibi subito motibus ibit amor
>
> Derived from the Greek *palindromos*, running back again.

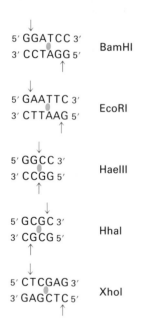

Figure 6-2
Specificities of some restriction endonucleases. The base-pair sequences that are recognized by these enzymes contain a twofold axis of symmetry. The two strands in these regions are related by a 180-degree rotation around the axis marked by the green symbol. The cleavage sites are denoted by red arrows. The abbreviated name of each restriction enzyme is given at the right of the sequence that it recognizes.

This chapter also introduces some of the benefits of these new methods. For example, the discovery of restriction enzymes led to the development of techniques for the rapid sequencing of DNA. A wealth of information concerning gene architecture, the control of gene expression, and protein structure has come from the sequencing of millions of bases in DNA molecules from viruses, bacteria, and higher organisms. DNA molecules can also be synthesized de novo. The automated solid-phase synthesis of DNA provides highly specific probes and synthetic tailor-made genes. The final part of this chapter deals with the construction and cloning of novel combinations of genes. New genes can be efficiently expressed by host cells, as exemplified by the production of human insulin by bacteria. Moreover, specific mutations can be made in vitro to engineer proteins in designed ways.

RESTRICTION ENZYMES SPLIT DNA INTO SPECIFIC FRAGMENTS

Restriction enzymes, also called restriction endonucleases, recognize specific base sequences in double-helical DNA and cleave both strands of the duplex at specific places. To biochemists, these exquisitely precise scalpels are marvelous gifts of nature. They are indispensable for analyzing chromosome structure, sequencing very long DNA molecules, isolating genes, and creating new DNA molecules that can be cloned. Restriction enzymes were discovered by Werner Arber, Hamilton Smith, and Daniel Nathans.

Restriction enzymes are found in a wide variety of procaryotes. Their biological role is to cleave foreign DNA molecules. The cell's own DNA is not degraded because the sites recognized by its own restriction enzymes are methylated. The interplay between modification and the action of endonucleases will be considered in a later chapter (Chapter 34). The important point here is that many of them recognize specific sequences of four to eight base pairs and hydrolyze a phosphodiester bond in each strand in this region. A striking characteristic of most of these cleavage sites is that they possess *twofold rotational symmetry*. In other words, the recognized sequence is *palindromic* and the cleavage sites are symmetrically positioned. For example, the sequence recognized by a restriction enzyme from *Streptomyces achromogenes* is

```
                          Cleavage site
                              ↓
        5'  C—C—G—C—G—G  3'
            :  :  :  :  :  :
        3'  G—G—C—G—C—C  5'
                  ↑
              Cleavage    Symmetry axis
              site
```

In each strand, the enzyme cleaves the CG phosphodiester bond on the 3' side of the symmetry axis.

More than ninety restriction enzymes have been purified and characterized. Their names consist of a three-letter abbreviation for the host organism (e.g., Eco for *E. coli*, Hin for *Haemophilus influenzae*, Hae for *H. aegyptius*) followed by a strain designation (if needed) and a Roman numeral (if more than one restriction enzyme is produced). The specificities of several of these enzymes are shown in Figure 6-2. Note that the cuts may be staggered or even.

Restriction enzymes are used to cleave DNA molecules into specific fragments that are more readily analyzed and manipulated than the parent molecule. For example, the 5.1-kilobase (kb) circular duplex DNA of the tumor-producing SV40 virus is cleaved at one site by EcoRI, four sites by HpaI, and eleven sites by HindIII. A piece of DNA produced by the action of one restriction enzyme can be specifically cleaved into smaller fragments by another restriction enzyme. The pattern of such fragments can serve as a *fingerprint* of a DNA molecule, as will be discussed shortly. Indeed, complex chromosomes containing hundreds of millions of base pairs can be mapped by using a series of restriction enzymes.

> *Kilobase (kb)*—
> A unit of length equal to 1000 base pairs of a double-stranded nucleic acid molecule (or 1000 bases of a single-stranded molecule).
> One kilobase of double-stranded DNA has a contour length of 0.34 μm and a mass of about 660 kd.

RESTRICTION FRAGMENTS CAN BE SEPARATED BY GEL ELECTROPHORESIS AND VISUALIZED

Small differences between related DNA molecules can be readily detected because their restriction fragments can be separated and displayed by gel electrophoresis. In many types of gels, the electrophoretic mobility of a DNA fragment is inversely proportional to the logarithm of the number of base pairs up to a certain limit. Polyacrylamide gels are used to separate fragments containing up to about 1000 base pairs, whereas more porous agarose gels are used to resolve mixtures of larger fragments (up to about 20 kb). An important feature of these gels is their high resolving power. In certain kinds of gels, fragments differing in length by just one nucleotide out of several hundred can be distinguished. Moreover, entire chromosomes containing millions of nucleotides can now be separated on agarose gels by applying pulsed electric fields in different directions (p. 46). Bands or spots of radioactive DNA in gels can be visualized by autoradiography. Alternatively, a gel can be stained with ethidium bromide, which fluoresces an intense orange when it has been bound to double-helical DNA. A band containing only 50 ng of DNA can readily be seen (Figure 6-3).

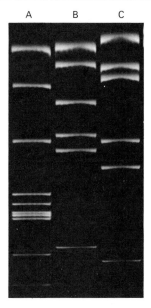

Figure 6-3
Gel electrophoresis pattern showing the fragments produced by cleaving SV40 DNA with each of three restriction enzymes. These fragments were made fluorescent by staining the gel with ethidium bromide. [Courtesy of Dr. Jeffrey Sklar.]

A restriction fragment containing a specific base sequence can be identified by hybridizing it with a labeled complementary DNA strand (Figure 6-4). A mixture of restriction fragments is separated by electro-

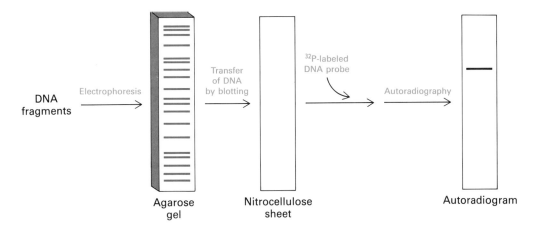

Figure 6-4
Southern blotting. A DNA fragment containing a specific sequence can be identified by separating a mixture of fragments by electrophoresis, transferring them to nitrocellulose, and hybridizing with a ^{32}P-labeled probe complementary to the sequence. The fragment containing the sequence is then visualized by autoradiography.

Restriction-fragment-length polymorphism (RFLP)—Southern blotting can be used to follow the inheritance of selected genes. Mutations within restriction sites change the sizes of restriction fragments and hence the positions of bands in Southern blot analyses. The existence of genetic diversity in a population is termed polymorphism. The detected mutation may itself cause disease or it may be closely linked to one that does. Genetic diseases such as sickle-cell anemia (p. 169), cystic fibrosis, and Huntington's chorea can be detected by RFLP analyses.

phoresis through an agarose gel, denatured to form single-stranded DNA, and transferred to a nitrocellulose sheet. The positions of the DNA fragments in the gel are preserved in the nitrocellulose sheet, where they can be hybridized with a ^{32}P-labeled single-stranded *DNA probe*. Autoradiography then reveals the position of the restriction fragment with a sequence complementary to that of the probe. A particular fragment in the midst of a million others can readily be identified in this way, like finding a needle in a haystack. This powerful technique is known as *Southern blotting* because it was devised by E. M. Southern. Likewise, RNA molecules can be separated by gel electrophoresis, and specific sequences can be identified by hybridization following transfer to nitrocellulose. This analogous technique for the analysis of RNA has been whimsically termed *Northern blotting*. A further play on words accounts for the term *Western blotting*, which refers to a technique for detecting a particular protein by staining with specific antibody (p. 63). Southern, Northern, and Western blotting are also known as DNA, RNA, and protein blotting.

DNA CAN BE SEQUENCED BY SPECIFIC CHEMICAL CLEAVAGE (MAXAM-GILBERT METHOD)

The analysis of DNA structure and its relation to gene expression has also been markedly facilitated by the development of powerful techniques for the sequencing of DNA molecules. The *chemical cleavage method* devised by Allan Maxam and Walter Gilbert starts with a DNA that is labeled at one end of one strand with ^{32}P. Polynucleotide kinase is usually used to add ^{32}P at the 5′-hydroxyl terminus. The labeled DNA is then broken preferentially at one of the four nucleotides. The conditions are chosen so that an average of one break is made per chain. In the reaction mixture for a given base, then, each broken chain yields a radioactive fragment extending from the ^{32}P label to one of the positions of that base, and such fragments are produced for every position of the base. For example, if the sequence is

$$5'-^{32}\text{P-GCTACGTA-}3'$$

the *radioactive* fragments produced by specific cleavage on the 5′ side of each of the four bases would be

Cleavage at A: ^{32}P-GCT
                   ^{32}P-GCTACGT

Cleavage at G: ^{32}P-GCTAC

Cleavage at C: ^{32}P-G
                   ^{32}P-GCTA

Cleavage at T: ^{32}P-GC
                   ^{32}P-GCTACG

The fragments in each mixture are then separated by polyacrylamide-gel electrophoresis, which can resolve DNA molecules differing in length by just one nucleotide. The next step is to look at an autoradiogram of this gel. In our example (Figure 6-5), the lowest band would be in the C lane, and the next one up in the T lane, followed by one in the A lane. Hence, the sequence of the first three nucleotides is 5′-CTA-3′ (the identity of the G at the 5′ end is not revealed). Reading all seven bands in ascending order gives the sequence 5′-CTACGTA-3′.

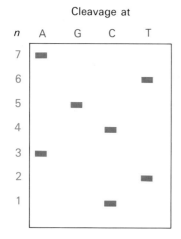

Figure 6-5
Schematic diagram of a gel showing the radioactive fragments formed by specific cleavage of 5′-^{32}P-GCTACGTA-3′ at each of the four bases (A, G, C, and T lanes). The number of nucleotides (*n*) in a fragment is shown on the left. The base sequence is read in ascending order.

Thus, the autoradiogram of a gel produced from four different chemical cleavages displays a pattern of bands from which the sequence can be read directly. In practice, DNA is specifically cleaved by reagents that modify and then remove certain bases from their sugars (Figure 6-6). Purines are damaged by *dimethylsulfate*, which methylates guanine at N-7 and adenine at N-3. The glycosidic bond of a methylated purine is readily broken by heating at neutral pH, which leaves the sugar without a base. Then the backbone is cleaved and the sugar unit eliminated by heating in alkali. When the resulting end-labeled fragments are resolved in a lane on a polyacrylamide gel, a pattern of dark and light bands is seen on the autoradiogram. The dark bands correspond to fragments formed by breakage at guanine, because this base is methylated much more rapidly than is adenine. The lane containing this sample is called the G lane because nearly all the cleavages are at G. In contrast, the glycosidic bond of methylated adenosine is less stable to dilute acid than that of methylated guanosine. Hence, treatment with dilute acid after methylation causes cleavages at both A and G, giving rise to an A + G lane. Comparison of this lane of a gel with a parallel G lane reveals whether cleavage occurred at A or G (Figure 6-7). Cytosine and thymine are split by *hydrazine*. The backbone is then cleaved by *piperidine*, which displaces the products of the hydrazine reaction and catalyzes elimination of the phosphates. The resulting mixture gives rise to a C + T lane, a series of bands of about equal intensity from cleavages at cytosines and thymines. These pyrimidines are distinguished by preparing a C lane, for which hydrazinolysis is carried out in the presence of 2 M NaCl. (This suppresses the reaction with thymine.)

The DNA sequence is then read from the autoradiogram of the gel by comparing the G, A + G, C + T, and C lanes (Figure 6-7). The shortest fragment has the highest electrophoretic mobility and so the 5' end of the sequence is at the bottom of the gel. Sequences of more than 250 bases can be readily determined by running several gels for different time intervals to resolve all of the labeled fragments.

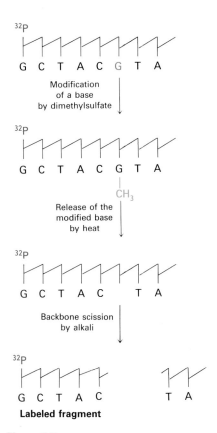

Figure 6-6
Strategy of the chemical-cleavage method for the sequencing of DNA. This particular procedure would produce the fragments visualized in the G lane of a set of gels.

DNA CAN BE SEQUENCED BY CONTROLLED INTERRUPTION OF REPLICATION (SANGER DIDEOXY METHOD)

DNA can also be sequenced by generating fragments through the *controlled interruption of enzymatic replication*, a method developed by Frederick Sanger and his associates. DNA polymerase I is used to copy a particular sequence of a single-stranded DNA. The synthesis is primed by a complementary fragment, which may be obtained from a restriction enzyme digest or synthesized chemically. In addition to the four deoxyribonucleoside triphosphates (radioactively labeled), the incubation mixture contains a *2',3'-dideoxy analog* of one of them. The incorporation of this analog blocks further growth of the new chain because it lacks the 3'-hydroxyl terminus needed to form the next phosphodiester bond. Hence, fragments of various lengths are produced in which the

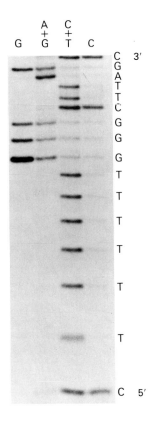

Figure 6-7
Autoradiogram of a gel showing labeled fragments produced by chemical cleavage. The 5' end of the DNA strand was labeled with ^{32}P. The shortest nucleotide is at the bottom of the gel. Hence, the base sequence is 5'-CTTTTTTGGGCTTAGC-3'. [Courtesy of Dr. David Dressler.]

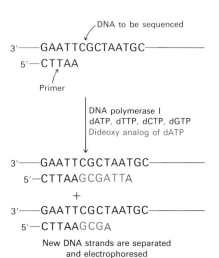

Figure 6-8
Strategy of the chain-termination method for the sequencing of DNA. Fragments are produced by adding the 2′,3′-dideoxy analog of a dNTP to each of four polymerization mixtures. For example, the addition of the dideoxy analog of dATP (shown in red) results in fragments ending in A.

Figure 6-9
Fluorescence detection of oligonucleotide fragments produced by the dideoxy method. Each of the four chain-terminating mixtures is primed with a tag that fluoresces at a different wavelength (e.g., blue for A). The sequence determined by fluorescence measurements at four wavelengths is shown at the bottom of the figure. [From L. M. Smith, J. Z. Sanders, R. J. Kaiser, P. Hughes, C. Dodd, C. R. Connell, C. Heiner, S. B. H. Kent, and L. E. Hood. *Nature* 321(1986):674.]

dideoxy analog is at the 3′ end (Figure 6-8). Four such sets of *chain-terminated fragments* (one for each dideoxy analog) are then electrophoresed, and the base sequence of the new DNA is read from the autoradiogram of the four lanes.

The complete sequence of the 5386 bases in φX174 DNA was determined in this way by Sanger and co-workers in 1977, just a quarter century after Sanger's pioneering elucidation of the amino acid sequence of a protein. This accomplishment is a landmark in molecular biology because it revealed the total information content of a DNA genome. This tour de force was followed several years later by another one, the determination of the sequence of human mitochondrial DNA, a double-stranded circular DNA molecule containing 16,569 base pairs. It encodes two ribosomal RNAs, 22 transfer RNAs, and 13 proteins. This achievement was followed by the sequencing of the 48,513 base pairs of the DNA of λ phage, a virus that infects *E. coli*. The wealth of information derived from these remarkable accomplishments will be considered in detail in later chapters.

About 5×10^6 bases of DNA have been sequenced in laboratories around the world since the introduction of the Maxam-Gilbert and Sanger methods. All of these studies have used autoradiographic images of gels to ascertain the lengths of DNA fragments generated by chemical cleavage and controlled interruption of replication. Recently, a variant of the dideoxy method has been devised. A fluorescent tag is attached to the oligonucleotide primer—a differently colored one in each of the four chain-terminating reaction mixtures (e.g., a blue emitter for termination at A and a red one for termination at C). The reaction mixtures are combined and electrophoresed together. The separated bands of DNA are then detected by their fluorescence as they pass out the bottom of the tube, and the sequence of their colors directly yields the base sequence (Figure 6-9). Sequences of up to 500 bases can now be determined in this way. An attractive feature of this fluorescence detection method is that it can readily be automated. The sequencing of the en-

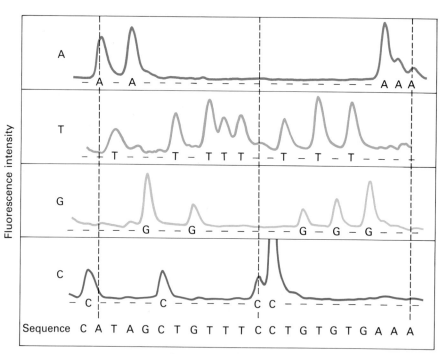

tire *E. coli* genome (3×10^6 base pairs) has now become feasible. We can even begin to think about determining the sequence of extensive stretches of the human genome, which contains 3×10^9 base pairs.

DNA PROBES AND GENES CAN BE SYNTHESIZED BY AUTOMATED SOLID-PHASE METHODS

DNA strands, like polypeptides (p. 66), can be synthesized by the sequential addition of activated monomers to a growing chain that is linked to an insoluble support. The activated monomers are protonated *deoxyribonucleoside 3'-phosphoramidites* (Figure 6-10). In step 1, the

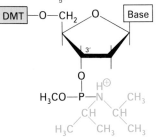

Protonated phosphoramidite
(The 5'-hydroxyl is blocked by a dimethoxytrityl protecting group.)

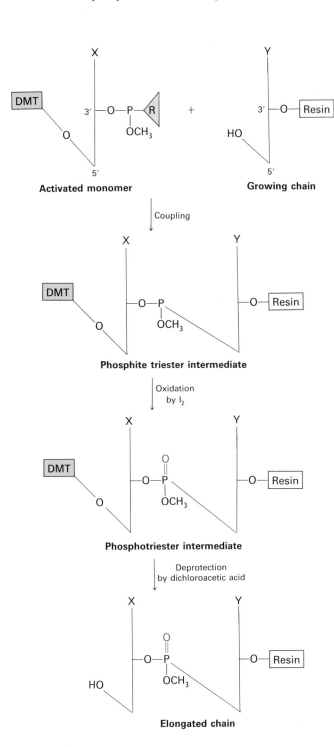

Figure 6-10
Solid-phase synthesis of a DNA chain by the phosphite triester method. The activated monomer added to the growing chain is a deoxyribonucleoside 3'-phosphoramidite containing a DMT (dimethoxytrityl) blocking group on its 5'-oxygen atom.

3′-phosphorus atom of this incoming unit becomes joined to the 5′-oxygen of the growing chain to form a *phosphite triester*. The 5′-OH of the activated monomer is unreactive because it is blocked by a dimethoxytrityl (DMT) protecting group. Likewise, amino groups on the purine and pyrimidine bases are blocked. Coupling is carried out under anhydrous conditions because water reacts with phosphoramidites. In step 2, the phosphite triester (in which P is trivalent) is oxidized by iodine to form a *phosphotriester* (in which P is pentavalent). In step 3, the DMT protecting group on the 5′-OH of the growing chain is removed by addition of dichloroacetic acid, which leaves other protecting groups intact. The DNA chain is now elongated by one unit and ready for another cycle of addition. Each monomer addition cycle takes only about ten minutes and elongates more than 98% of the chains.

This solid-phase approach is ideal for the synthesis of DNA, as it is for polypeptides, because the desired product stays on the insoluble support until the final release step. All of the reactions occur in a single vessel, and excess soluble reagents can be added to drive reactions to completion. At the end of each step, soluble reagents and by-products are washed away from the glass beads that bear the growing chains.

After assembly of the desired DNA chain, the methyl groups protecting the phosphates are removed by addition of thiophenol. The DNA strand is then released from the glass bead by cleavage of the ester bond between the 3′-OH of the terminal nucleoside and the resin that links it to the glass support. This bond is hydrolyzed by the addition of concentrated ammonium hydroxide. Finally, the benzoyl and isobutyryl groups protecting the bases are removed by heating the DNA in ammonium hydroxide. Because elongation is never 100% complete, the new DNA chains are of diverse lengths—the desired chain is the longest one. The sample can be purified by high-performance liquid chromatography or by electrophoresis on polyacrylamide gels. DNA chains up to 100 nucleotides long can readily be synthesized by this automated method.

The ability to rapidly synthesize DNA chains of any selected sequence opens many experimental avenues. For example, an oligonucleotide labeled at one end with ^{32}P can be used to search for a complementary sequence in a very long DNA molecule or even in a genome consisting of many chromosomes. The use of labeled oligonucleotides as *DNA probes* is powerful and general. For example, a DNA probe that is base-paired to a known complementary sequence in a chromosome can serve as the starting point of an exploration of adjacent uncharted DNA. For example, the probe can be used as a *primer* to initiate the replication of neighboring DNA by DNA polymerase. One of the most exciting applications of the solid-phase approach is the *synthesis of new tailor-made genes*. New proteins with novel properties can now be produced in abundance by expressing synthetic genes. *Protein engineering* has become a reality. Moreover, regulatory sequences in DNA can be changed at will to control gene expression.

NEW GENOMES CAN BE CONSTRUCTED, CLONED, AND EXPRESSED

The pioneering work of Paul Berg, Herbert Boyer, and Stanley Cohen in the early 1970s led to the development of recombinant DNA technology, which has revolutionized biochemistry. New combinations of unre-

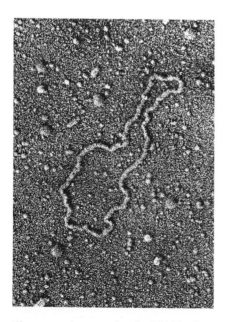

Electron micrograph of pSC101, the first plasmid vector used in the cloning of DNA. [Courtesy of Dr. Stanley N. Cohen.]

lated genes can be constructed in the laboratory by applying recombinant DNA techniques. These novel combinations can be cloned, amplified manyfold, by introducing them into suitable cells, where they are replicated by the DNA synthesizing machinery of the host. The inserted genes are often transcribed and translated in their new setting. What is most striking is that the genetic endowment of the host can be permanently altered in a designed way.

The major steps in the cloning of DNA are (Figure 6-11):

1. *Construction of a recombinant molecule.* A DNA fragment of interest is covalently joined to a DNA *vector*. The essential feature of a vector is that it can replicate autonomously in an appropriate host. Plasmids (naturally occurring circles of DNA that act as accessory chromosomes) and λ phage (a virus) are choice vectors for cloning in *E. coli*.

2. *Introduction into host cells.* Many bacterial and eucaryotic cells take up naked DNA molecules from the medium. The efficiency of uptake is low (about 1 of 10^6 DNA molecules), but an appreciable proportion of cells can be transformed under appropriate experimental conditions. Mutant bacteria that do not rapidly degrade foreign DNA are often used as host cells. DNA molecules can also be injected into many animal and plant cells. Alternatively, target cells can be infected with virus particles reassembled to harbor the recombinant DNA molecule. In this synthetic viral genome, a gene of interest replaces a segment of viral DNA that is not essential for replication.

Chimeric DNA—
A recombinant DNA molecule containing unrelated genes. From *chimera*, a mythological creature with the head of a lion, the body of a goat, and the tail of a serpent.

" . . . a thing of immortal make, not human, lion-fronted and snake behind, a goat in the middle, and snorting out the breath of the terrible flame of bright fire. . . ."

Iliad (6.179)

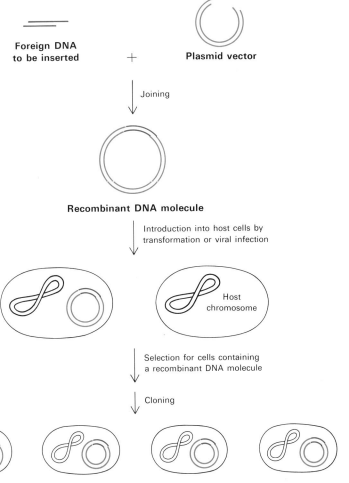

Figure 6-11
Construction and cloning of recombinant DNA molecules.

3. *Selection.* The next step is to determine which cells harbor the recombinant DNA molecule containing the gene of interest. The desired cells can be selected by the presence either of the vector or of the inserted gene itself. For example, some plasmid vectors confer resistance to an antibiotic, which can be used to eliminate the undesired cells. Another approach is to culture individual cells, then test a sample from each clone for the desired DNA sequence by Southern blotting with a labeled complementary probe. An expressed protein can be detected by Western blotting with a specific antibody, or by assaying for a functional property such as the appearance of functional enzymatic activity. Many cell lines containing recombinant DNA are genetically stable.

RESTRICTION ENZYMES AND DNA LIGASE ARE KEY TOOLS IN FORMING RECOMBINANT DNA MOLECULES

Let us begin by seeing how novel DNA molecules can be constructed in the laboratory. The vector in a recombinant DNA experiment can be prepared for splicing by cleaving it at a single specific site with a restriction enzyme. For example, the plasmid pSC101 (a 9.9-kb double-helical circular DNA molecule) is split at a unique site by the EcoRI restriction enzyme. The staggered cuts made by this enzyme produce *complementary single-stranded ends*, which have specific affinity for each other and hence are known as *cohesive ends*. Any DNA fragment can be inserted into this plasmid if it has the same cohesive ends. Such a fragment can be prepared from a larger piece of DNA by using the same restriction enzyme as was used to open the plasmid DNA. The single-stranded ends of the fragment are then complementary to those of the cut plasmid. The DNA fragment and the cut plasmid can be annealed and then joined by *DNA ligase*, which catalyzes the formation of a phosphodiester bond between two DNA chains (Figure 6-12). DNA ligase requires a free OH group at the 3' end of one DNA chain and a phosphate group at the 5' end of the other. Furthermore, the chains joined by ligase must belong to double-helical DNA molecules. An energy source, such as ATP or NAD^+ is required for the joining reaction, which will be discussed in Chapter 27.

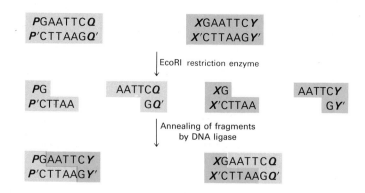

Figure 6-12
Joining of DNA molecules by the cohesive-end method. One of the parental DNA molecules (shown in green) carries genes *P* and *Q* separated by a restriction site, and the other (shown in red) carries *X* and *Y*. One of the recombinant molecules contains *P* and *Y*, and the other contains *Q* and *X*.

This cohesive-end method for joining DNA molecules can be made general by using a *short, chemically synthesized DNA linker* that can be cleaved by restriction enzymes. First, the linker is covalently joined to the ends of a DNA fragment or vector. For example, the 5' ends of a decameric linker and a DNA molecule are phosphorylated by polynucleotide kinase and then joined by the ligase from T4 phage (Figure 6-13). This ligase can form a covalent bond between blunt-ended (flush-ended) double-helical DNA molecules. Cohesive ends are produced when these terminal extensions are cut by an appropriate restriction enzyme. Thus, *cohesive ends corresponding to a particular restriction enzyme can be added to virtually any DNA molecule.* We see here the fruits of combining enzymatic and synthetic chemical approaches in crafting new DNA molecules.

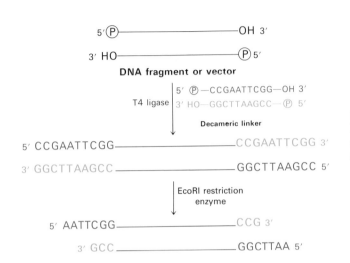

Figure 6-13
Formation of cohesive ends by the addition and cleavage of a chemically synthesized linker.

PLASMIDS AND LAMBDA PHAGE ARE CHOICE VECTORS FOR DNA CLONING IN BACTERIA

Many plasmids and bacteriophages have been ingeniously modified to enhance the delivery of recombinant DNA molecules into bacteria and to facilitate the selection of bacteria harboring them. *Plasmids* are naturally occurring circular duplex DNA molecules ranging in size from two kilobases to several hundred kilobases. They carry genes for the inactivation of antibiotics, the production of toxins, and the breakdown of natural products. These *accessory chromosomes* can replicate independently of the host chromosome. In contrast with the host genome, they are dispensable under certain conditions. A bacterial cell may have no plasmids at all or it may house as many as twenty copies of a plasmid.

One of the most useful plasmids for cloning is pBR322, which contains genes for resistance to tetracycline and ampicillin (an antibiotic like penicillin). This plasmid can be cleaved at a variety of unique sites by different endonucleases, and DNA fragments inserted. Insertion of DNA at the EcoRI restriction site does not alter either of the genes for antibiotic resistance (Figure 6-14). However, insertion at the HindIII, SaII, or BamHI restriction site inactivates the gene for tetracycline resistance, an effect called *insertional inactivation*. Cells containing pBR322 with a DNA insert at one of these restriction sites are resistant to ampicillin but sensitive to tetracycline, and so they can be readily *selected*. Cells that failed to take up the vector are sensitive to both antibiotics, whereas cells containing pBR322 without a DNA insert are resistant to both.

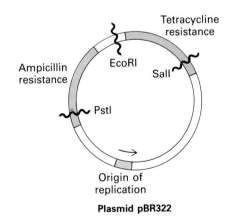

Figure 6-14
Genetic map of pBR322, a plasmid with two genes for antibiotic resistance.

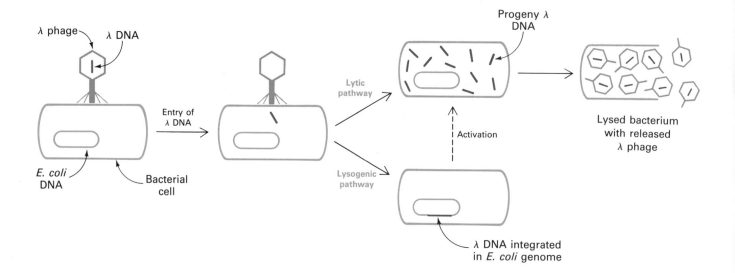

Figure 6-15
Lambda phage can multiply within a host and lyse it (lytic pathway) or its DNA can become integrated into the host genome (lysogenic pathway), where it is dormant until activated.

Lambda (λ) *phage* is another widely-used vector (Figure 6-15). This bacteriophage enjoys a choice of life styles: it can destroy its host or it can become part of its host (Chapter 34). In the *lytic pathway*, viral functions are fully expressed: viral DNA and proteins are quickly produced and packaged into virus particles, which leads to the lysis (destruction) of the host cell and the sudden appearance of about 100 progeny virus particles, or *virions*. In the *lysogenic pathway*, the phage DNA becomes inserted into the host-cell genome and can be replicated together with host-cell DNA for many generations, remaining inactive. Certain environmental changes can trigger the expression of this dormant viral DNA, which leads to the formation of progeny virus and lysis of the host. Large segments of the 48-kb DNA of λ phage are not essential for productive infection and can be replaced by foreign DNA.

Mutant λ phages designed for the cloning of DNA have been constructed. One of the mutants, called λgt-λβ, contains only two EcoRI cleavage sites instead of the five normally present (Figure 6-16). After

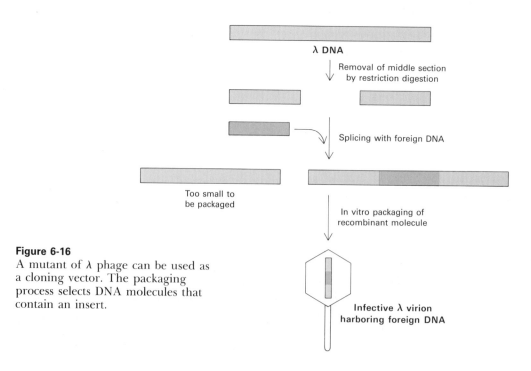

Figure 6-16
A mutant of λ phage can be used as a cloning vector. The packaging process selects DNA molecules that contain an insert.

cleavage, the middle segment of this λ DNA molecule can be removed. The two remaining pieces of DNA have a combined length equal to 72% of a normal genome length. This amount of DNA is too little to be packaged into a λ particle. The range of lengths that can be readily packaged is from 75% to 105% of a normal genome length. However, *a suitably long DNA insert (such as 10 kb) between the two ends of λ DNA enables such a recombinant DNA molecule (93% of normal length) to be packaged.* Nearly all infective λ particles formed in this way will contain an inserted piece of foreign DNA. Another advantage of using these modified viruses as vectors is that they enter bacteria much more easily than do plasmids. A variety of λ mutants have been constructed for use as cloning vectors. One of them, called a *cosmid*, can serve as a vector for large DNA inserts (up to about 45 kb).

M13 phage is another very useful vector for cloning DNA. This filamentous virus is 900 nm long and only 9 nm wide (Figure 6-17). Its 6.4-kb single-stranded circle of DNA is protected by a coat of 2710 identical protein subunits. M13 enters *E. coli* through the bacterial sex pilus, a protein appendage. The single-stranded DNA in the virus particle (called the + strand) is replicated by a double-stranded replicative form (RF) containing + and − strands, much as in φX174 (p. 85). Only the + strand is packaged into new virus particles. About a thousand progeny M13 are produced per generation. A striking feature of M13 is that it does not kill its bacterial host. Consequently, large quantities of M13 can be grown and easily harvested (1 g from 10 liters of culture fluid).

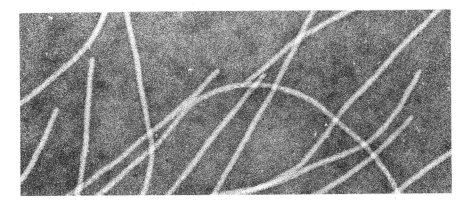

Figure 6-17
Electron micrograph of M13 filamentous phage. [Courtesy of Dr. Robley Williams.]

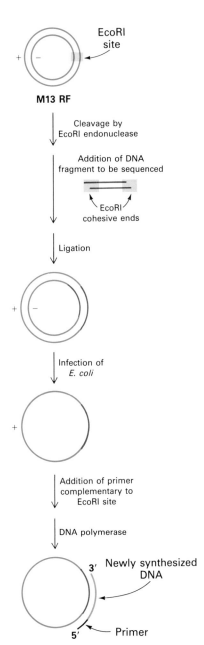

Figure 6-18
Sequencing by the dideoxy method of a DNA fragment inserted into M13 phage DNA. Synthesis of a new strand is primed by an oligonucleotide that is complementary to the restriction sequence adjacent to the inserted DNA.

M13 is prepared for cloning by cutting its circular double-stranded RF at a single site with a restriction enzyme. A double-stranded foreign DNA fragment produced by cleavage with the same restriction enzyme is then ligated to the cut RF (Figure 6-18). The foreign DNA can be inserted into the RF in two different orientations because the ends of both DNA molecules are the same. Hence, half of the new + strands packaged into virus particles will contain one of the strands of the foreign DNA, and half will contain the other strand. Infection of *E. coli* by a single virus particle will yield a large amount of single-stranded M13 DNA containing the same strand of the foreign DNA. The sequence in M13 DNA adjacent to the inserted DNA is known because it is the target for cleavage by the restriction enzyme. Consequently, a synthetic

SPECIFIC GENES CAN BE CLONED FROM A DIGEST OF GENOMIC DNA

Ingenious cloning and selection methods have made feasible the isolation of a specific DNA segment several kilobases long out of a genome containing more than 3×10^6 kb. Let us see how a gene that occurs just once in a human genome can be cloned. A sample of the total genomic DNA is first mechanically sheared or partly digested by restriction enzymes into large fragments (Figure 6-19). This nearly random population of overlapping DNA molecules is then separated by gel electrophoresis into a set of fragments about 20 kb long. Synthetic linkers are attached to the ends of these fragments, cohesive ends are formed, and the fragments are then inserted into a vector, such as λ phage DNA, prepared with the same cohesive ends. The in vitro packaging of DNA into virus particles selects for recombinant DNA molecules that contain a large insert. *E. coli* are then infected with these recombinant phages. The resulting lysate contains fragments of human DNA housed in a large number of virus particles. These constitute a *genomic library* because they contain fragments of the entire human genome. Phage can be propagated indefinitely, and so the library can be used repeatedly over long periods.

This genomic library is then screened to find the very small proportion of phage harboring the gene of interest. A calculation shows that a 99% probability of success requires screening about 500,000 clones; hence, a very rapid and efficient screening process is essential. This can be accomplished by DNA hybridization.

A dilute suspension of the recombinant phage is first plated on a lawn of bacteria (Figure 6-20). Where each phage particle has landed and infected a bacterium, a plaque develops on the plate. Then a replica of this "master" plate is made by applying a sheet of nitrocellulose. Infected bacteria and phage DNA released from lysed cells adhere to the sheet in a pattern of spots corresponding to the plaques. Intact bacteria on this sheet are lysed with NaOH, which also serves to denature the DNA so that it becomes accessible for hybridization with a ^{32}P-labeled probe. *The presence of a specific DNA sequence in a single spot on the replica can be detected by using a radioactive complementary DNA or RNA molecule as a probe.* Autoradiography then reveals the positions of spots harboring recombinant DNA. The corresponding plaques are picked out of the intact master plate and grown. A million clones can readily be screened in a day by a single investigator.

This method makes it possible to isolate virtually any gene, *provided that a probe is available*. How does one obtain a specific probe? One approach is to *start with the corresponding mRNA from cells in which it is abundant*. For example, precursors of red blood cells contain large amounts of mRNAs for hemoglobin, and plasma cells are rich in mRNAs for antibody molecules. mRNAs from these cells can be fractionated by size to enrich for the one of interest. As will be described shortly, a DNA complementary to this mRNA can be synthesized in vitro and cloned to produce a highly specific probe.

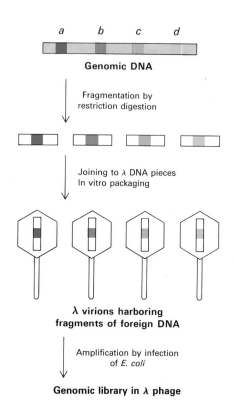

Figure 6-19
Creating a genomic library from a digest of a whole eucaryotic genome.

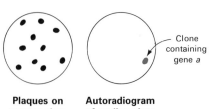

Figure 6-20
Screening a genomic library for a specific gene. Here, a plate is tested for plaques containing gene *a* of Figure 6-19.

Alternatively, *a probe for a gene can be prepared if part of the amino acid sequence of the protein encoded by the gene is known.* A problem arises because a given peptide sequence can be encoded by a number of oligonucleotides. Thus, for this purpose, peptide sequences containing tryptophan and methionine are preferred, because these amino acids are specified by a single codon, whereas other amino acid residues have between two and six codons (p. 107). The strategy is to choose a peptide that has a high proportion of tryptophan and methionine. For example, even for the pentapeptide

Trp-Tyr-Met-Cys-Met

there are four possible DNA coding sequences because tyrosine and cysteine can each be specified by either of two codons.

A mixture of all the coding DNA sequences (or their complements) is synthesized by the solid-phase method and made radioactive by phosphorylating their 5′ ends with ^{32}P-orthophosphate. The replica plate is exposed to these probes and autoradiographed to identify any clone with a complementary DNA sequence. Among these, the ones containing the desired gene can be identified by sequencing the recombinant DNA from their plaques to determine whether it matches the known amino acid sequence of the protein of interest.

A typical genomic DNA library consists of DNA fragments about 20 kb long. How can we obtain information about longer stretches of DNA, say 300 kb long? Recall that the fragments in the library are produced by random cleavage of many DNA molecules and so some of the fragments overlap one another. Suppose that a fragment containing region A selected by hybridization with a complementary probe A′ also contains region B (Figure 6-21). A new probe B′ can be prepared by cleaving this fragment and subcloning region B. If the library is screened again with probe B′, new fragments containing region B will be found. Some will contain a contiguous region C. Hence, we now have information about a segment of DNA encompassing regions A, B, and C. This process of subcloning and rescreening is called *chromosome walking*. Long stretches of DNA can be analyzed in this way provided that each of the new probes is complementary to a unique region.

3′ ACC-ATA-TAC-ACA-TAC 5′
ACC-ATG-TAC-ACA-TAC
ACC-ATA-TAC-ACG-TAC
ACC-ATG-TAC-ACG-TAC
Trp-Tyr-Met-Cys-Met

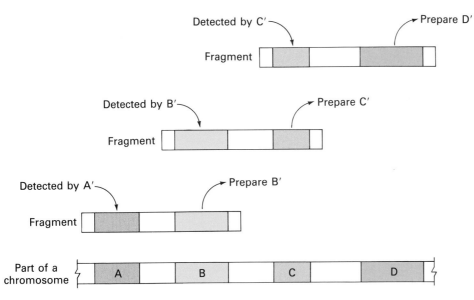

Figure 6-21
Chromosome walking. Long regions of unknown DNA can be explored starting with a known base sequence by subcloning and rescreening.

COMPLEMENTARY DNA (cDNA) PREPARED FROM mRNA CAN BE EXPRESSED IN HOST CELLS

Can mammalian DNA be cloned and expressed by *E. coli*? Recall that most mammalian genes are mosaics of introns and exons. These interrupted genes cannot be expressed by bacteria, which lack the machinery to splice introns out of the primary transcript. However, this difficulty can be circumvented by causing bacteria to take up recombinant DNA that is complementary to mRNA. For example, proinsulin, a precursor of insulin, is synthesized by bacteria harboring plasmids that contain DNA complementary to mRNA for proinsulin (Figure 6-22). Indeed, much of the insulin used today by millions of diabetics is produced by bacteria.

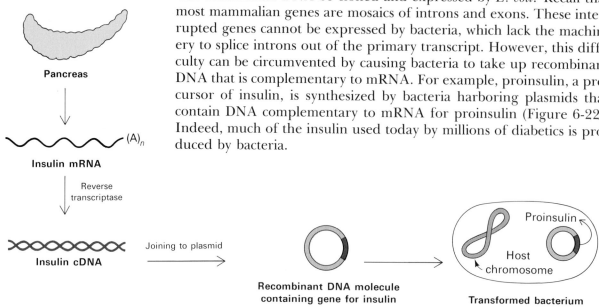

Figure 6-22
Synthesis of proinsulin, a precursor of insulin, by transformed clones of *E. coli*.

The key to forming complementary DNA (cDNA) is the enzyme *reverse transcriptase*. As was discussed earlier (p. 87), retroviruses use this enzyme to form a DNA-RNA hybrid in the replication of their genomic RNA. Reverse transcriptase synthesizes a DNA strand complementary to an RNA template if it is provided with a primer that is base-paired to the RNA and contains a free 3'-OH group. We can use this enzyme to synthesize DNA from mRNA by providing an oligo-dT primer that pairs with the poly-A sequence at the 3' end of most eucaryotic mRNA molecules (Figure 6-23). The rest of the cDNA strand is then synthesized in the presence of the four deoxyribonucleoside triphosphates. The RNA strand of this RNA-DNA hybrid is subsequently hydrolyzed by raising the pH. Unlike RNA, DNA is resistant to alkaline hydrolysis. The 3'-end of the newly-synthesized DNA strand then forms a hairpin loop and primes the synthesis of the opposite DNA strand. The hairpin loop is removed by digestion with S1 nuclease, which recognizes unpaired nucleotides. Synthetic linkers can be added to this double-helical DNA for ligation to a suitable vector.

cDNA molecules can be inserted into vectors that favor their efficient expression in hosts such as *E. coli*. Such plasmids or phages are called *expression vectors*. To maximize transcription, the cDNA is inserted into

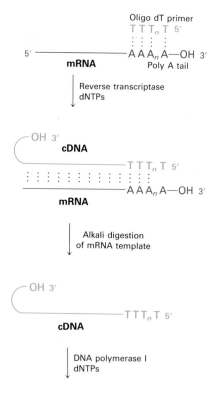

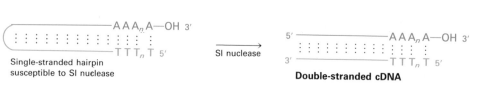

Figure 6-23
Forming a cDNA duplex from mRNA by using reverse transcriptase.

the vector in the correct reading frame near a strong bacterial promoter. In addition, these vectors assure efficient translation by encoding a ribosome-binding site on the mRNA near the initiation codon. *cDNA clones can be screened on the basis of their capacity to direct the synthesis of a foreign protein in bacteria.* A radioactive antibody specific for the protein of interest can be used to identify the colonies of bacteria that harbor the corresponding cDNA vector (Figure 6-24). As before, spots of bacteria on a replica plate are lysed to release proteins, which bind to an applied nitrocellulose filter. A ^{125}I-labeled antibody is added, and autoradiography reveals the location of the desired colonies on the master plate. This *immunochemical screening* approach can be used whenever a protein is expressed and corresponding antibody is available.

NEW GENES INSERTED INTO EUCARYOTIC CELLS CAN BE EFFICIENTLY EXPRESSED

Bacteria are ideal hosts for the amplification of DNA molecules. They can also serve as factories for the production of a wide range of procaryotic and eucaryotic proteins. However, posttranslational modifications such as specific cleavages of polypeptides and attachment of carbohydrate units are not carried out by bacteria, because they lack the necessary enzymes. Thus, many eucaryotic genes can be correctly expressed only in eucaryotic host cells. Another motivation for introducing recombinant DNA molecules into cells of higher organisms is to gain insight into how their genes are organized and expressed: How are genes turned on and off in embryological development? How does a fertilized egg give rise to an organism with highly differentiated cells that are organized in space and time? These central questions of biology can now be fruitfully approached because it has become feasible to express foreign genes in mammalian cells as well as in bacteria.

Recombinant DNA molecules can be introduced into animal cells in several ways. In one, foreign DNA molecules precipitated by *calcium phosphate* are taken up by animal cells. A small fraction of the imported DNA becomes stably integrated into the chromosomal DNA. The efficiency of incorporation is low but the method is useful because it is easy to apply. In another method, DNA is injected into cells. A fine-tipped (0.1 μm diameter) glass micropipet containing a solution of foreign DNA is inserted into a nucleus (Figure 6-25). A skilled investigator can inject hundreds of cells per hour. About 2% of injected mouse cells are viable and contain the new gene. In a third method, *viruses* can be used to bring new genes into animal cells. The most effective vectors are

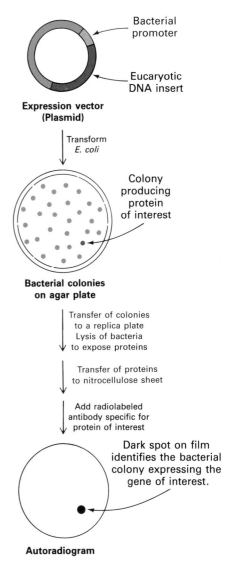

Figure 6-24
Screening of expression vector products by staining with specific antibody.

Micropipet with DNA solution

Fertilized mouse egg

Holding pipet

Figure 6-25
Microinjection of cloned plasmid DNA into the male pronucleus of a fertilized mouse egg.

retroviruses (RNA tumor viruses) (Figure 6-26). As discussed earlier (p. 87), these viruses replicate through DNA intermediates, the reverse of the normal flow of information (hence the prefix *retro*). A striking feature of the life cycle of a retrovirus is that the double-helical DNA form of its genome, produced by the action of reverse transcriptase, becomes randomly incorporated into host chromosomal DNA. This DNA version of the viral genome, called proviral DNA, can be efficiently expressed by the host cell and replicated along with normal cellular DNA (Chapter 34). Retroviruses do not usually kill their hosts. Foreign genes have been efficiently introduced into mammalian cells by infecting them with vectors derived from *Maloney murine leukemia virus*, which can accept inserts as long as 6 kb. Some genes introduced by this retroviral vector into the genome of a transformed host cell are efficiently expressed.

Genetically engineered giant mice (Figure 6-27) illustrate the expression of foreign genes in mammalian cells. *Growth hormone* (somatotropin), a 21-kd protein, is normally synthesized by the pituitary gland. A deficiency of this hormone produces dwarfism and an excess leads to gigantism. The gene for rat growth hormone was placed next to the mouse metallothionein promoter on a plasmid (Figure 6-28). This promoter is normally located on a chromosome, where it controls the transcription of metallothionein, a cysteine-rich protein that has high affinity for heavy metals. The synthesis of this protective protein by the liver is induced by heavy-metal ions such as cadmium. Several hundred copies of this plasmid were microinjected into the male pronucleus of a fertilized mouse egg, which was then inserted into the uterus of a foster-mother mouse. A number of mice that developed from these microinjected eggs contained the gene for rat growth hormone, as shown by Southern blots of their DNA. These *transgenic mice*, containing multiple copies (~30 per cell) of the rat growth hormone gene, grew much more rapidly than did control mice. The level of growth hormone in these mice was 500 times as high as in normal mice and their body weight at maturity was twice normal. The foreign DNA had been transcribed and its five introns correctly spliced out to form functional mRNA. *These experiments strikingly demonstrate that a foreign gene under the control of a new promoter can be integrated and efficiently expressed in mammalian cells.*

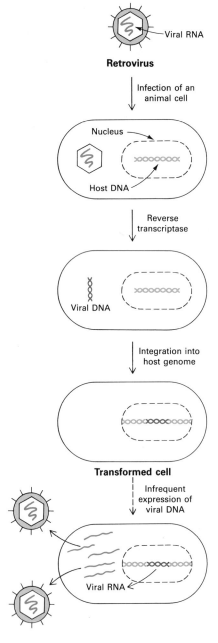

Figure 6-26
Life cycle of a retrovirus.

Figure 6-27
Injection of the gene for growth hormone into a fertilized mouse egg gave rise to giant mouse (left), about twice the weight of his sibling (right). [Courtesy of Dr. Ralph Brinster.]

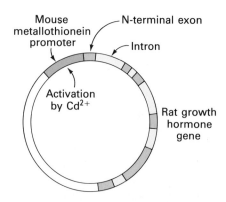

Figure 6-28
The gene for rat growth hormone was inserted into a plasmid next to the metallothionein promoter, which is activated by the addition of heavy metals, such as cadmium ion.

TUMOR-INDUCING (Ti) PLASMIDS CAN BE USED TO BRING NEW GENES INTO PLANT CELLS

The common soil bacterium *Agrobacterium tumefaciens* infects plants and introduces foreign genes into them (Figure 6-29). A lump of tumor tissue called a *crown gall* grows at the site of infection. Crown galls synthesize opines, a group of amino acid derivatives that are metabolized by the infecting bacteria. In essence, the metabolism of the plant cell is diverted to satisfy the highly distinctive appetite of the intruder. The instructions for the synthesis of opines and the switch to the tumor state come from *Ti plasmids* (tumor-inducing plasmids) that are carried by *Agrobacterium*. A small portion of the Ti plasmid becomes integrated into the genome of infected plant cells; this 20-kb segment is called *T-DNA* (transferred DNA).

Ti plasmid derivatives can be used as vectors to deliver foreign genes into plant cells (Figure 6-30). First, a segment of foreign DNA is inserted into the T-DNA region of a small plasmid by restriction enzymes and ligases. This synthetic plasmid is added to *Agrobacterium* colonies harboring naturally occurring Ti plasmids. By recombination, Ti plasmids containing the foreign gene are formed. These Ti vectors hold great promise for exploring the genomes of plant cells and modifying plants to improve their agricultural value and crop yield. However, they are not suitable for transforming all types of plants. Ti-plasmid transfer works with dicots (broad-leaved plants such as grapes) and a few kinds of monocots but not with economically important cereal monocots.

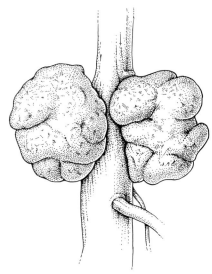

Figure 6-29
Crown gall, a plant tumor, is caused by a bacterium (*Agrobacterium tumefaciens*) that carries a tumor-inducing plasmid (Ti plasmid).

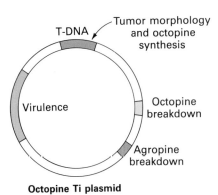

Figure 6-30
Agrobacteria containing Ti plasmids can deliver foreign genes into some plant cells. [After M. Chilton. A vector for new genes into plant. Copyright © 1983 by Scientific American, Inc. All rights reserved.]

Foreign DNA has recently been introduced into cereal monocots as well as dicots by applying intense electric fields, a technique called *electroporation* (Figure 6-31). First, the cellulose wall surrounding plant cells is removed by adding cellulases; this produces *protoplasts*, plant cells with exposed plasma membranes. Electric pulses then are applied to a suspension of protoplasts and plasmid DNA. Because high electric fields make membranes transiently permeable to large molecules, plasmid DNA molecules enter the cells. The cell wall is then allowed to reform, which results in viable plant cells. Maize cells and carrot cells have been stably transformed in this way with plasmid DNA that includes genes for resistance to antibiotics. Moreover, the plasmid DNA is efficiently expressed by the transformed cells.

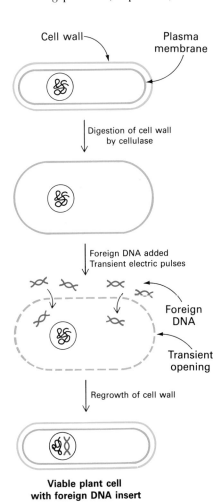

Figure 6-31
Foreign DNA can be introduced into plant cells by electroporation, applying intense electric fields to make their plasma membranes transiently permeable.

NOVEL PROTEINS CAN BE ENGINEERED BY SITE-SPECIFIC MUTAGENESIS

Much has been learned about genes and proteins by selecting genes from the repertoire offered by nature. In the classic genetic approach, mutations are generated randomly throughout the genome and those exhibiting a particular phenotype are selected. Analysis of these mutants then reveals which genes are altered, and DNA sequencing identifies the precise nature of the changes. *Recombinant DNA technology now makes it feasible to create specific mutations in vitro.* We can construct new genes with designed properties by four kinds of directed changes: *deletion, insertion, transposition,* and *substitution.*

A specific deletion can be produced by cleaving a plasmid at two sites with a restriction enzyme and religating to form a smaller circle (Figure 6-32). This simple approach usually removes a large block of DNA. A smaller deletion can be made by cutting a plasmid at a single site. The ends of the linear DNA are then digested with an exonuclease that removes nucleotides from both strands. The shortened piece of DNA is then religated to form a circle that is missing a short length of DNA about the restriction site.

Mutant proteins with single amino acid substitutions can readily be produced by *oligonucleotide-directed mutagenesis* (Figure 6-33). Suppose that we want to replace a particular serine residue with cysteine. This mutation can be made if (1) we have a plasmid containing the gene or cDNA for this protein and (2) we know the base sequence in the vicinity of the site to be altered. If the serine of interest is encoded by TCT, we need to change the C to a G to get cysteine, which is encoded by TGT. The key to this mutation is to prepare an oligonucleotide primer that is complementary to this region of the gene except that it contains TGT instead of TCT. The two strands of the plasmid are separated and the primer is then annealed to the complementary strand. (The mismatch of one base pair out of fifteen is tolerable if the annealing is carried out at an appropriate temperature. An attractive feature of nucleic acid hybridization is that its *stringency*—the required closeness of the match—can be experimentally controlled by choice of temperature and ionic strength.) The primer is then elongated by DNA polymerase, and the double-stranded circle is closed by adding DNA ligase. Subsequent replication of this duplex yields two kinds of progeny plasmid, half with the original TCT sequence and half with the mutant TGT sequence. Expression of the plasmid containing the new TGT sequence will produce a protein with the desired substitution of serine for cysteine at a unique site. We will encounter many examples of the use of oligonucleotide-directed mutagenesis to precisely alter regulatory regions of genes and to produce proteins with tailor-made features.

Novel proteins can also be created by splicing gene segments that encode domains that are not associated in nature. For example, a gene for an antibody can be joined to a gene for an enzyme to produce a

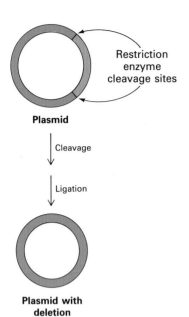

Figure 6-32
Deletions of parts of genes can be produced by cutting a plasmid at a pair of sites with a restriction enzyme (or a pair of enzymes) and religating. A set of smaller deletions can be produced by cutting a plasmid at a single site, digesting the ends to different extents with an exonuclease, and religating.

Figure 6-33
Oligonucleotide-directed mutagenesis. A primer containing a mismatched nucleotide is used to produce a desired point mutation.

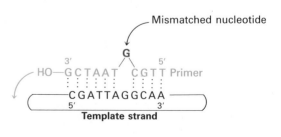

chimeric protein that could be useful as a therapeutic agent. Moreover, entirely new genes can be synthesized de novo by the solid-phase method. There is much interest in using them to direct the formation of synthetic vaccines that could be safer than conventional vaccines prepared by inactivating pathogenic viruses.

RECOMBINANT DNA TECHNOLOGY HAS OPENED NEW VISTAS

The analysis of the molecular basis of life has been revolutionized by recombinant DNA technology. Complex chromosomes are rapidly being mapped and dissected into units that can be manipulated and deciphered. The amplification of genes by cloning has provided abundant quantities of DNA for sequencing. Genes are now open books that can be read. New insights are emerging, as exemplified by the discovery of introns in eucaryotic genes. Central questions of biology, such as the molecular basis of development, are now being fruitfully explored. The reading of the genome opens a new record of evolution. Biochemists now move back and forth between gene and protein and feel at home in both areas of inquiry.

Analyses of genes and cDNA can reveal the existence of previously unknown proteins, which can be isolated and purified (Figure 6-34A). Conversely, purification of a protein can be the starting point for the isolation and cloning of its gene or cDNA (Figure 6-34B). Very small amounts of protein or nucleic acid suffice because of the sensitivity of recently developed microchemical techniques and the amplification afforded by gene cloning. The powerful techniques of protein chemistry, nucleic acid chemistry, immunology, and molecular genetics are highly synergistic.

New kinds of proteins can be created by altering genes in specific ways. Site-specific mutagenesis opens the door to understanding how proteins fold, recognize other molecules, catalyze reactions, and process information. Large amounts of proteins can be obtained by expressing cloned genes or cDNAs in bacteria. Hormones such as insulin and antiviral agents such as interferon are being produced by bacteria. A

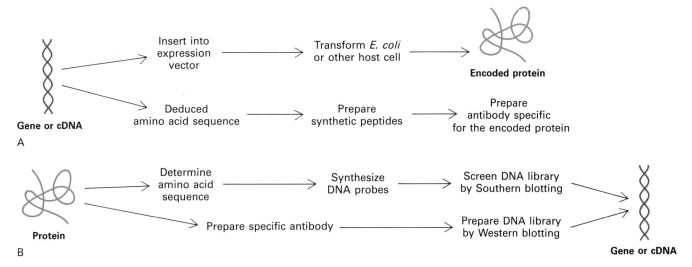

Figure 6-34
The techniques of protein chemistry and nucleic acid chemistry are mutually reinforcing. (A) From DNA (or RNA) to protein, and (B) from protein to DNA.

new pharmacology that will profoundly alter medicine has come into being. Recombinant DNA technology is also providing highly specific diagnostic reagents, such as DNA probes for the detection of genetic diseases, infections, and cancers. Retroviruses are being used to deliver missing genes into mice with genetic diseases, and human gene therapy will probably be initiated in the near future. Agriculture, too, will almost certainly benefit in the near future from the capacity to construct new genes and introduce them into eucaryotic cells. The genetic engineering of plants is likely to lead to crops that are more resistant to drought and have greater nutritional value.

SUMMARY

The recombinant DNA revolution in biology is rooted in the repertoire of enzymes that act on nucleic acids. Restriction enzymes, a group of key reagents, are endonucleases that recognize specific base sequences in double-helical DNA and cleave both strands of the duplex. Specific fragments of DNA formed by the action of restriction enzymes can be separated and displayed by gel electrophoresis. The pattern of these restriction fragments is a fingerprint of a DNA molecule. A DNA fragment containing a particular sequence can be identified by hybridizing it with a labeled single-stranded DNA probe (Southern blotting). Analysis of DNA molecules has also advanced by the development of rapid sequencing techniques. DNA can be sequenced by chemical cleavage at particular bases (Maxam-Gilbert method) or by controlled interruption of replication (Sanger dideoxy method). The fragments produced by either method are separated by gel electrophoresis and visualized by autoradiography of a ^{32}P label at the 5' end or by fluorescent tags. DNA probes for hybridization reactions, as well as new genes, can be synthesized by the sequential addition of deoxyribonucleoside 3'-phosphoramidites to a growing chain that is linked to an insoluble support. DNA chains a hundred nucleotides long can readily be synthesized by this automated solid-phase method.

New genomes can be constructed in the laboratory, introduced into host cells, and expressed. Novel DNA molecules are made by joining fragments that have complementary cohesive ends produced by the action of a restriction enzyme. DNA ligase joins the ends of DNA chains if they are within a double helix. Plasmids (circular accessory chromosomes) and λ phage are choice vectors for cloning DNA in bacteria. Specific genes can be cloned from a digest of DNA. This genomic library can be screened with a complementary DNA or RNA probe. Alternatively, one can form cDNA from mRNA by the action of reverse transcriptase. cDNA inserted into expression vectors that contain strong promoters is efficiently expressed by bacteria. Foreign DNA can be carried into mammalian cells by a retrovirus or be directly injected. The production of large mice by injecting the gene for rat growth hormone into fertilized mouse eggs shows that mammalian cells can be genetically altered in a designed way. New DNA can be brought into plant cells by the soil bacterium *Agrobacterium tumefaciens*, which harbors Ti (tumor-inducing) plasmids. DNA can also be introduced into cells by applying intense electric fields, which render the cells transiently permeable to very large molecules.

Novel proteins can be engineered by generating specific mutations in vitro. A mutant protein with a single amino acid substitution can be produced by priming DNA replication with an oligonucleotide encoding the new amino acid. The techniques of protein and nucleic acid chemistry are highly synergistic. Investigators now move back and forth between gene and protein with great facility.

SELECTED READINGS

Where to start

Berg, P., 1981. Dissections and reconstructions of genes and chromosomes. *Science* 213:296–303.

Gilbert, W., 1981. DNA sequencing and gene structure. *Science* 214:1305–1312.

Nathans, D., 1979. Restriction endonucleases, simian virus 40, and the new genetics. *Science* 206:903–909.

Sanger, F., 1981. Determination of nucleotide sequences in DNA. *Science* 214:1205–1210.

Books on recombinant DNA technology

Watson, J. D., Tooze, J., and Kurtz, D. T., 1983. *Recombinant DNA: A Short Course.* Scientific American Books. [A highly readable and lucid introduction.]

Old, R. W., and Primrose, S. B., 1985. *Principles of Gene Manipulation: An Introduction to Genetic Engineering.* Blackwell Scientific Publications.

Mantell, S. H., Matthews, J. A., and McKee, R. A., 1985. *Principles of Plant Biotechnology: An Introduction to Genetic Engineering in Plants.* Blackwell Scientific Publications.

Inouye, M., (ed.), 1983. *Experimental Manipulation of Gene Expression.* Academic Press.

Maniatis, T., Fritsch, E. F., and Sambrook, J., 1982. *Molecular Cloning: A Laboratory Manual.* Cold Spring Harbor Laboratory.

Automated synthesis of DNA

Caruthers, M. H., 1985. Gene synthesis machines: DNA chemistry and its uses. *Science* 230:281–285.

Itakura, K., Rossi, J. J., and Wallace, R. B., 1984. Synthesis and use of synthetic oligonucleotides. *Ann. Rev. Biochem.* 53:323–356.

Hunkapiller, M., Kent, S., Caruthers, M., Dreyer, W., Firca, J., Giffin, C., Horvath, S., Hunkapiller, T., Tempst, P., and Hood, L., 1984. A microchemical facility for the analysis and synthesis of genes and proteins. *Nature* 310:105–111.

DNA sequencing

Sanger, F., Nicklen, S., and Coulson, A. R., 1977. DNA sequencing with chain-terminating inhibitors. *Proc. Nat. Acad. Sci.* 74:5463–5467.

Maxam, A. M., and Gilbert, W., 1977. A new method for sequencing DNA. *Proc. Nat. Acad. Sci.* 74:560–564.

Smith, L. M., Sanders, J. Z., Kaiser, R. J., Hughes, P., Dodd, C., Connell, C. R., Heiner, C., Kent, S. B. H., and Hood, L. E., 1986. Fluorescence detection in automated DNA sequence analysis. *Nature* 321:674–679.

Origins of DNA cloning

Jackson, D. A., Symons, R. H., and Berg, P., 1972. Biochemical method for inserting new genetic information into DNA of simian virus 40: circular SV40 DNA molecules containing lambda phage genes and the galactose operon of *Escherichia coli. Proc. Nat. Acad. Sci.* 69:2904–2909.

Lobban, P. E., and Kaiser, A. D., 1973. Enzymatic end-to-end joining of DNA molecules. *J. Mol. Biol.* 78:453–471.

Cohen, S. N., Chang, A., Boyer, H., and Helling, R., 1973. Construction of biologically functional bacterial plasmids in vitro. *Proc. Nat. Acad. Sci.* 70:3240–3244.

Cohen, S. N., 1985. DNA cloning: historical perspectives. *Biogenetics of Neurohormonal Peptides,* pp. 3–14. Academic Press.

Watson, J. D., and Tooze, J., 1981. *The DNA Story.* W. H. Freeman.

Protein production by recombinant bacteria

Pestka, S., 1983. The purification and manufacture of human interferons. *Sci. Amer.* 249(2):36.

Gilbert, W., and Villa-Komaroff, L., 1980. Useful proteins from recombinant bacteria. *Sci. Amer.* 242(4):74–94. [Available as *Sci. Amer.* Offprint 1466.]

Johnson, I. S., 1983. Human insulin from recombinant DNA technology. *Science* 219:632–637.

Introduction of genes into animal cells

Brinster, R. L., and Palmiter, R. D., 1986. Introduction of genes into the germ lines of animals. *Harvey Lectures* 80:1–38. [The production of large mice by injection of the growth hormone gene is discussed in this review.]

Karlsson, S., Humphries, R. K., Gluzman, Y., and Nienhuis, A. W., 1985. Transfer of genes into hematopoietic cells using recombinant DNA viruses. *Proc. Nat. Acad. Sci.* 82:158–162.

Anderson, W. F., 1984. Prospects for human gene therapy. *Science* 226:401–409.

Cepko, C. L., Roberts, B. E., and Mulligan, R. C., 1984. Construction and applications of a highly transmissible murine retrovirus shuttle vector. *Cell* 37:1053–1062.

Friedman, R. L., 1985. Expression of human adenosine deaminase using a transmissible murine retrovirus vector system. *Proc. Nat. Acad. Sci.* 82:703–707.

Introduction of genes into plant cells

Chilton, M-D., 1983. A vector for introducing new genes into plants. *Sci. Amer.* 248(6):50.

Hooykaas, P. J. J., and Schilperoort, R. A., 1985. The Ti-plasmid of *Agrobacterium tumefaciens*: a natural genetic engineer. *Trends Biochem. Sci.* 10:307–309.

Fromm, M. E., Taylor, L. P., and Walbot, V., 1986. Stable transformation of maize after gene transfer by electroporation. *Nature* 319:791–793.

Site-specific mutagenesis

Botstein, D., and Shortle, D., 1985. Strategies and applications of in vitro mutagenesis. *Science* 229:1193–1201.

Myers, R. M., Lerman, L. S., and Maniatis, T., 1985. A general method for saturation mutagenesis of cloned DNA fragments. *Science* 229:242–247.

PROBLEMS

1. An autoradiogram of a gel containing four lanes of DNA fragments produced by chemical cleavage is shown in Figure 6-35. The DNA contained a ^{32}P label at its 5′ end. What is its sequence?

   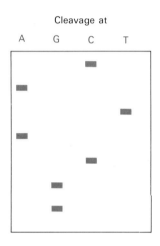

   Figure 6-35
   Schematic diagram of a gel showing the radioactive fragments formed by specific cleavage of an oligonucleotide.

2. Suppose that you determined the DNA sequence of 5′-GCCATTGCA-3′ by the Sanger dideoxy method. Sketch the gel pattern that revealed the sequence of this oligonucleotide.

3. Ovalbumin is the major protein of egg white. The chicken ovalbumin gene contains eight exons separated by seven introns. Should one use ovalbumin cDNA or ovalbumin genomic DNA to form the protein in *E. coli*? Why?

4. Suppose that a human genomic library were prepared by exhaustive digestion of human DNA with the EcoRI restriction enzyme. Fragments averaging about 4 kb in length would be generated.
   (a) Would this procedure be desirable for cloning large genes? Why?
   (b) Would this procedure be desirable for mapping extensive stretches of the genome by chromosome walking? Why?

5. Sickle-cell anemia arises from a mutation in the gene for the β chain of human hemoglobin. The change from GAG to GTG in the mutant eliminates a cleavage site for the restriction enzyme MstII, which recognizes the target sequence CCTGAGG. These findings form the basis of a valuable diagnostic test for the presence of the sickle-cell gene. Propose a rapid diagnostic procedure for distinguishing between the normal and the mutant gene. Would a positive result with your test prove that the mutant contains GTG in place of GAG?

6. Thomas Cech showed that the ribosomal RNA (rRNA) precursor in the ciliate protozoan *Tetrahymena* can self-splice without binding any *Tetrahymena* protein. He cloned in a plasmid a region of *Tetrahymena* DNA consisting of the intron and flanking sequences present in the precursor RNA. Suggest how this plasmid was used to establish that *Tetrahymena* proteins are not required for the splicing of the ribosomal RNA precursor.

7. The introduction into a bacterial cell of a single M13 RF DNA molecule containing an inserted fragment of foreign DNA yields M13 progeny DNA containing only one strand of the foreign DNA.
   (a) Why is it important to establish which DNA strand is contained in a particular virus particle?
   (b) Suppose that the foreign DNA strand was inserted into M13 DNA using restriction enzyme 1 and that the foreign DNA has an internal site for restriction enzyme 2. How could one determine which strand of foreign DNA is contained in the virus particle?
   (c) Agarose gel electrophoresis can distinguish between single-stranded and double-stranded DNA. How would this method be used to determine whether two phage particles contain the same strand or different strands of foreign DNA? How could an enzyme be used to make this determination?

8. Suppose that a colleague purified from cardiac muscle a 30-residue peptide that modulates calcium transport across membranes. The peptide shows promise as a drug, but unfortunately it is rapidly degraded by proteolytic enzymes in blood. You are intrigued by this problem and decide that it would be worthwhile to use site-specific mutagenesis to engineer a series of modified peptides in the hope that one of them will retain biological activity and be resistant to proteolytic degradation. What information is essential before you can begin to prepare new peptides by molecular genetic techniques? What other information would be helpful in this project?

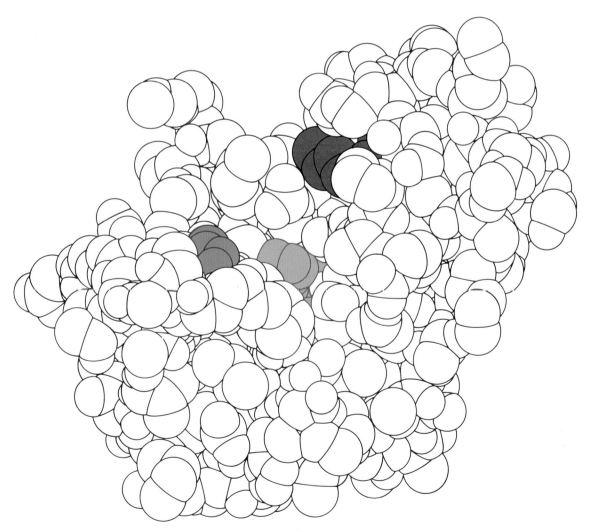

Model of ribonuclease S, an enzyme that hydrolyzes ribonucleic acids. Amino acid residues at the active site that are critical for catalysis are shown in color. The three-dimensional structure of this enzyme was solved by Frederic Richards and Harold Wyckoff. [After a drawing kindly provided by Dr. Frederic Richards, Steven Anderson; and Arthur Perlo.]

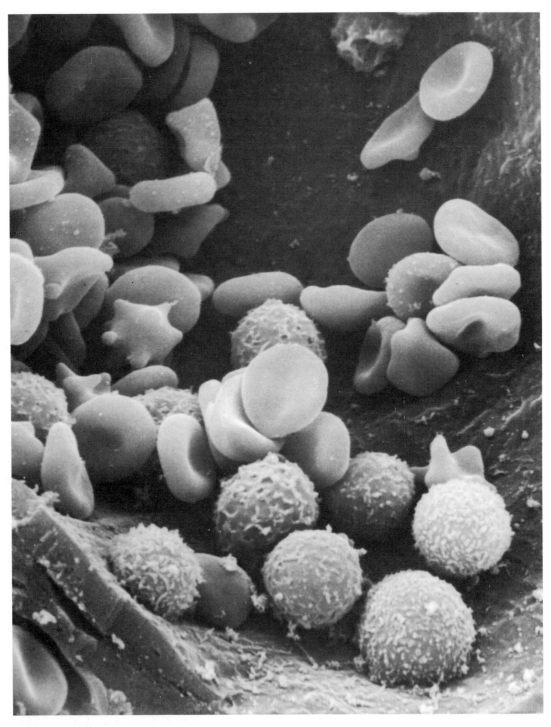

Scanning electron micrograph showing erythrocytes (biconcave-shaped) and leucocytes (rounded) in a small blood vessel. [From R. G. Kessel and R. H. Kardon. *Tissues and Organs*, W. H. Freeman. Copyright © 1979.]

CHAPTER 7

Oxygen-transporting Proteins: Myoglobin and Hemoglobin

The transition from anaerobic to aerobic life was a major step in evolution because it uncovered a rich reservoir of energy. Eighteen times as much energy is extracted from glucose in the presence of oxygen as in its absence. Vertebrates have evolved two principal mechanisms for supplying their cells with a continuous and adequate flow of oxygen. The first is a circulatory system that actively delivers oxygen to cells. The second is the use of *oxygen-carrying molecules* to overcome the limitation imposed by the low solubility of oxygen in water. The oxygen carriers in vertebrates are the proteins *hemoglobin* and *myoglobin*. Hemoglobin, which is contained in red blood cells, serves as the oxygen carrier in blood and also plays a vital role in the transport of carbon dioxide and hydrogen ion. Myoglobin, which is located in muscle, serves as a reserve supply of oxygen and facilitates the movement of oxygen within muscle.

Part II of this book begins with myoglobin and hemoglobin because they illustrate many important principles of protein conformation, dynamics, and function. Their three-dimensional structures, known in atomic detail, reveal much about how proteins fold, bind other molecules, and integrate information. The binding of O_2 by hemoglobin is regulated by H^+, CO_2, and organic phosphates. These regulators greatly affect the oxygen-binding properties of hemoglobin by binding to sites on the protein far from where O_2 is bound. Indeed, interactions between separate sites, termed *allosteric interactions*, occur in many proteins. Allosteric effects play a critical role in controlling and integrating molecular events in biological systems. Hemoglobin is the best understood allosteric protein, so that examining its normal structure and function in some detail is rewarding. Furthermore, the discovery of mutant hemoglobins has revealed that disease can arise from a change

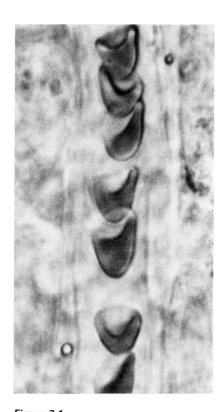

Figure 7-1
Erythrocytes flowing through a small blood vessel. [From P. I. Brånemark. *Intravascular Anatomy of Blood Cells in Man* (Basel: S. Karger AG, 1971).]

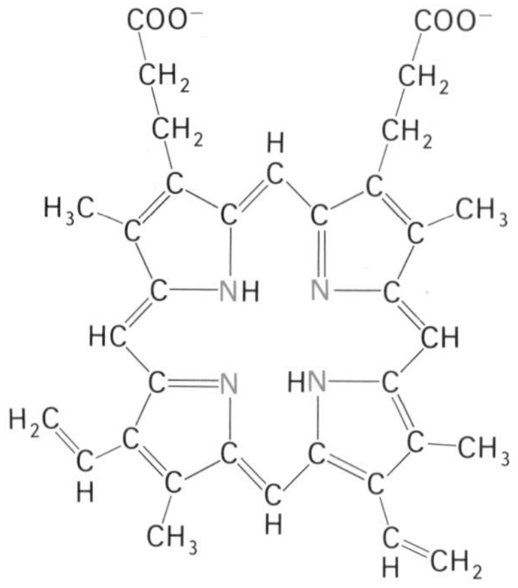

Part II
PROTEIN CONFORMATION, DYNAMICS, AND FUNCTION

Protoporphyrin IX

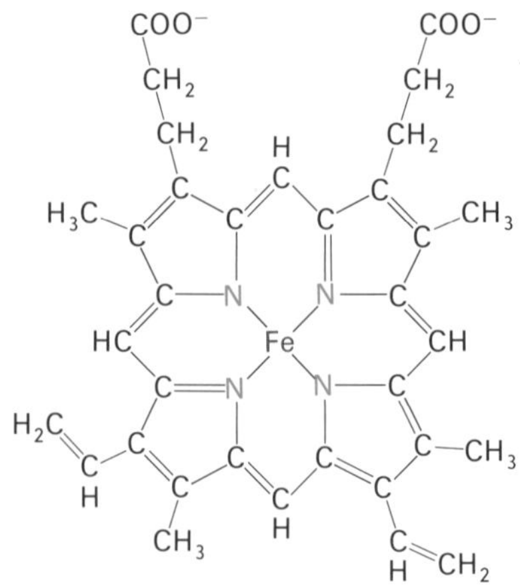

Heme
(Fe-protoporphyrin IX)

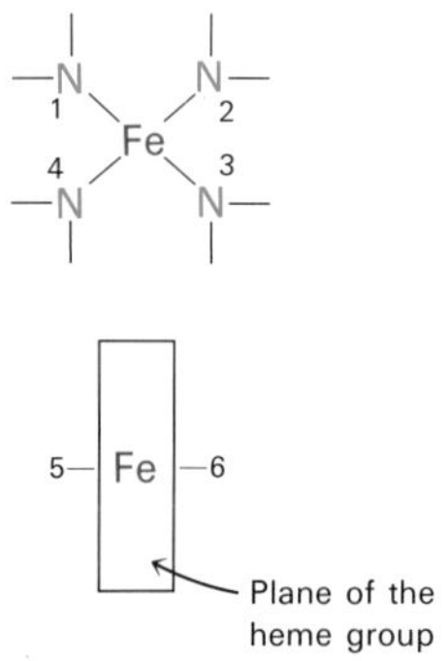

Figure 7-2
The iron atom in heme can form six bonds.

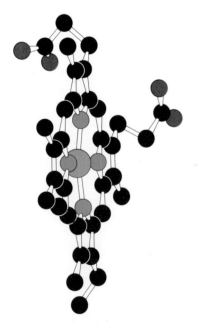

Figure 7-3
The heme group in myoglobin (yellow for Fe, blue for N, red for O, and black for C).

of a single amino acid in a protein. The concept of molecular disease, now an integral part of medicine, came from studies of the abnormal hemoglobin causing sickle-cell anemia. Hemoglobin has also been a rich source of insight into the molecular basis of evolution.

OXYGEN BINDS TO A HEME PROSTHETIC GROUP

The capacity of myoglobin or hemoglobin to bind oxygen depends on the presence of a nonpolypeptide unit, namely, a *heme group*. The heme also gives myoglobin and hemoglobin their distinctive color. Indeed, many proteins require tightly bound, specific nonpolypeptide units for their biological activities. Such a unit is called a *prosthetic group*. A protein without its characteristic prosthetic group is termed an *apoprotein*.

The heme consists of an organic part and an iron atom. The organic part, *protoporphyrin*, is made up of four *pyrrole* rings. The four pyrroles are linked by methene bridges to form a tetrapyrrole ring. Four methyl, two vinyl, and two propionate side chains are attached to the tetrapyrrole ring. These substituents can be arranged in fifteen different ways. Only one of these isomers, called protoporphyrin IX, is present in biological systems.

The iron atom in heme binds to the four nitrogens in the center of the protoporphyrin ring (Figures 7-2 and 7-3). The iron can form two additional bonds, one on either side of the heme plane. These bonding sites are termed the fifth and sixth coordination positions. The iron atom can be in the ferrous (+2) or the ferric (+3) oxidation state, and the corresponding forms of hemoglobin are called *ferrohemoglobin* and *ferrihemoglobin*. Ferrihemoglobin is also called methemoglobin. Only ferrohemoglobin, the +2 oxidation state, can bind oxygen. The same nomenclature applies to myoglobin.

MYOGLOBIN WAS THE FIRST PROTEIN TO BE SEEN AT ATOMIC RESOLUTION

The elucidation of the three-dimensional structure of myoglobin by John Kendrew and of hemoglobin by Max Perutz are landmarks in molecular biology. These studies came to fruition in the late 1950s and showed that x-ray crystallography (p. 60) can reveal the structure of molecules as large as proteins. The largest structure solved before then had been vitamin B_{12}, which is an order of magnitude smaller than myoglobin and hemoglobin. The determination of the three-dimensional structures of these proteins was a great stimulus to the field of protein crystallography. Kendrew chose myoglobin for x-ray analysis because it is relatively small, easily prepared in quantity, and readily crystallized. It had the additional advantage of being closely related to hemoglobin, which was already being studied by his colleague Perutz. Myoglobin from the skeletal muscle of the sperm whale was selected because it is stable and forms excellent crystals. The skeletal muscle of diving mammals, such as whale, seal, and porpoise, is particularly rich in myoglobin, which serves as a store of oxygen during a dive.

In 1957, Kendrew and his colleagues saw what no one had ever seen before: a three-dimensional picture of a protein molecule in all its complexity. The model derived from their Fourier synthesis at 6-Å resolution contained a set of high-density rods of just the dimensions expected for a polypeptide chain. The molecule appeared very compact. Closer examination showed that it consisted of a complicated and intertwining set of these rods, going straight for a distance, then turning a corner and going off in a new direction (Figure 7-4). The location of the iron atom of the heme was also evident because it contains many more electrons than does any other atom in the structure.

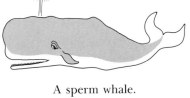

A sperm whale.

MYOGLOBIN HAS A COMPACT STRUCTURE AND A HIGH CONTENT OF ALPHA HELICES

The high-resolution electron-density map of myoglobin, obtained two years later, contained a wealth of structural detail. The positions of 1200 of the 1260 nonhydrogen atoms were clearly defined to a precision of better than 0.3 Å. The course of the main chain and the position of the heme group are shown in Figure 7-5. Some important features of myoglobin are:

1. Myoglobin is *extremely compact*. The overall dimensions are about $45 \times 35 \times 25$ Å. There is rather little empty space inside.

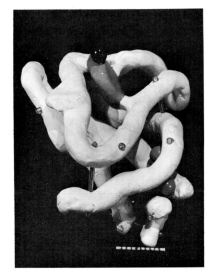

Figure 7-4
Model of myoglobin at low resolution. [Courtesy of Dr. John Kendrew.]

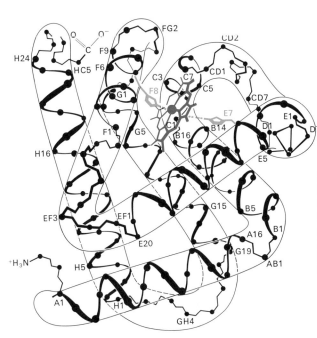

Figure 7-5
Model of myoglobin at high resolution. Only the α-carbon atoms are shown. The heme group is shown in red. [After R. E. Dickerson. In *The Proteins*, H. Neurath, ed., 2nd ed., (Academic Press, 1964), vol. 2, p. 634.]

Val-	Leu-	Ser-	Glu-	Gly-	Glu-	Trp-	Gln-	Leu-	Val-
NA1	NA2	A1	A2	A3	A4	A5	A6	A7	A8

10

Leu-	His-	Val-	Trp-	Ala-	Lys-	Val-	Glu-	Ala-	Asp-
A9	A10	A11	A12	A13	A14	A15	A16	AB1	B1

20

Val-	Ala-	Gly-	His-	Gly-	Gln-	Asp-	Ile-	Leu-	Ile-
B2	B3	B4	B5	B6	B7	B8	B9	B10	B11

30

Arg-	Leu-	Phe-	Lys-	Ser-	His-	Pro-	Glu-	Thr-	Leu-
B12	B13	B14	B15	B16	C1	C2	C3	C4	C5

40

Glu-	Lys-	Phe-	Asp-	Arg-	Phe-	Lys-	His-	Leu-	Lys-
C6	C7	CD1	CD2	CD3	CD4	CD5	CD6	CD7	CD8

50

Thr-	Glu-	Ala-	Glu-	Met-	Lys-	Ala-	Ser-	Glu-	Asp-
D1	D2	D3	D4	D5	D6	D7	E1	E2	E3

60

Leu-	Lys-	Lys-	His-	Gly-	Val-	Thr-	Val-	Leu-	Thr-
E4	E5	E6	E7	E8	E9	E10	E11	E12	E13

70

Ala-	Leu-	Gly-	Ala-	Ile-	Leu-	Lys-	Lys-	Lys-	Gly-
E14	E15	E16	E17	E18	E19	E20	EF1	EF2	EF3

80

His-	His-	Glu-	Ala-	Glu-	Leu-	Lys-	Pro-	Leu-	Ala-
EF4	EF5	EF6	EF7	EF8	F1	F2	F3	F4	F5

90

Gln-	Ser-	His-	Ala-	Thr-	Lys-	His-	Lys-	Ile-	Pro-
F6	F7	F8	F9	FG1	FG2	FG3	FG4	FG5	G1

100

Ile-	Lys-	Tyr-	Leu-	Glu-	Phe-	Ile-	Ser-	Glu-	Ala-
G2	G3	G4	G5	G6	G7	G8	G9	G10	G11

110

Ile-	Ile-	His-	Val-	Leu-	His-	Ser-	Arg-	His-	Pro-
G12	G13	G14	G15	G16	G17	G18	G19	GH1	GH2

120

Gly-	Asn-	Phe-	Gly-	Ala-	Asp-	Ala-	Gln-	Gly-	Ala-
GH3	GH4	GH5	GH6	H1	H2	H3	H4	H5	H6

130

Met-	Asn-	Lys-	Ala-	Leu-	Glu-	Leu-	Phe-	Arg-	Lys-
H7	H8	H9	H10	H11	H12	H13	H14	H15	H16

140

Asp-	Ile-	Ala-	Ala-	Lys-	Tyr-	Lys-	Glu-	Leu-	Gly-
H17	H18	H19	H20	H21	H22	H23	H24	HC1	HC2

150

Tyr-	Gln-	Gly
HC3	HC4	HC5

153

2. About 75% of the main chain is in an *α-helical conformation*. The eight major helical segments, all right-handed, are referred to as A, B, C, . . . , H. The first residue in helix A is designated A1, the second A2, and so forth (Figure 7-6). Five nonhelical segments lie between helices (named CD, e.g., if located between the C and D helices). Myoglobin has two other nonhelical regions: two residues at the amino-terminal end (named NA1 and NA2) and five residues at the carboxyl-terminal end (named HC1 through HC5).

3. Four of the helices are terminated by a *proline* residue, whose five-membered ring simply does not fit within a straight stretch of α helix.

4. The main-chain peptide groups are *planar*, and the carbonyl group of each is *trans* to the NH. Also, the bond angles and distances are like those in dipeptides and other organic compounds.

5. The inside and outside are well defined. *The interior consists almost entirely of nonpolar residues* such as leucine, valine, methionine, and phenylalanine. In contrast, glutamic and aspartic acids, glutamine, asparagine, lysine, and arginine are absent from the interior of the protein. Residues that have both a polar and a nonpolar part, such as threonine, tyrosine, and tryptophan, are oriented so that their nonpolar portions point inward. The only polar residues inside myoglobin are two histidines, which have a critical function at the binding site. The outside of the protein has both polar and nonpolar residues.

Figure 7-6
Amino acid sequence of sperm-whale myoglobin. The label below each residue in the sequence refers to its position in an α-helical region or a nonhelical region. For example, B4 is the fourth residue in the B helix; EF7 is the seventh residue in the nonhelical region between the E and F helices. [After A. E. Edmundson. *Nature* 205(1965):883; H. C. Watson. *Prog. Stereochem.* 4(1969):299–333.]

OXYGEN BINDS WITHIN A CREVICE IN MYOGLOBIN

The heme group is located in a crevice in the myoglobin molecule. The highly polar propionate side chains of the heme are on the surface of the molecule. At physiological pH, these carboxylic acid groups are ionized. The rest of the heme is inside the molecule, where it is surrounded by nonpolar residues except for two histidines. The iron atom of the heme is directly bonded to one of these histidines, namely, residue F8 (Figures 7-7 and 7-8). This histidine, which occupies the fifth coordination position, is called the proximal histidine. The iron atom is about 0.3 Å out of the plane of the porphyrin, on the same side as histidine F8. *The oxygen-binding site is on the other side of the heme plane, at the sixth coordination position*. A second histidine residue (E7), termed the distal histidine, is near the heme but not bonded to it. A section of an electron-density map displaying the heme is shown in Figure 7-9.

The conformations of the three physiologically pertinent forms of myoglobin—deoxymyoglobin, oxymyoglobin, and ferrimyoglobin—are very similar except at the sixth coordination position (Table 7-1). In deoxymyoglobin, it is empty; in oxymyoglobin, it is occupied by O_2; in ferrimyoglobin, it is occupied by water. The axis of the bound O_2 is at an angle to the iron-oxygen bond (Figure 7-10).

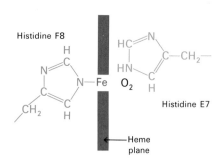

Figure 7-7
Schematic diagram of the oxygen-binding site in myoglobin.

Chapter 7
OXYGEN TRANSPORTERS

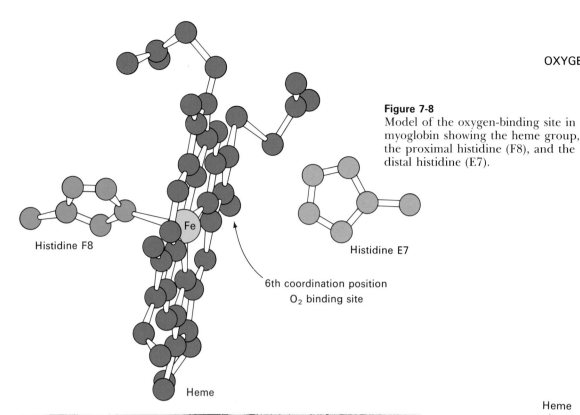

Figure 7-8
Model of the oxygen-binding site in myoglobin showing the heme group, the proximal histidine (F8), and the distal histidine (E7).

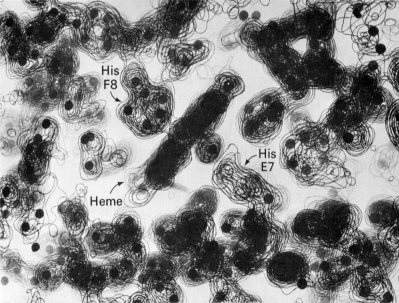

Figure 7-9
Section of an electron-density map of myoglobin near the oxygen-binding site. Electron density extending across the lower part of the map is the E helix. [Courtesy of Dr. John Kendrew.]

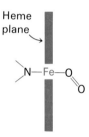

Figure 7-10
Bent, end-on orientation of O_2 in oxymyoglobin. The angle between the O_2 axis and the Fe–O bond is 121 degrees.

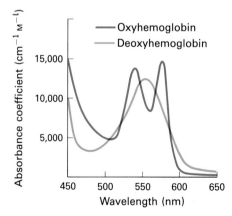

Figure 7-11
Visible absorption spectra of myoglobin and hemoglobin change markedly on binding oxygen. Myoglobin and hemoglobin have very similar visible absorption spectra.

Table 7-1
Heme environment

Form	Oxidation state of Fe	Occupant	
		5th coordination position	6th coordination position
Deoxymyoglobin	+2	His F8	Empty
Oxymyoglobin	+2	His F8	O_2
Ferrimyoglobin	+3	His F8	H_2O

A HINDERED HEME ENVIRONMENT IS ESSENTIAL FOR REVERSIBLE OXYGENATION

The oxygen-binding site comprises only a small fraction of the volume of the myoglobin molecule. Indeed, oxygen is directly bonded only to the iron atom of the heme. Why is the polypeptide portion of myoglobin needed for oxygen transport and storage? The answer lies in the oxygen-binding properties of an isolated heme group. In water, a free ferrous heme group can bind oxygen, but it does so for only a fleeting moment. The reason is that O_2 very rapidly oxidizes the ferrous heme to ferric heme, which cannot bind oxygen. A complex of O_2 sandwiched between two hemes is an intermediate in this reaction. *In myoglobin, the heme group is much less susceptible to oxidation because two myoglobin molecules cannot readily associate to form a heme-O_2-heme complex.* The formation of this sandwich is sterically blocked by the distal histidine and other residues surrounding the sixth coordination site.

The strongest evidence for the importance of steric factors in determining the rate of oxidation of heme comes from studies of synthetic model compounds. James Collman synthesized *picket-fence iron porphyrin complexes* (Figure 7-12) that mimic the oxygen-binding sites of myoglobin and hemoglobin. These compounds have a protective enclosure for binding O_2 on one side of the porphyrin ring, whereas the other side is left unhindered so that it can bind a base. In fact, when the base is a substituted imidazole (like histidine), the oxygen affinity of the picket-fence compound (Figure 7-13) is like that of myoglobin. Furthermore, the picket fence stabilizes the ferrous form of this iron porphyrin and thus enables it to reversibly bind oxygen for long periods. The critical difference between this model compound and free heme is the presence of the picket fence, which blocks the formation of the sandwich dimer.

Thus, myoglobin has created a special microenvironment that confers distinctive properties on its prosthetic group. In general, *the function of a prosthetic group is modulated by its polypeptide environment.* For example, the same heme group has quite a different function in cytochrome *c*, a protein in the terminal oxidation chain in the mitochondria of all aerobic organisms. In cytochrome *c*, the heme is a reversible carrier of electrons rather than of oxygen. Heme has yet another function in the enzyme catalase, where it catalyzes the conversion of hydrogen peroxide into water and oxygen.

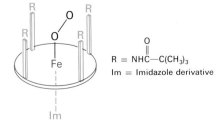

Figure 7-12
Schematic diagram of a picket-fence iron porphyrin with a bound O_2. The picket fence prevents two of these porphyrins from coming together to form the major intermediate in oxidation.

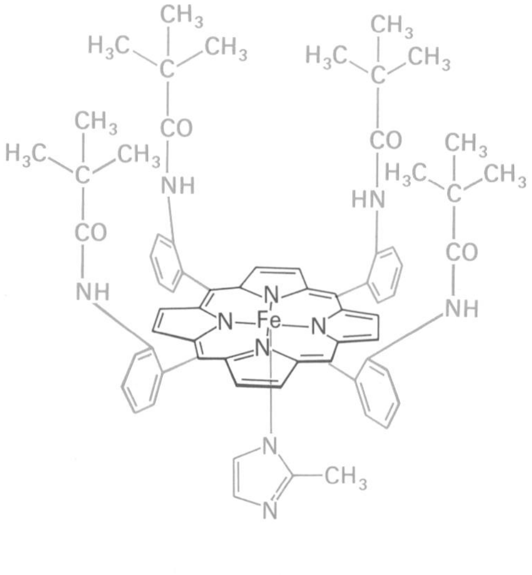

Figure 7-13
Structural formula of a picket-fence iron porphyrin. [After J. Collman, J. I. Brauman, E. Rose, and K. S. Suslick. *Proc. Nat. Acad. Sci.* 75(1978):1053.]

CARBON MONOXIDE BINDING IS DIMINISHED BY THE PRESENCE OF THE DISTAL HISTIDINE

Carbon monoxide is a poison because it combines with ferromyoglobin and ferrohemoglobin and thereby blocks oxygen transport. An isolated heme in solution typically binds CO some 25,000 times as strongly as O_2. However, the binding affinity of myoglobin and hemoglobin for CO is only about 200 times as great as for O_2. How do these proteins suppress the innate preference of heme for carbon monoxide? The answer comes from x-ray crystallographic and infra-red spectroscopic studies of complexes of CO and O_2 with myoglobin and model iron porphyrins. In the very tightly bound complexes of CO with isolated iron porphyrins, the Fe, C, and O atoms are in a linear array (Figure 7-14). In carbonmonoxymyoglobin, by contrast, the CO axis is at an angle to the Fe–C bond. The linear binding of CO is prevented mainly by the steric hindrance of the distal histidine. On the other hand, the O_2 axis is at an angle to the Fe–O bond both in oxymyoglobin and in model compounds. *Thus, the protein forces CO to bind at an angle rather than in line. This bent geometry in the globins weakens the interaction of CO with the heme.*

The decreased affinity of myoglobin and hemoglobin for CO is biologically important. Carbon monoxide was a potential hazard long before the emergence of industrialized societies because it is produced endogenously (within cells) in the breakdown of heme (p. 597). The level of endogenously formed carbon monoxide is such that about 1% of the sites in myoglobin and hemoglobin are blocked by CO, a tolerable degree of inhibition. However, endogenously produced CO would cause massive poisoning if the affinity of these proteins for CO was like that of isolated iron porphyrins. This challenge was solved by the evolution of heme proteins that discriminate between O_2 and CO by sterically imposing a bent and hence weaker mode of binding for CO.

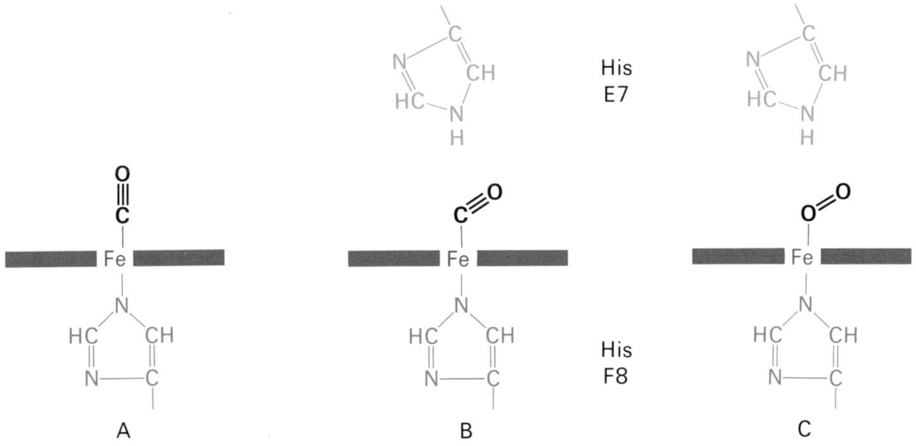

Figure 7-14
Structural basis of the diminished affinity of myoglobin and hemoglobin for carbon monoxide: (A) linear mode of binding of CO to isolated iron porphyrins; (B) bent mode of binding of CO to myoglobin and hemoglobin, in which the distal histidine (E7) prevents CO from binding linearly and so the affinity for CO is markedly reduced; (C) bent mode of binding of O_2 in myoglobin and hemoglobin. Isolated iron porphyrins also bind O_2 in a bent mode.

THE CENTRAL EXON OF MYOGLOBIN ENCODES A FUNCTIONAL HEME-BINDING UNIT

As was discussed in Chapter 5, most eucaryotic genes are mosaics of exons (coding sequences) and introns (noncoding intervening sequences). It was mentioned that exons encode discrete structural and functional units of proteins (p. 112). Myoglobin illustrates this general principle nicely. The gene for myoglobin consists of three exons: an amino-terminal one encoding residues 1 to 30 (NA1 to B2), a central one encoding residues 31 to 105 (B3 to G6), and a C-terminal one encoding residues 106 to 153 (G7 to HC5). A look at the structure of native myoglobin shows that nearly all of the heme-binding site is specified by the central exon (Figure 7-15). The regions encoded by the other two exons make very few contacts with the heme. What are the functional properties of an isolated polypeptide consisting of residues 31 to 105? The answer to this question is not yet known but the properties of another fragment are revealing. Digestion of apomyoglobin with clostripain, an arginine-specific protease, yields a polypeptide containing residues 32 to 139, which corresponds to the central exon and part of the C-terminal one. This fragment binds heme. Furthermore, this complex, called *mini-myoglobin*, binds O_2 and CO reversibly. Indeed, the rates of association and dissociation are nearly the same as those of the intact protein.

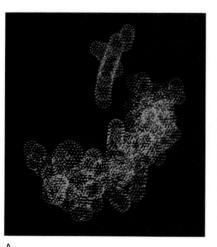
A

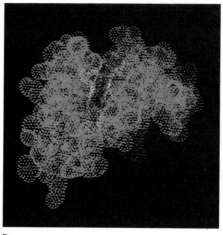
B

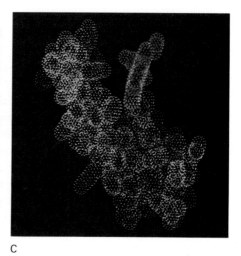
C

Figure 7-15
Region of myoglobin encoded by (A) exon 1, (B) exon 2, and (C) exon 3. The central exon encodes a functional oxygen-binding module.

These findings indicate that the conformation of mini-myoglobin is very similar to that of native myoglobin. Because the amino-terminal exon and the last fourteen residues of the C-terminal exon are not essential for reversible oxygenation, it is evident that the central exon contains much of the information for binding heme and maintaining the native fold of an oxygen-binding protein. The exon-intron organization of the constituent chains of hemoglobin is very similar to that of myoglobin. Amino acid sequence comparisons suggest that the genes for myoglobin and hemoglobin diverged some 700 million years ago (p. 841). Thus, the central exon of these oxygen carriers is an ancient piece of DNA that encoded a functional heme-binding module eons ago.

HEMOGLOBIN CONSISTS OF FOUR POLYPEPTIDE CHAINS

We turn now to hemoglobin, the oxygen transporter in erythrocytes. Vertebrate hemoglobins consist of four polypeptide chains, two of one kind and two of another (Table 7-2). The four chains are held together by noncovalent attractions. Each contains a heme group and a single oxygen-binding site. Hemoglobin A, the principal hemoglobin in

Table 7-2
Subunits of human hemoglobins

Embryonic hemoglobins

Hb Gower 1	$\zeta_2\epsilon_2$
Hb Gower 2	$\alpha_2\epsilon_2$
Hb Portland	$\zeta_2\gamma_2$

Fetal hemoglobin

Hb F	$\alpha_2\gamma_2$

Adult hemoglobins

Hb A	$\alpha_2\beta_2$
Hb A_2	$\alpha_2\delta_2$

adults, consists of two alpha (α) chains and two beta (β) chains. Adults also have a minor hemoglobin (~2% of the total hemoglobin) called hemoglobin A$_2$, which contains delta (δ) chains in place of the β chains of hemoglobin A. Thus, the subunit composition of hemoglobin A is $\alpha_2\beta_2$, and that of hemoglobin A$_2$ is $\alpha_2\delta_2$.

Fetuses have distinctive hemoglobins. Shortly after conception, fetuses synthesize zeta (ζ) chains (which are α-like chains) and epsilon (ϵ) chains (which are β-like). In the course of fetal life, ζ is replaced by α, and ϵ is replaced by gamma (γ), which is replaced by β (Figure 7-16). The major hemoglobin during the latter two-thirds of fetal life, hemoglobin F, has the subunit composition $\alpha_2\gamma_2$. The α and ζ chains contain 141 residues; the β, γ, and δ chains contain 146 residues. Why do hemoglobins consist of two kinds of chains? Why do fetuses have distinctive hemoglobins? Some answers to these intriguing questions will be given later in this chapter.

X-RAY ANALYSIS OF HEMOGLOBIN

Perutz's elucidation of the three-dimensional structure of hemoglobin, a monumental accomplishment, began in 1936. He left Austria that year to pursue graduate work in England and decided on hemoglobin as his thesis subject. The largest structure that had been solved then was the dye phthalocyanin, which contains 58 atoms. In choosing to tackle a molecule one hundred times as large, as Perutz wrote years later, it was little wonder that "my fellow students regarded me with a pitying smile. . . . Fortunately, the examiners of my doctoral thesis did not insist on a determination of the structure, otherwise I should have had to remain a graduate student for twenty-three years." Fortunately, too, Lawrence Bragg became director of the Cavendish Laboratory in Cambridge at this time and supported the project. In 1912, he and his father were the first to use x-ray crystallography to solve structures. Bragg wrote, "I was frank about the outlook. It was like multiplying a zero probability that success would be achieved by an infinity of importance if the structure came out; the result of this mathematic operation was anyone's guess." Success came in 1959, when Perutz obtained a low-resolution electron-density image of horse oxyhemoglobin, followed several years later by high-resolution maps of both human and horse oxyhemoglobin and deoxyhemoglobin. The three-dimensional structures of human and horse hemoglobins are very similar.

The hemoglobin molecule is nearly spherical, with a diameter of 55 Å. The four chains are packed together in a tetrahedral array (Figure 7-17). The heme groups are located in crevices near the exterior of the

Figure 7-16
Expression of hemoglobin genes in human development. (A) α and ζ genes. (B) β, γ, and δ genes.

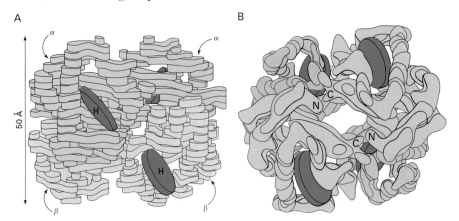

Figure 7-17
Model of hemoglobin at low resolution. The α chains in this model are yellow, the β chains blue, and the heme groups red. View (A) is at right angles to view (B). [After M. F. Perutz. The hemoglobin molecule. Copyright © 1964 by Scientific American, Inc. All rights reserved.]

molecule, one in each subunit. The four oxygen-binding sites are far apart; the distance between the two closest iron atoms is 25 Å. Each α chain is in contact with both β chains. In contrast, there are few interactions between the two α chains or between the two β chains.

THE ALPHA AND BETA CHAINS OF HEMOGLOBIN CLOSELY RESEMBLE MYOGLOBIN

The three-dimensional structures of myoglobin and the α and β chains of human hemoglobin are strikingly similar (Figure 7-18). This close resemblance in the folding of their main chains was unexpected because their amino acid sequences are rather different. In fact, these three chains are identical at only 24 of 141 positions. Hence, quite different amino acid sequences can specify very similar three-dimensional structures (Figure 7-19).

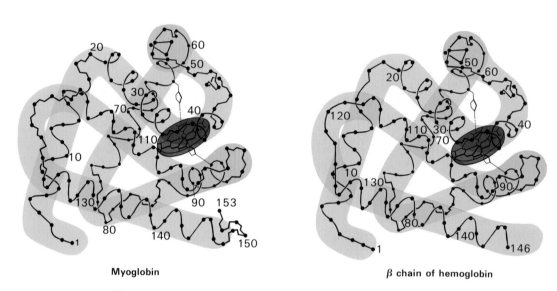

Myoglobin **β chain of hemoglobin**

Figure 7-18
Comparison of the conformations of the main chain of myoglobin and the β chain of hemoglobin. The similarity of their conformations is evident. [From M. F. Perutz. The hemoglobin molecule. Copyright © 1964 by Scientific American, Inc. All rights reserved.]

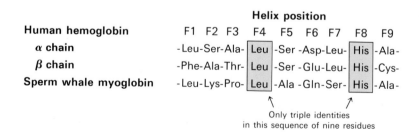

Figure 7-19
Comparison of the amino acid sequences of sperm whale myoglobin and the α and β chains of human hemoglobin, for residues F1 to F9. The amino acid sequences of these three polypeptide chains are much less alike than are their three-dimensional structures.

It is evident that the three-dimensional form of sperm whale myoglobin and of the α and β chains of human hemoglobin has broad biological significance. In fact, this motif, called the *globin fold*, is common to all known vertebrate myoglobins and hemoglobins. *The intricate folding of the polypeptide chain, first discovered in myoglobin, is nature's fundamental design for an oxygen carrier: it places the heme in an environment that enables it to carry oxygen reversibly.* The genes for myoglobin and for the α, β, and other chains of hemoglobin are variations on a fundamental theme. This family of genes almost certainly arose by gene duplication and diversification.

CRITICAL RESIDUES IN THE AMINO ACID SEQUENCE

The amino acid sequences of hemoglobins from more than sixty species (ranging from lamprey eels to humans) are known. A comparison of these sequences shows considerable variability at most positions. However, nine positions have the same residue in all or nearly all species studied thus far (Table 7-3). These invariant residues are especially important for the function of the hemoglobin molecule. Several of them directly affect the oxygen-binding site. Another invariant residue is tyrosine HC2, which stabilizes the molecule by forming a hydrogen bond between the H and F helices. Glycine B6 is invariant because of its small size: a side chain larger than a hydrogen atom would not allow the B and E helices to approach each other as closely as they do (Figure 7-20). Proline C2 may be essential because it defines one end of the C helix.

The amino acid residues in the interior of hemoglobins vary considerably. However, the change is always of one nonpolar residue for another (as from alanine to isoleucine). Thus, *the striking nonpolar character of the interior of the molecule is conserved.* The nonpolar core is also important in stabilizing the three-dimensional structure of hemoglobin.

In contrast, residues on the surface of the molecule are highly variable. Indeed, few are consistently positively or negatively charged. It might be thought that proline residues would be preserved because of their role as helix-breakers. However, this is not so. Only one proline is invariant, yet the lengths and directions of the helices in all globins are very similar. Obviously, there are other ways of terminating or bending α helices. For example, an α helix can be disrupted by the interaction of the OH group of serine or threonine with a main-chain carbonyl group.

Hemo—
A prefix from the Greek word meaning "blood"

Myo—
A prefix from the Greek word "muscle"

Globin—
A protein belonging to the myoglobin-hemoglobin family

Table 7-3
Invariant amino acid residues in hemoglobins

Position	Amino acid	Role
F8	Histidine	Proximal heme-linked histidine
E7	Histidine	Distal histidine near the heme
CD1	Phenylalanine	Heme contact
F4	Leucine	Heme contact
B6	Glycine	Allows the close approach of the B and E helices
C2	Proline	Helix termination
HC2	Tyrosine	Cross-links the H and F helices
C4	Threonine	Uncertain
H10	Lysine	Uncertain

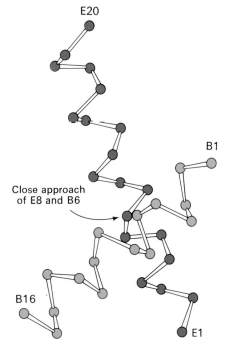

Figure 7-20
Crossing of the B and E helices in myoglobin. Residue B6 is almost invariably glycine because there is no space for a larger side chain.

HEMOGLOBIN IS AN ALLOSTERIC PROTEIN

The α and β subunits of hemoglobin have the same structural design as myoglobin. However, new properties of profound biological importance emerge when different subunits come together to form a tetramer. Hemoglobin is a much more intricate and sentient molecule than is myoglobin. Hemoglobin transports H^+ and CO_2 in addition to O_2. Furthermore, the oxygen-binding properties of hemoglobin are regulated by interactions between separate, nonadjacent sites. Hemoglobin is an allosteric protein, whereas myoglobin is not. This difference is expressed in three ways:

1. The binding of O_2 to hemoglobin enhances the binding of additional O_2 to the same hemoglobin molecule. In other words, O_2 binds cooperatively to hemoglobin. In contrast, the binding of O_2 to myoglobin is not cooperative.

2. The affinity of hemoglobin for oxygen depends on pH, whereas that of myoglobin is independent of pH. The CO_2 molecule also affects the oxygen-binding characteristics of hemoglobin.

3. The oxygen affinity of hemoglobin is further regulated by organic phosphates such as 2,3-bisphosphoglycerate (BPG). The result is that hemoglobin has a lower affinity for oxygen than does myoglobin.

Torr—
A unit of pressure equal to that exerted by a column of mercury 1 mm high at 0°C and standard gravity (1 mm Hg).
Named after Evangelista Torricelli (1608–1647), the inventor of the mercury barometer.

OXYGEN BINDS COOPERATIVELY TO HEMOGLOBIN

The saturation Y is defined as the fractional occupancy of the oxygen-binding sites. The value of Y can range from 0 (all sites empty) to 1 (all sites filled). A plot of Y versus pO_2, the partial pressure of oxygen, is called an *oxygen dissociation curve*. The oxygen dissociation curves of myoglobin and hemoglobin differ in two ways (Figures 7-21 and 7-22). For any given pO_2, Y is higher for myoglobin than for hemoglobin. This means that *myoglobin has a higher affinity for oxygen than does hemoglobin*. Oxygen affinity can be characterized by a quantity called P_{50}, which is the partial pressure of oxygen at which 50% of sites are filled (i.e., at which $Y = 0.5$). For myoglobin, P_{50} is typically 1 torr, whereas for hemoglobin, P_{50} is 26 torrs.

The second difference is that *the oxygen dissociation curve of myoglobin is hyperbolic, whereas that of hemoglobin is sigmoidal*. Let us consider these curves in quantitative terms, starting with the one for myoglobin be-

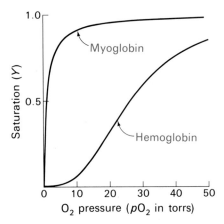

Figure 7-21
Oxygen dissociation curves of myoglobin and hemoglobin. Saturation of the oxygen-binding sites is plotted as a function of the partial pressure of oxygen surrounding the solution.

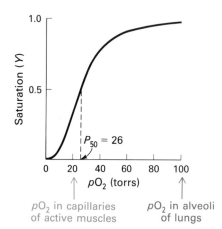

Figure 7-22
Oxygen dissociation curve of hemoglobin. Typical values for pO_2 in the capillaries of active muscle and in the alveoli of the lung are marked on the horizontal axis. Note that P_{50} for hemoglobin under physiological conditions lies between these values.

cause it is simpler. The binding of oxygen to myoglobin (Mb) can be described by a simple equilibrium:

$$MbO_2 \rightleftharpoons Mb + O_2 \quad (1)$$

The equilibrium constant K for the dissociation of oxymyoglobin is

$$K = \frac{[Mb][O_2]}{[MbO_2]} \quad (2)$$

in which $[MbO_2]$ is the concentration of oxymyoglobin, $[Mb]$ is the concentration of deoxymyoglobin, and $[O_2]$ is the concentration of uncombined oxygen, all in moles per liter. Then the saturation Y is

$$Y = \frac{[MbO_2]}{[MbO_2] + [Mb]} \quad (3)$$

Substitution of equation 2 into equation 3 yields

$$Y = \frac{[O_2]}{[O_2] + K} \quad (4)$$

Because oxygen is a gas, it is convenient to express its concentration in terms of pO_2, the partial pressure of oxygen (in torrs) in the atmosphere surrounding the solution. Equation 4 then becomes

$$Y = \frac{pO_2}{pO_2 + P_{50}} \quad (5)$$

Equation 5 plots as a hyperbola. In fact, the oxygen dissociation curve calculated from equation 5, taking P_{50} to be 1 torr, closely matches the experimentally observed curve for myoglobin.

In contrast, the sigmoidal curve for hemoglobin cannot be matched by any curve described by equation 5. In 1913, Archibald Hill showed that the curve obtained from the oxygen-binding data for hemoglobin agrees with the equation derived for the *hypothetical* equilibrium:

$$Hb(O_2)_n \rightleftharpoons Hb + n\,O_2 \quad (6)$$

This expression yields

$$Y = \frac{(pO_2)^n}{(pO_2)^n + (P_{50})^n} \quad (7)$$

which can be rearranged to give

$$\frac{Y}{1 - Y} = \left(\frac{pO_2}{P_{50}}\right)^n \quad (8)$$

This equation states that the ratio of oxyheme (Y) to deoxyheme ($1 - Y$) is equal to the nth power of the ratio of pO_2 to P_{50}. Taking the logarithms of both sides of equation 8 gives

$$\log \frac{(Y)}{1 - Y} = n \log pO_2 - n \log P_{50} \quad (9)$$

A plot of $\log[Y/(1-Y)]$ versus $\log pO_2$, called a *Hill plot*, approximates a straight line. Its slope n at the midpoint of the binding ($Y = 0.5$) is called the *Hill coefficient*. The value of n increases with the degree of cooperativity; the maximum possible value of n is equal to the number of binding sites.

Myoglobin gives a linear Hill plot with $n = 1.0$, whereas $n = 2.8$ for hemoglobin (Figure 7-23). The slope of 1.0 for myoglobin means that O_2 molecules bind independently of each other, as indicated in equa-

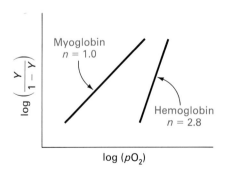

Figuire 7-23
Hill plot for the binding of O_2 to myoglobin and hemoglobin. The slope of 2.8 for hemoglobin indicates that it binds oxygen cooperatively, in contrast with myoglobin, which has a slope of 1.0.

tion 1. In contrast, the Hill coefficient of 2.8 for hemoglobin means that *the binding of oxygen in hemoglobin is cooperative*. Binding at one heme facilitates the binding of oxygen at the other hemes on the same tetramer. Conversely, the unloading of oxygen at one heme facilitates the unloading of oxygen at the others. In other words, the heme groups of a hemoglobin molecule communicate with each other. The mechanism of cooperative binding of oxygen by hemoglobin, sometimes called *heme-heme interaction*, will be discussed shortly.

THE COOPERATIVE BINDING OF OXYGEN BY HEMOGLOBIN ENHANCES OXYGEN TRANSPORT

The cooperative binding of oxygen by hemoglobin makes hemoglobin a more efficient oxygen transporter. The oxygen saturation of hemoglobin changes more rapidly with changes in the partial pressure of O_2 than it would if the oxygen-binding sites were independent of each other. Let us consider a specific example. Assume that the alveolar pO_2 is 100 torrs, and that the pO_2 in the capillary of an active muscle is 20 torrs. Let $P_{50} = 26$ torrs, and take $n = 2.8$. Then Y in the alveolar capillaries will be 0.98, and Y in the muscle capillaries will be 0.32. The oxygen delivered will be proportional to the difference in Y, which is 0.66. Let us now make the same calculation for a hypothetical oxygen carrier for which P_{50} is also 26 torrs, but in which the binding of oxygen is not cooperative ($n = 1$). Then $Y_{alveoli} = 0.79$, and $Y_{muscle} = 0.43$, and so the difference in Y is equal to 0.36. Thus, *the cooperative binding of oxygen by hemoglobin enables it to deliver 1.83 times as much oxygen under typical physiological conditions as it would if the sites were independent.*

H^+ AND CO_2 PROMOTE THE RELEASE OF O_2 (THE BOHR EFFECT)

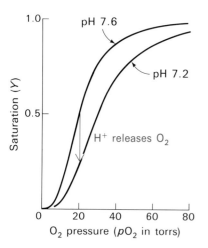

Figure 7-24
Effect of pH on the oxygen affinity of hemoglobin. Lowering the pH from 7.6 to 7.2 results in the release of O_2 from oxyhemoglobin.

Myoglobin shows no change in oxygen binding over a broad range of pH, nor does CO_2 have an appreciable effect. In hemoglobin, however, acidity enhances the release of oxygen. In the physiological range, a lowering of pH shifts the oxygen dissociation curve to the right, so that the oxygen affinity is decreased (Figure 7-24). Increasing the concentration of CO_2 (at constant pH) also lowers the oxygen affinity. In rapidly metabolizing tissue, such as contracting muscle, much CO_2 and acid are produced. *The presence of higher levels of CO_2 and H^+ in the capillaries of such metabolically active tissue promotes the release of O_2 from oxyhemoglobin.* This important mechanism for meeting the higher oxygen needs of metabolically active tissues was discovered by Christian Bohr in 1904. The reciprocal effect, discovered ten years later by J. S. Haldane, occurs in the alveolar capillaries of the lungs. The high concentration of O_2 there unloads H^+ and CO_2 from hemoglobin, just as the high concentration of H^+ and CO_2 in active tissues drives off O_2. These linkages between the binding of O_2, H^+, and CO_2 are known as the *Bohr effect* (Figure 7-25).

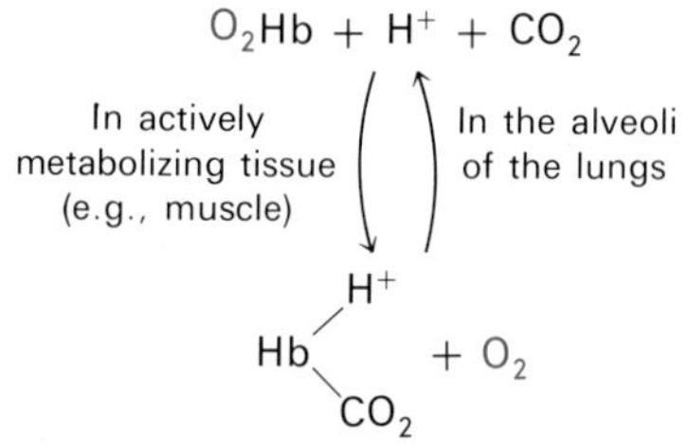

BPG LOWERS THE OXYGEN AFFINITY OF HEMOGLOBIN

Figure 7-25
Summary of the Bohr effect. The actual mechanism and stoichiometry are more complex than indicated in this diagram.

The oxygen affinity of hemoglobin within red cells is lower than that of hemoglobin in free solution. As early as 1921, Joseph Barcroft wondered, "Is there some third substance present . . . which forms an integral part of the oxygen-hemoglobin complex?" Indeed there is. Reinhold Benesch and Ruth Benesch showed in 1967 that 2,3-bisphosphoglycerate (BPG, also known as 2,3-diphosphoglycerate, DPG) binds to hemoglobin and has a large effect on its affinity for oxygen. This highly anionic organic phosphate is present in human red cells at about the same molar concentration as hemoglobin. In the absence of BPG, the P_{50} of hemoglobin is 1 torr, like that of myoglobin. In its presence, P_{50} becomes 26 torrs (Figure 7-26). Thus, *BPG lowers the oxygen affinity of hemoglobin by a factor of 26, which is essential in enabling hemoglobin to unload oxygen in tissue capillaries.* BPG diminishes the oxygen affinity of hemoglobin by binding to deoxyhemoglobin but not to the oxygenated form.

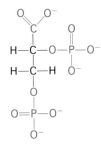

2,3-Bisphosphoglycerate
(2,3-Diphosphoglycerate, DPG)

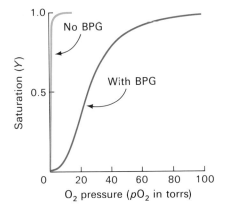

Figure 7-26
2,3-Bisphosphoglycerate (BPG) decreases the oxygen affinity of hemoglobin.

FETAL HEMOGLOBIN HAS A HIGHER OXYGEN AFFINITY THAN MATERNAL HEMOGLOBIN

Fetuses have their own kind of hemoglobin, called hemoglobin F ($\alpha_2\gamma_2$), which differs from adult hemoglobin A ($\alpha_2\beta_2$), as mentioned previously. An important property of hemoglobin F is that it has a higher oxygen affinity under physiological conditions than does hemoglobin A (Figure 7-27). The higher oxygen affinity of hemoglobin F optimizes the transfer of oxygen from the maternal to the fetal circulation. Hemoglobin F is oxygenated at the expense of hemoglobin A on the other side of the placental circulation. The higher oxygen affinity of fetal blood was known for many years, but an understanding of its basis could come only after the discovery of BPG. *Hemoglobin F binds BPG less strongly than does hemoglobin A and consequently has a higher oxygen affinity.* In the absence of BPG, the oxygen affinity of hemoglobin F is actually lower than that of hemoglobin A.

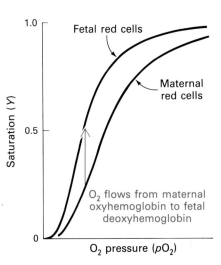

Figure 7-27
Fetal red blood cells have a higher oxygen affinity than do maternal red blood cells. The reason is that, in the presence of BPG, the oxygen affinity of fetal hemoglobin is higher than that of maternal hemoglobin.

SUBUNIT INTERACTIONS ARE REQUIRED FOR ALLOSTERIC EFFECTS

Let us now consider the structural basis of these allosteric effects. Hemoglobin can be dissociated into its constituent chains. The properties of the isolated α chain are very much like those of myoglobin. The α chain by itself has a high oxygen affinity, a hyperbolic oxygen dissociation curve, and oxygen-binding characteristics that are insensitive to pH, CO_2 concentration, and BPG level. The isolated β chain readily forms a tetramer, β_4, which is called hemoglobin H. Like the α chain

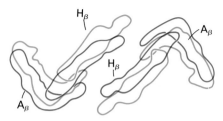

Figure 7-28
Projection of part of the electron-density maps of oxyhemoglobin (shown in red) and deoxyhemoglobin (shown in blue) at a resolution of 5.5 Å. The A and H helices of the two β chains of hemoglobin are shown here. The center of the diagram corresponds to the central cavity of the molecule. It shows one of the conformational changes accompanying oxygenation—a movement of the H helices toward each other. [After M. F. Perutz. *Nature* 228(1970):738.]

and myoglobin, β_4 entirely lacks the allosteric properties of hemoglobin and has a high oxygen affinity. In short, *the allosteric properties of hemoglobin arise from interactions between its subunits. The functional unit of hemoglobin is a tetramer consisting of two kinds of polypeptide chains.*

THE QUATERNARY STRUCTURE OF HEMOGLOBIN CHANGES MARKEDLY ON OXYGENATION

X-ray crystallographic studies have shown that oxy- and deoxyhemoglobin differ markedly in quaternary structure (Figure 7-28). The oxygenated molecule is more compact. For instance, the distance between the iron atoms of the β chains decreases from 40 to 33 Å on oxygenation. The changes in the contacts between the α and β chains are of special interest. There are two kinds of contact regions between the α and β chains (Figure 7-29). The $\alpha_1\beta_1$ contact (and the identical $\alpha_2\beta_2$

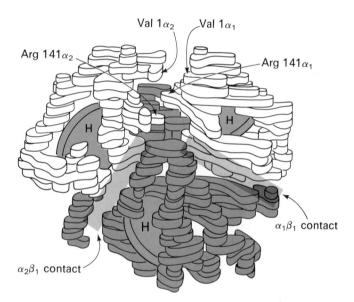

Figure 7-29
Model of oxyhemoglobin at low resolution showing the two kinds of interfaces between α and β chains. The α chains are white, the β chains black. Three hemes (red) can be seen in this view of the molecule. The $\alpha_2\beta_1$ contact region is shown in blue, the $\alpha_1\beta_1$ contact region in yellow. [After M. F. Perutz and L. F. TenEyck. *Cold Spring Harbor Symp. Quant. Biol.* 36(1971):296.]

contact) is one type. The other type is the $\alpha_1\beta_2$ contact (and the identical $\alpha_2\beta_1$ contact). In the transition from oxy- to deoxyhemoglobin, large structural changes take place at the $\alpha_1\beta_2$ contact but only small ones at the $\alpha_1\beta_1$ contact. Furthermore, the $\alpha_1\beta_1$ pair rotates relative to the other pair by 15 degrees (Figure 7-30). Some atoms at the interface between these pairs shift by as much as 6 Å.

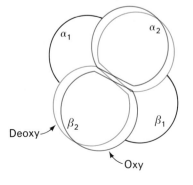

Figure 7-30
Schematic diagram showing the change in quaternary structure on oxygenation. One pair of $\alpha\beta$ subunits shifts with respect to the other by a rotation of 15 degrees and a translation of 0.8 Å. The oxy form of the rotated $\alpha\beta$ subunit is shown in red and the deoxy form in blue. [After J. Baldwin and C. Chothia. *J. Mol. Biol.* 129(1979):192. Copyright by Academic Press, Inc. (London) Ltd.]

In fact, the $\alpha_1\beta_2$ contact region is designed to act as a switch between two alternative structures. The two forms of this dove-tailed interface are stabilized by different sets of hydrogen bonds (Figure 7-31). This interface is closely connected to the heme groups, and so structural changes in it can be expected to affect the hemes. The importance of the interface is reinforced by the finding that most residues in it are the same in all species. Also, almost all mutations in this interface region diminish heme-heme interaction, whereas mutations in the $\alpha_1\beta_1$ interface do not.

DEOXYHEMOGLOBIN IS CONSTRAINED BY SALT LINKS BETWEEN DIFFERENT CHAINS

In oxyhemoglobin, the carboxyl-terminal residues of all four chains have almost complete freedom of rotation. In deoxyhemoglobin, by contrast, these terminal groups are anchored (Figure 7-32), and all of them, as well as the side chains of the C-terminal residues, participate in salt links (electrostatic interactions). Deoxyhemoglobin is a tauter, more constrained molecule than oxyhemoglobin because of these eight salt links. *The quaternary structure of deoxyhemoglobin is termed the T (tense or taut) form; that of oxyhemoglobin, the R (relaxed) form.* The designations R and T are generally used to describe alternative quaternary structures of an allosteric protein, the T form having a lower affinity for the substrate (p. 239).

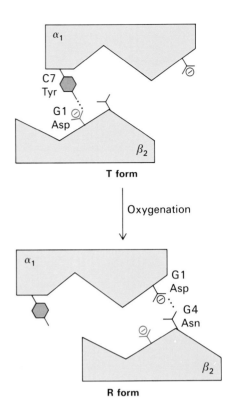

Figure 7-31
The $\alpha_1\beta_2$ interface switches from the T to the R form on oxygenation. The dove-tailed construction of this interface allows the subunits to readily adopt either of the two forms.

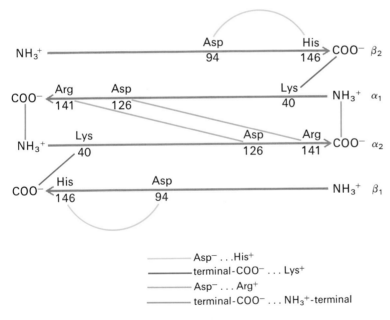

Figure 7-32
Salt links between different subunits in deoxyhemoglobin. These noncovalent, electrostatic cross-links are disrupted on oxygenation.

OXYGENATION MOVES THE IRON ATOM INTO THE PLANE OF THE PORPHYRIN

The conformational changes discussed thus far take place at some distance from the heme. Now let us see what happens at the heme group itself on oxygenation. In deoxyhemoglobin, the iron atom is about 0.4

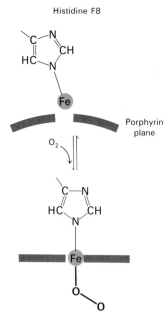

Figure 7-33
The iron atom moves into the plane of the heme on oxygenation. The proximal histidine (F8) is pulled along with the iron atom and becomes less tilted.

Å out of the heme plane toward the proximal histidine, so that the heme group is domed (convex) in the same direction (Figure 7-33). On oxygenation, the iron atom moves into the plane of the porphyrin to form a strong bond with O_2, and the heme becomes more planar. Structural studies of many synthetic iron porphyrins have shown that the iron atom is out of plane in five-coordinated compounds, whereas it is in plane or nearly so in six-coordinated complexes.

MOVEMENT OF THE IRON ATOM IS TRANSMITTED TO OTHER SUBUNITS BY THE PROXIMAL HISTIDINE

How does the movement of the iron atom into the plane of the heme favor the switch in quaternary structure from T to R? The proximal histidine is thought to play a key role in transmitting structural changes from one heme to another. When the iron atom moves into the plane of the heme on oxygenation, it pulls the proximal histidine with it. This movement of histidine F8 results in shifts of the F helix, the EF corner, and the FG corner (Figure 7-34). These conformational changes are transmitted to the subunit interfaces, where they rupture interchain salt links. Consequently, oxygenation shifts the equilibrium between the two quaternary structures to the R form. Thus, *a structural change (oxygenation) within a subunit is translated into structural changes at the interfaces between subunits. The binding of oxygen at one heme site is thereby communicated to parts of the molecule that are far away.*

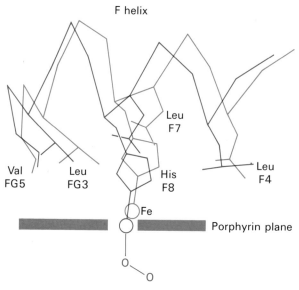

Figure 7-34
Conformational changes induced by the movement of the iron atom on oxygenation. The oxygenated structure is shown in red and the deoxygenated structure in blue. [After J. Baldwin and C. Chothia. *J. Mol. Biol.* 129(1979):192.]

MECHANISM OF THE COOPERATIVE BINDING OF OXYGEN

The fourth molecule of O_2 binds to hemoglobin some 300 times as tightly as the first one. Why? A postage-stamp analogy is helpful in considering this question (Figure 7-35). Deoxyhemoglobin is a taut mol-

Figure 7-35
Postage-stamp analogy of the oxygenation of hemoglobin. Two perforated edges must be torn to remove the first stamp. Only one perforated edge must be torn to remove the second stamp, and one edge again to remove the third stamp. The fourth stamp is then free.

ecule, constrained by the eight salt links between its four subunits. Oxygenation does not readily occur unless some of these salt links are broken, enabling the iron atom to move easily into the plane of the heme group. The number of salt links that need to be broken for the binding of an O_2 molecule depends on whether it is the first, second, third, or fourth to be bound. More salt links must be broken to permit the entry of the first O_2 than of subsequent ones. Because energy is required to break salt links, the binding of the first O_2 is energetically less favored than that of subsequent oxygen molecules. In this scheme, the binding affinity for the second and third O_2 molecules is intermediate between that for the first and last. This sequential increase in oxygen affinity would give the observed sigmoidal oxygen-binding curve.

BPG DECREASES OXYGEN AFFINITY BY CROSS-LINKING DEOXYHEMOGLOBIN

BPG binds specifically to deoxyhemoglobin in the ratio of one BPG per hemoglobin tetramer. This is an unusual stoichiometry—an $\alpha_2\beta_2$ protein would be expected to have at least two binding sites for a small molecule. The finding of one binding site immediately suggested that BPG binds on the symmetry axis of the hemoglobin molecule in the central cavity, where the four subunits are near each other (Figure 7-36). Indeed, x-ray analysis confirmed this proposal and showed that the binding site for BPG is constituted by three positively charged residues on *each* β chain: the α-amino group, lysine EF6, and histidine H21. These groups interact with the strongly negatively charged BPG, which carries nearly four negative charges at physiological pH. BPG is stereochemically complementary to this constellation of six positively charged groups facing the central cavity of the hemoglobin molecule (Figure 7-37). BPG binds more weakly to fetal hemoglobin than to hemoglobin A, because residue H21 in fetal hemoglobin is serine instead of histidine.

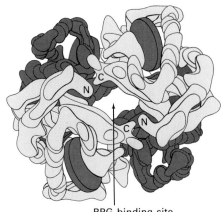

Figure 7-36
The binding site for BPG is in the central cavity of deoxyhemoglobin. [After M. F. Perutz. The hemoglobin molecule. Copyright © 1964 by Scientific American, Inc. All rights reserved.]

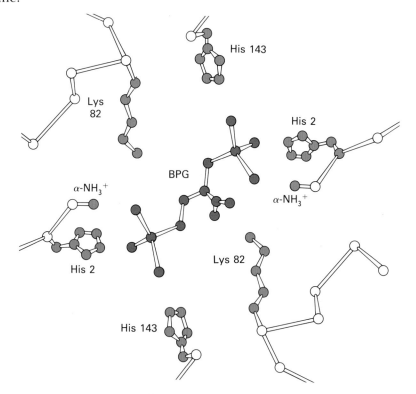

Figure 7-37
Mode of binding of BPG to human deoxyhemoglobin. BPG interacts with three positively charged groups on each β chain. [After A. Arnone. *Nature* 237(1972):148.]

On oxygenation, BPG is extruded because the central cavity becomes too small. Specifically, the gap between the H helices of the β chains becomes narrowed (see Figure 7-28). Also, the distance between the α-amino groups increases from 16 to 20 Å, which prevents them from simultaneously binding the phosphates of a BPG molecule.

The reason why BPG decreases oxygen affinity is now evident. *BPG stabilizes the deoxyhemoglobin quaternary structure by cross-linking the β chains.* In other words, BPG shifts the equilibrium toward the T form. As mentioned previously, the carboxyl-terminal residues of deoxyhemoglobin form eight salt links that must be broken for oxygenation to occur. The binding of BPG contributes additional cross-links that must be broken, and so the oxygen affinity of hemoglobin is diminished.

CO₂ BINDS TO THE TERMINAL AMINO GROUPS OF HEMOGLOBIN AND LOWERS ITS OXYGEN AFFINITY

In aerobic metabolism, about 0.8 equivalents of CO_2 are produced per O_2 consumed. Most of the CO_2 is transported as *bicarbonate*, which is formed within red cells by the action of *carbonic anhydrase*:

$$CO_2 + H_2O \rightleftharpoons HCO_3^- + H^+$$

Much of the H^+ generated by this reaction is taken up by deoxyhemoglobin as part of the Bohr effect. The CO_2 is carried by hemoglobin in the form of *carbamate*, because the un-ionized form of the α-amino groups of hemoglobin can react reversibly with CO_2:

$$R-NH_2 + CO_2 \rightleftharpoons R-\underset{\underset{O}{\|}}{\underset{|}{N}}-\overset{H}{\underset{}{C}}-O^- + H^+$$

The bound carbamates form salt bridges that stabilize the T form. *Hence, the binding of CO_2 lowers the oxygen affinity of hemoglobin.*

MECHANISM OF THE BOHR EFFECT

About 0.5 H^+ is taken up by hemoglobin for each molecule of O_2 that is released. This uptake of H^+ helps to buffer the pH in active tissues. It indicates that deoxygenation increases the affinities of some sites for H^+. Specifically, the pKs of some groups must be *raised* in the transition from oxy- to deoxyhemoglobin; an increase in pK means stronger binding of H^+. Which groups have their pKs raised? The potential sites and typical pK values are given in Figure 7-38. The side-chain carboxylates of glutamate and aspartate normally have pKs of about 4. It is unlikely that their pKs would be raised to at least 7 or 8, which would be necessary for a group to participate in the Bohr effect. On the other hand, the normal pKs of the side chains of tyrosine, lysine, and arginine are usually more than 10, and so it seems unlikely that they could be lowered enough to allow uptake of H^+. Hence, the most plausible candidates are the terminal amino group and the side chains of histidine and cysteine, which normally have pK values near 7.

X-ray and chemical studies suggest that three groups account for much of the Bohr effect: the side chains of histidines 146β and 122α and the α-amino group of the α chain. In oxyhemoglobin, histidine 146β rotates freely, whereas, in deoxyhemoglobin, this terminal resi-

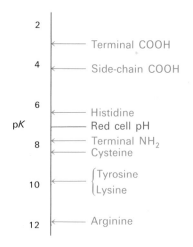

Figure 7-38
Typical pKs of some functional groups in proteins. The environment of a particular group can make its actual pK higher or lower than the value given here.

due participates in a number of interactions. Of particular significance is the interaction of its imidazole ring with the negatively charged aspartate 94 in the FG corner of the same β chain. The close proximity of this negatively charged group enhances the likelihood that the imidazole group binds a proton (Figure 7-39). In other words, the proximity of aspartate 94 raises the pK of histidine 146. Thus, *in the transition from oxy- to deoxyhemoglobin, histidine 146 acquires a greater affinity for H^+ because its local environment becomes more negatively charged.* The immediate environments of the other two groups implicated in the Bohr effect also are more negatively charged in deoxyhemoglobin because of changes of quaternary structure induced by the release of O_2.

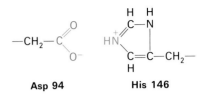

Figure 7-39
Aspartate 94 raises the pK of histidine 146 in deoxyhemoglobin but not in oxyhemoglobin. The proximity of the negative charge on aspartate 94 favors protonation of histidine 146 in deoxyhemoglobin.

COMMUNICATION WITHIN A PROTEIN MOLECULE

We have seen that the binding of O_2, H^+, CO_2, and BPG by hemoglobin are linked. These molecules are bound to spatially distinct sites that communicate with each other by means of conformational changes within the protein. The binding sites are separate because these molecules are structurally very different. *The interplay between these different sites is mediated by changes in quaternary structure.* In fact, every known allosteric protein consists of multiple polypeptide chains. The contact region between two chains can amplify and transmit conformational changes from one subunit to another. An allosteric protein does not have fixed properties. Rather, its functional characteristics are regulated by specific molecules in its environment. Consequently, allosteric interactions have immense importance in cellular function. *In the evolutionary transition from myoglobin to hemoglobin, a macromolecule capable of perceiving information from its environment has emerged.*

SICKLE-SHAPED RED BLOOD CELLS IN A CASE OF SEVERE ANEMIA

In 1904, James Herrick, a Chicago physician, examined a twenty-year-old black college student who had been admitted to the hospital because of a cough and fever. The patient felt weak and dizzy and had a headache. For about a year he had been experiencing palpitation and shortness of breath. On physical examination, the patient appeared rather well developed physically. There was a tinge of yellow in the whites of his eyes, and the visible mucous membranes were pale. His heart was distinctly enlarged. Examination of the blood showed that the patient was markedly anemic. The number of red cells was half of what is normal.

The patient's blood smear contained unusual red cells, which were described by Herrick in these terms: "The shape of the red cells was very irregular, but *what especially attracted attention was the large number of thin, elongated, sickle-shaped and crescent-shaped forms.*" The treatment was supportive, consisting of rest and nourishing food. The patient left the hospital four weeks later, less anemic and feeling much better. However, his blood still exhibited a "tendency to the peculiar crescent-shape in the red corpuscles though this was by no means as noticeable as before."

Herrick was puzzled by the clinical picture and laboratory findings. Indeed, he waited six years before publishing the case history and then candidly asserted that "not even a definite diagnosis can be made." He

Figure 7-40
Red cells from the blood of a patient with sickle-cell anemia, as viewed under a light microscope. [Courtesy of Dr. Frank Bunn.]

noted the chronic nature of the disease, and the diversity of abnormal physical and laboratory findings: cardiac enlargement, a generalized swelling of lymph nodes, jaundice, anemia, and evidence of kidney damage. He concluded that the disease could not be explained on the basis of an organic lesion in any one organ. He singled out the abnormal blood picture as the key finding and titled his case report *Peculiar Elongated and Sickle-Shaped Red Blood Corpuscles in a Case of Severe Anemia.* Herrick suggested that "some unrecognized change in the composition of the corpuscle itself may be the determining factor."

SICKLE-CELL ANEMIA IS A GENETICALLY TRANSMITTED, CHRONIC, HEMOLYTIC DISEASE

Other cases of this disease, called *sickle-cell anemia*, were found soon after the publication of Herrick's description. Indeed, sickle-cell anemia is not a rare disease. It is a significant public health problem wherever there is a substantial black population. The incidence of sickle-cell anemia among blacks is about four per thousand. In the past, it has usually been a fatal disease, often before age thirty, as a result of infection, renal failure, cardiac failure, or thrombosis. Sickled red cells become trapped in the small blood vessels, which impairs the circulation and leads to damage of multiple organs. Sickled cells are more fragile than normal ones. They hemolyze readily and consequently have a shorter life than normal cells, which leads to a severe anemia. The chronic course of the disease is punctuated by crises in which the proportion of sickled cells is especially high. During such a crisis, the patient may go into shock.

Sickle-cell anemia is genetically transmitted. *Patients with sickle-cell anemia are homozygous for an abnormal gene* located on an autosomal chromosome. Offspring who receive the abnormal gene from one parent but its normal allele from the other have *sickle-cell trait. Such heterozygous people are usually not symptomatic.* Only 1% of the red cells in a heterozygote's venous circulation are sickled, in contrast with about 50% in a homozygote. However, sickle-cell trait, which occurs in about one of ten blacks, is not entirely benign. Vigorous physical activity at high altitude, air travel in unpressurized planes, and anesthesia are potentially hazardous to people with sickle-cell trait. The reason will be evident shortly.

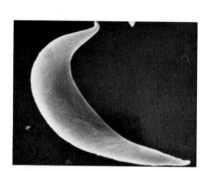

Figure 7-41
Scanning electron micrograph of an erythrocyte from a patient with sickle-cell anemia. [Courtesy of Dr. Jerry Thornwaite and Dr. Robert Leif.]

THE SOLUBILITY OF DEOXYGENATED SICKLE HEMOGLOBIN IS ABNORMALLY LOW

Herrick correctly surmised the location of the defect in sickle-cell anemia. Red cells from a patient with this disease will sickle on a microscope slide in vitro if the concentration of oxygen is reduced. In fact, the hemoglobin in these cells is itself defective. Deoxygenated sickle-cell hemoglobin has an abnormally low solubility, only 4% of that of normal deoxygenated hemoglobin. A fibrous precipitate is formed when a concentrated solution of sickle-cell hemoglobin is deoxygenated. This precipitate deforms red cells and gives them their sickle shape. Sickle-cell hemoglobin is commonly referred to as *hemoglobin S* (Hb S) to distinguish it from hemoglobin A (Hb A), the normal adult hemoglobin.

HEMOGLOBIN S HAS AN ABNORMAL ELECTROPHORETIC MOBILITY

In 1949, Linus Pauling and his associates examined the physical-chemical properties of hemoglobin from normal people and from those with sickle-cell trait or sickle-cell anemia. Their experimental approach was to search for differences between these hemoglobins by electrophoresis. They found that the isoelectric point (p. 45) of sickle-cell hemoglobin is higher than that of normal hemoglobin in both the oxygenated and the deoxygenated state (Figure 7-42):

	Normal	Sickle-cell anemia	Difference
Oxyhemoglobin	6.87	7.09	0.22
Deoxyhemoglobin	6.68	6.91	0.23

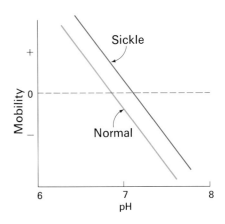

Figure 7-42
Electrophoretic mobility of sickle-cell hemoglobin and of normal hemoglobin as a function of pH. The isoelectric point of a molecule is the pH at which its mobility is 0.

These observations suggested that *there is a difference in the number or kind of ionizable groups in the two hemoglobins*. An estimate was made from the acid-base titration curve of hemoglobin in the neighborhood of pH 7. A change of one pH unit in the hemoglobin solution produces a change of about thirteen charges. The difference in isoelectric pH of 0.23 therefore corresponds to about three charges per hemoglobin molecule. It was concluded that *sickle-cell hemoglobin has between two and four more net positive charges per molecule than does normal hemoglobin*.

Patients with sickle-cell anemia (who are homozygous for the sickle gene) have hemoglobin S but no hemoglobin A. In contrast, people with sickle-cell trait (who are heterozygous for the sickle gene) have both kinds of hemoglobin in approximately equal amounts (Figure 7-43). Thus, Pauling's study revealed "*a clear case of a change produced in a protein molecule by an allelic change in a single gene.*" This was the first demonstration of a *molecular disease*.

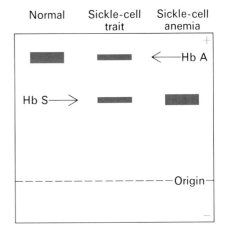

Figure 7-43
Gel electrophoresis pattern at pH 8.6 of hemoglobin isolated from a normal person, from a person with sickle-cell trait, and from a person with sickle-cell anemia.

A SINGLE AMINO ACID IN THE BETA CHAIN IS ALTERED IN SICKLE-CELL HEMOGLOBIN

The precise difference between normal and sickle-cell hemoglobin was identified in 1954, when Vernon Ingram devised a new technique for detecting amino acid substitutions in proteins. The hemoglobin molecule was split into smaller units for analysis, because it was anticipated that it would be easier to detect an altered amino acid in a peptide containing about 20 residues than in a protein ten times as large. Trypsin was used to specifically cleave hemoglobin on the carboxyl side of its lysine and arginine residues. Because an αβ half of hemoglobin contains a total of 27 lysine and arginine residues, *tryptic digestion* produced 28 different peptides. The next step was to separate the peptides. This was accomplished by a *two-dimensional procedure* (Figure 7-44). The mix-

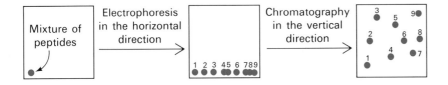

Figure 7-44
Fingerprinting. A mixture of peptides produced by proteolytic cleavage is resolved by electrophoresis in the horizontal direction followed by chromatography in the vertical direction.

Chromatography—
A term introduced by Mikhail Tswett in 1906 to describe the separation of a mixture of leaf pigments on a calcium carbonate column, like "light rays in the spectrum." Derived from the Greek words *chroma*, color, and *graphein*, to draw or write.

ture of peptides was placed in a spot at one corner of a large sheet of filter paper. *Electrophoresis* was first carried out in one direction, followed by *chromatography* in the perpendicular direction. Finally, the peptide spots were made visible by staining the filter paper with ninhydrin. This sequence of steps—selective cleavage of a protein into small peptides, followed by their separation in two dimensions—is called *fingerprinting*.

The fingerprints of hemoglobins A and S were highly revealing (Figure 7-45). When they were compared, *all but one of the peptide spots matched*. The one spot that was different was eluted from each fingerprint and shown to be a single peptide consisting of eight amino acids. Amino acid analysis indicated that this peptide in hemoglobin S differed from the one in hemoglobin A by a single amino acid. Ingram determined the sequence of this peptide and showed that *hemoglobin S contains valine instead of glutamate at position 6 of the β chain*:

Hemoglobin A   Val-His-Leu-Thr-Pro-Glu-Glu-Lys-
Hemoglobin S   Val-His-Leu-Thr-Pro-Val-Glu-Lys-
                 β1   2   3   4   5   6   7   8

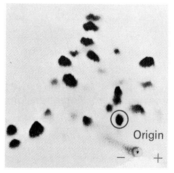
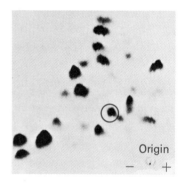

Figure 7-45
Comparison of the ninhydrin-stained fingerprints of hemoglobin A and hemoglobin S. The position of the peptide that is different in these hemoglobins is encircled in red. [Courtesy of Dr. Corrado Baglioni.]

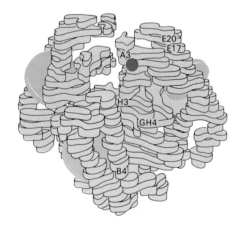

Figure 7-46
The position of the amino acid change in hemoglobin S (glutamate to valine at β6) is marked in red in this model of deoxyhemoglobin. Note that this site is at the surface of the protein. The α chains are shown in yellow, the β chains in blue. [After J. T. Finch, M. F. Perutz, J. F. Bertles, and J. Dobler. *Proc. Nat. Acad. Sci.* 70(1973):721.]

SICKLE HEMOGLOBIN HAS STICKY PATCHES ON ITS SURFACE

The side chain of valine is distinctly nonpolar, whereas that of glutamate is highly polar. The substitution of valine for glutamate at position 6 of the β chains places a nonpolar residue on the outside of hemoglobin S (Figure 7-46). *This alteration markedly reduces the solubility of deoxygenated hemoglobin S but has little effect on the solubility of oxygenated hemoglobin S.* This fact is crucial to an understanding of the clinical picture of sickle-cell anemia and sickle-cell trait.

The molecular basis of sickling can be visualized as follows:

1. The substitution of valine for glutamic acid gives hemoglobin S a sticky patch on the outside of each of its β chains (Figure 7-47). This sticky patch is present on both oxy- and deoxyhemoglobin S and is missing from hemoglobin A.

2. In the EF corner of each β chain of deoxyhemoglobin S is a hydrophobic site that is complementary to the sticky patch (Figure 7-47). The complementary site on one deoxyhemoglobin S molecule can bind to the sticky patch on another deoxyhemoglobin S molecule, which results in the formation of long fibers that distort the red cell.

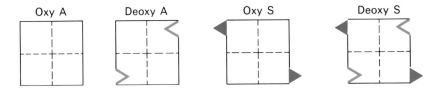

Figure 7-47
The red triangle represents the sticky patch that is present on both oxy- and deoxyhemoglobin S but not on either form of hemoglobin A. The complementary site is represented by an indentation that can accommodate the triangle. This complementary site is present in deoxyhemoglobin S and is probably also present in deoxyhemoglobin A.

3. In oxyhemoglobin S, the complementary site is masked. The sticky patch is present, but it cannot bind to another oxyhemoglobin S because the complementary site is unavailable.

4. Thus, *sickling occurs when there is a high concentration of the deoxygenated form of hemoglobin S* (Figure 7-48).

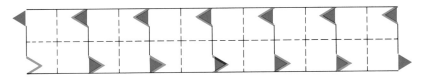

Figure 7-48
Interaction of sticky patches on deoxyhemoglobin S molecules with the complementary sites on other deoxyhemoglobin S molecules forms long strands.

These facts account for several clinical characteristics of sickle-cell anemia. A vicious cycle is set up when sickling occurs in a small blood vessel. The blockage of the vessel creates a local region of low oxygen concentration. Hence, more hemoglobin goes into the deoxy form and so more sickling occurs. A person with sickle-cell trait is usually asymptomatic because not more than half of his hemoglobin is hemoglobin S. This is too low a concentration for extensive sickling at normal oxygen levels. However, if the oxygen level is unusually low (as at high altitude), sickling can occur in such a person.

DEOXYHEMOGLOBIN S FORMS LONG HELICAL FIBERS

As described previously, deoxyhemoglobin S forms fibrous precipitates that deform red cells and give them their sickle shape. The main precipitate seen by electron microscopy consists of fibers having a diameter of 215 Å (Figure 7-49). Each fiber appears to be a fourteen-stranded

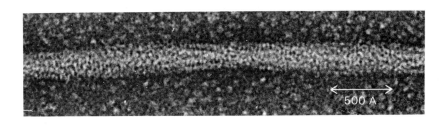

Figure 7-49
Electron micrograph of a negatively stained fiber of deoxyhemoglobin S. [From G. Dykes, R. H. Crepeau, and S. J. Edelstein. *Nature* 272(1978):509.]

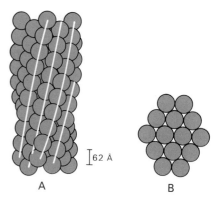

Figure 7-50
Fourteen-stranded helical model of the deoxyhemoglobin S fiber: (A) axial view; (B) cross-sectional view. Each circle represents a hemoglobin S molecule. [After G. Dykes, R. H. Crepeau, and S. J. Edelstein. *Nature* 272(1978):509.]

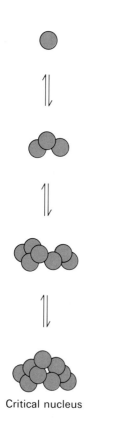

Figure 7-51
Nucleation phase in the formation of deoxyhemoglobin S fibers. The assembly of these nuclei is much slower than their subsequent growth. [After J. Hofrichter, P. D. Ross, and W. A. Eaton. *Proc. Nat. Acad. Sci.* 71(1974):4864.]

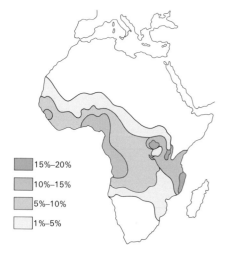

Figure 7-52
Frequency of the sickle-cell gene in Africa. High frequencies are restricted to regions where malaria is a major cause of death. [After A. C. Allison. Sickle cells and evolution. Copyright © 1956 by Scientific American, Inc. All rights reserved.]

helix (Figure 7-50). A significant feature of this helix is that each hemoglobin S molecule makes contact with at least eight others. It is evident that the fiber is stabilized by multiple interactions.

What determines whether a red cell becomes sickled during its passage through the capillary circulation, which takes about a second? The most important determinant is the concentration of deoxyhemoglobin S. *The striking experimental finding is that the rate of fiber formation is proportional to the tenth power of the effective concentration of deoxyhemoglobin S. Thus, fiber formation is a highly concerted reaction.* These kinetic data indicate that nucleation is the rate-limiting phase in fiber formation. The fiber grows rapidly once a critical cluster of about ten deoxyhemoglobin S molecules has been formed (Figure 7-51). Such a nucleus may correspond to a major part of one turn of the fourteen-stranded helix. These findings have important clinical implications. They demonstrate that *kinetic*, as well as thermodynamic, factors are important in sickling. A red cell that is supersaturated with deoxyhemoglobin S will not sickle if the lag time for fiber formation is longer than the transit time from the peripheral capillaries to the alveoli of the lungs, where reoxygenation occurs. The very strong dependence of the polymerization rate on the concentration of deoxyhemoglobin S accounts for the fact that people with sickle-cell trait are usually asymptomatic. The concentration of deoxyhemoglobin S in the red cells of these heterozygotes is about half of that in homozygotes, and so their rate of fiber formation is about a thousandfold slower ($2^{10} = 1024$).

HIGH INCIDENCE OF THE SICKLE GENE IS DUE TO THE PROTECTION CONFERRED AGAINST MALARIA

The frequency of the sickle gene is as high as 40% in certain parts of Africa. Until recently, most homozygotes have died before reaching adulthood, and so there must have been strong selective pressures to maintain the high incidence of the gene. James Neel proposed that the heterozygote enjoys advantages not shared by either the normal homozygote or the sickle-cell homozygote. In fact, Anthony Allison found that people with sickle-cell trait are protected against the most lethal form of malaria. The incidence of malaria and the frequency of the sickle gene in Africa are definitely correlated (Figure 7-52). This is a clear-cut example of *balanced polymorphism*—the heterozygote is protected against malaria and does not suffer from sickle-cell disease, whereas the normal homozygote is vulnerable to malaria.

FETAL DNA CAN BE ANALYZED FOR THE PRESENCE OF THE SICKLE-CELL GENE

The substitution of valine for glutamate at β6 in hemoglobin S results from a change in a single base, T for A. This mutation can readily be detected by cleaving DNA with a restriction enzyme that recognizes the sequence in this immediate region. The target for the restriction endonuclease MstII is the palindromic sequence CCTNAGG (where N denotes any base), which is present in the gene for the β chain of hemoglobin A ($β^A$ gene) but not in the one for hemoglobin S ($β^S$). Because of the absence of this target site in the $β^S$ gene, complete digestion of the gene by MstII produces a 1.3-kb fragment, corresponding to a 1.1-kb fragment from the $β^A$ gene (Figure 7-53A). The fragments in the digested sample of DNA are separated by gel electrophoresis and visualized by Southern blotting (p. 120) with a ^{32}P-labeled DNA probe that is complementary to the 1.1-kb fragment. The 1.3-kb fragment is also stained by this probe because it contains the 1.1-kb sequence. An autoradiogram reveals whether the $β^A$ gene, the $β^S$ gene, or both are present in the DNA sample (Figure 7-53B).

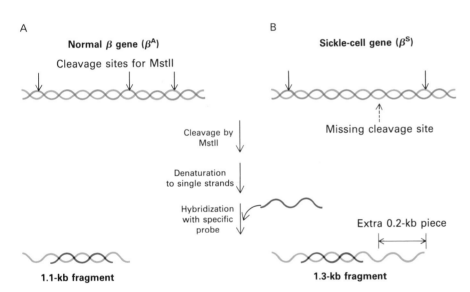
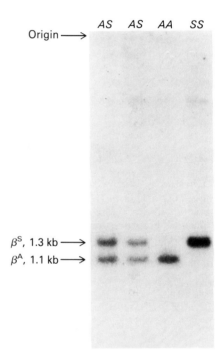

Figure 7-53
Restriction endonuclease method for detecting the sickle-cell gene. (A) Target site in the gene and fragments produced by digestion. (B) Electrophoresis pattern of a digest from parents who are heterozygous for the gene (lanes labeled *AS*), a normal child (*AA*), and a child with sickle-cell anemia (*SS*). [Part B is from Y. W. Kan. In *Medicine, Science, and Society*, K. J. Isselbacher, ed., (Wiley, 1984), p. 297.]

An attractive feature of this restriction enzyme method is that the DNA sample can come from any fetal cell. In contrast, a hemoglobin sample could be obtained only from red blood cells or their precursors. A sample of amniotic fluid is obtained from the fetus more readily and with less potential hazard than are blood-forming cells. DNA analyses can also be performed on biopsy samples of the chorionic villi obtained early in pregnancy, at about eight weeks gestation. These techniques for obtaining and analyzing genomic DNA are generally applicable. The number of genetic diseases that can be detected early in pregnancy by restriction enzyme cleavage of fetal DNA followed by Southern blot-

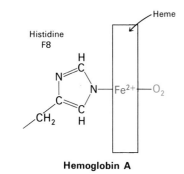

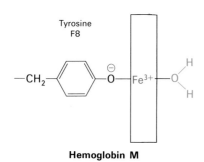

Figure 7-54
Substitution of tyrosine for the proximal histidine (F8) results in the formation of a hemoglobin M. The negatively charged oxygen atom of tyrosine is coordinated to the iron atom, which is in the ferric state. Water rather than O_2 is bound at the sixth coordination position.

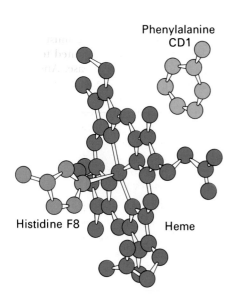

ting with highly specific probes is increasing rapidly. Parents of fetuses who are at risk can now make informed decisions as to whether to terminate pregnancy.

MOLECULAR PATHOLOGY OF HEMOGLOBIN

More than three hundred abnormal hemoglobins have been discovered by the examination of patients with clinical symptoms and by electrophoretic surveys of normal populations. In northern Europe, one of three hundred persons is heterozygous for a variant of hemoglobin A. The frequency of any one mutant allele is less than 10^{-4}, which is lower by several orders of magnitude than the frequency of the sickle allele in regions where malaria is endemic. In other words, most abnormal hemoglobins do not confer a selective advantage on the person. They are almost always neutral or harmful.

Abnormal hemoglobins are of several types:

1. *Altered exterior.* Nearly all substitutions on the surface of the hemoglobin molecule are harmless. Hemoglobin S is a striking exception.

2. *Altered active site.* The defective subunit cannot bind oxygen because of a structural change near the heme that directly affects oxygen binding. For example, substitution of tyrosine for the histidine proximal or distal to heme stabilizes the heme in the ferric form, which can no longer bind oxygen (Figure 7-54). The tyrosine side chain is ionized in this complex with the ferric ion of the heme. Mutant hemoglobins characterized by a permanent ferric state of two of the hemes are called *hemoglobin M*. The letter M signifies that the altered chains are in the *methemoglobin* (ferrihemoglobin) form. These patients are usually cyanotic. The disease has been seen only in heterozygous form, because homozygosity would almost certainly be lethal.

3. *Altered tertiary structure.* The polypeptide chain is prevented by the amino acid substitution from folding into its normal conformation. These hemoglobins are usually unstable. For example, in hemoglobin Hammersmith, phenylalanine at CD1, adjacent to the heme, is replaced by serine (Figure 7-55). The affinity of this hemoglobin for its heme groups is much lower than normal. Amino acid substitutions at sites far from the heme also can prevent hemoglobin from folding into its normal conformation. An instructive mutant is hemoglobin Riverdale-Bronx, which has arginine in place of glycine at B6. Recall that glycine occupies this position in all known normal myoglobins and hemoglobins (p. 153). This mutant hemoglobin does not fold normally because arginine at B6 is too large to fit into the narrow space between the B and E helices (see Figure 7-20).

4. *Altered quaternary structure.* Some mutations at subunit interfaces lead to the loss of allosteric properties. These hemoglobins usually have an abnormal oxygen affinity. The $\alpha_1\beta_2$ contact region, which changes markedly on oxygenation, is especially vulnerable to effects of mutation.

Figure 7-55
Location of phenylalanine CD1, in normal hemoglobin. The aromatic ring of this residue is in contact with the heme. In hemoglobin Hammersmith, serine replaces this phenylalanine residue; this markedly weakens the binding of heme.

THALASSEMIAS ARE GENETIC DISORDERS OF HEMOGLOBIN SYNTHESIS

The genetic diseases considered thus far are ones in which hemoglobin molecules are produced in essentially normal amounts but have impaired function because of the change of a single amino acid residue. The *thalassemias*, a different class of genetic disorders, are characterized by *defective synthesis of one or more hemoglobin chains*. The chain that is affected is denoted by a prefix, as in α-thalassemia. The name of this group of diseases comes from the Greek word "thalassa," meaning "sea," because of the high incidence of some forms of thalassemia among people living near the Mediterranean Sea. Indeed, some 20% of the population in parts of Italy are carriers of a gene for this disease. The geographic distribution of some thalassemia genes parallels that of malaria, which suggests that the heterozygote benefits from the presence of the gene, as in sickle-cell trait.

Thalassemias are produced by many different mutations that lead to the absence or deficiency of a globin chain in a variety of ways:

1. *The gene is missing.*

2. The gene is present but *RNA synthesis or processing is impaired*. For example, a mutation in the TATA box of the promoter site (p. 98) decreases the amount of RNA synthesized. Mutations within introns or near the exon-intron boundary can lead to aberrant splicing of the primary RNA transcript.

3. Globin mRNA is produced in normal amount but it encodes a grossly abnormal protein. For example, mutation of an amino acid codon to a stop codon (say TGG for tryptophan to TGA) will result in *premature termination* of protein synthesis. The deletion or addition of a nucleotide, a *frameshift mutation*, leads to an entirely different amino acid sequence on the distal side of the mutation. Most of these hemoglobin chains are very unstable and are rapidly degraded.

IMPACT OF THE DISCOVERY OF MOLECULAR DISEASES

Analyses of mutations affecting oxygen transport have had a major impact on molecular biology, medicine, and genetics. Their significance is threefold:

1. *They are sources of insight into relations between the structure and function of normal hemoglobin.* Mutations of a single amino acid residue are highly specific chemical modifications provided by nature. They shed light on facets of the protein that are critical for function.

2. *The discovery of mutant hemoglobins has revealed that disease can arise from a change of a single amino acid in one kind of polypeptide chain.* The concept of molecular disease, which is now an integral part of medicine, had its origins in the incisive studies of sickle-cell hemoglobin. The thalassemias have provided striking illustrations of the consequences of aberrant transcription, RNA splicing, and protein synthesis.

3. *The finding of mutant hemoglobins has enhanced our understanding of evolutionary processes.* Mutations are the raw material of evolution; the studies of sickle-cell anemia have shown that a mutation may be simultaneously beneficial and harmful.

Chapter 7
OXYGEN TRANSPORTERS

On disease and evolution— "Subjectively, to evolve must most often have amounted to suffering from a disease. And these diseases were of course molecular. The appearance of the concept of good and evil, interpreted by man as his painful expulsion from Paradise, was probably a molecular disease that turned out to be evolution."

From E. Zuckerkandl and L. Pauling. In *Horizons in Biochemistry*, M. Kasha and B. Pullman, eds. (Academic Press, 1964), pp. 189–225.

SUMMARY

Myoglobin and hemoglobin are the oxygen-carrying proteins in vertebrates. Myoglobin facilitates the transport of oxygen in muscle and serves as a reserve store of oxygen, whereas hemoglobin is the oxygen carrier in blood. These proteins contain tightly bound heme, a substituted porphyrin with a central iron atom. The ferrous (+2) state of heme binds O_2, whereas the ferric (+3) state does not.

Myoglobin, a single polypeptide chain of 153 residues (18 kd), has a compact shape. The inside of myoglobin consists almost exclusively of nonpolar residues. About 75% of the polypeptide chain is α helical. The single ferrous heme group is located in a nonpolar niche, which protects it from oxidation to the ferric form. The iron atom of the heme is directly bonded to a nitrogen atom of a histidine side chain. This proximal histidine occupies the fifth coordination position. The sixth coordination position on the other side of the heme plane is the binding site for O_2. The nearby distal histidine diminishes the binding of CO at the oxygen-binding site and inhibits the oxidation of heme to the ferric state. The central exon of the myoglobin gene encodes nearly all of the heme-binding site.

Hemoglobin consists of four polypeptide chains, each with a heme group. Hemoglobin A, the predominant hemoglobin in adults, has the subunit structure $\alpha_2\beta_2$. The three-dimensional structure of the α and β chains of hemoglobin is strikingly similar to that of myoglobin. Yet new properties appear in tetrameric hemoglobin that are not present in monomeric myoglobin. Hemoglobin transports H^+ and CO_2 in addition to O_2. Furthermore, the binding of these molecules is regulated by allosteric interactions, which are interactions between separate sites on the same protein. Indeed, hemoglobin is the best-understood allosteric protein.

Hemoglobin exhibits three kinds of allosteric effects. First, the oxygen-binding curve of hemoglobin is sigmoidal, which means that the binding of oxygen is cooperative. The binding of oxygen to one heme facilitates the binding of oxygen to the other hemes in the same molecule. Second, H^+ and CO_2 promote the release of O_2 from hemoglobin, an effect that is physiologically important in enhancing the release of O_2 in metabolically active tissues such as muscle. Conversely, O_2 promotes the release of H^+ and CO_2 in the alveolar capillaries of the lungs. These allosteric linkages between the bindings of H^+, CO_2, and O_2 are known as the Bohr effect. Third, the affinity of hemoglobin for O_2 is further regulated by 2,3-bisphosphoglycerate (BPG), a small molecule with a high density of negative charge. BPG can bind to deoxyhemoglobin but not to oxyhemoglobin. Hence, BPG lowers the oxygen affinity of hemoglobin. Fetal hemoglobin ($\alpha_2\gamma_2$) has a higher oxygen affinity than adult hemoglobin because it binds BPG less tightly.

The allosteric properties of hemoglobin arise from interactions between its α and β subunits. The T (tense) quaternary structure is constrained by salt links between different subunits, giving it a low affinity for O_2. These intersubunit salt links are absent from the R (relaxed) form, which has a high affinity for O_2. On oxygenation, the iron atom moves into the plane of the heme, pulling the proximal histidine with it. This motion cleaves some of the salt links, and the equilibrium is shifted from T to R. BPG stabilizes the deoxy state by binding to positively charged groups surrounding the central cavity of hemoglobin. Carbon dioxide, another allosteric effector, binds to the terminal amino groups of all four chains by forming readily reversible carbamate linkages. The hydrogen ions participating in the Bohr effect are bound to several pairs of sites that have a more negatively charged environment in the deoxy than in the oxy state.

Gene mutation resulting in the change of a single amino acid in a single protein can produce disease. The best-understood molecular disease is sickle-cell anemia. Hemoglobin S, the abnormal hemoglobin in this disease, consists of two normal α chains and two mutant β chains. In hemoglobin S, glutamate at residue 6 of the β chain is replaced by valine. This substitution of a nonpolar side chain for a polar one drastically reduces the solubility of deoxyhemoglobin S, which leads to the formation of fibrous precipitates that deform the red cell and give it a sickle shape. The resulting destruction of red cells produces a chronic hemolytic anemia. Sickle-cell anemia arises when a person is homozygous for the mutant sickle gene. The heterozygous condition, called sickle-cell trait, is relatively asymptomatic. About one of ten blacks in the United States is heterozygous for the sickle gene, and as many as four of ten in some parts of Africa. This high incidence of a gene that is harmful in homozygotes is due its beneficial effect in heterozygotes—people with sickle-cell trait are protected against the most lethal form of malaria. This is an example of balanced polymorphism. Sickle-cell anemia can be diagnosed in utero by restriction-endonuclease digestion of a sample of fetal DNA, followed by Southern blotting.

Several hundred mutant hemoglobins have been found by studies of the hemoglobin of patients with hematologic symptoms and by surveys of normal populations. Several classes of mutant hemoglobins are known: (1) Most substitutions on the surface of hemoglobin are harmless. Hemoglobin S is a striking exception. (2) Most substitutions near the heme impair the oxygen-binding site. For example, replacement of the proximal or distal histidine with tyrosine locks the heme in the ferric state, which cannot bind oxygen. (3) Many alterations in the interior of the molecule distort the tertiary structure and produce unstable hemoglobins. (4) Many changes at subunit interfaces lead to changes in oxygen affinity and the loss of allosteric properties. Thalassemias, a different class of genetic disorders, are characterized by the absence or defect of a globin chain. Some underlying causes are (1) absence of a globin gene, (2) impaired transcription or defective RNA processing, and (3) premature termination of protein synthesis or a shift in the reading frame.

SELECTED READINGS

Where to start

Kendrew, J. C., 1961. The three-dimensional structure of a protein molecule. *Sci. Amer.* 205(6):96–11. [Available as *Sci. Amer.* Offprint 121.]

Perutz, M. F., 1978. Hemoglobin structure and respiratory transport. *Sci. Amer.* 239(6):92–125. [Available as *Sci. Amer.* Offprint 1413.]

Dickerson, R. E., and Geis, I., 1983. *Hemoglobin: Structure, Function, Evolution and Pathology.* Benjamin/Cummings. [A beautifully illustrated account with a broad perspective.]

Structure of myoglobin and hemoglobin

Fermi, G., and Perutz, M. F., 1981. *Atlas of Molecular Structures in Biology. 2. Haemoglobin and myoglobin.* Clarendon Press, Oxford.

Takano, T., 1977. Structure of myoglobin refined at 2.0 Å resolution. *J. Mol. Biol.* 110:537–584.

Shaanan, B., 1983. Structure of human oxyhaemoglobin at 2.1 Å resolution. *J. Mol. Biol.* 171:31–59.

Fermi, G., Perutz, M. F., Shaanan, B., and Fourme, R., 1984. The crystal structure of human deoxyhaemoglobin at 1.74 Å resolution. *J. Mol. Biol.* 175:159–174.

Model systems

Collman, J. P., 1977. Synthetic models for the oxygen-binding hemoproteins. *Acc. Chem. Res.* 10:265–272. [A highly informative review of picket-fence porphyrins.]

Collman, J. P., Brauman, J. I., Halbert, T. R., and Suslick, K. S., 1976. Nature of O_2 and CO binding to metalloporphyrins and heme proteins. *Proc. Nat. Acad. Sci.* 73:3333–3337. [Presents the structural basis for the decreased binding of carbon monoxide by myoglobin and hemoglobin.]

Interaction of hemoglobin with H^+, CO_2, and BPG

Kilmartin, J. V., 1976. Interaction of haemoglobin with protons, CO_2, and 2,3-diphosphoglycerate. *Brit. Med. Bull.* 32:209–222.

Benesch, R., and Benesch, R. E., 1969. Intracellular organic phosphates as regulators of oxygen release by haemoglobin. *Nature* 221:618–622.

Tyuma, I., and Shimizu, K., 1970. Effect of organic phosphates on the difference in oxygen affinity between fetal and adult human hemoglobin. *Fed. Proc.* 29:1112–1114.

Allosteric mechanism of hemoglobin

Ho, C., (ed.), 1982. *Hemoglobin and Oxygen Binding.* Elsevier.

Perutz, M. F., 1980. Stereochemical mechanism of oxygen transport by haemoglobin. *Proc. Roy. Soc. Lond. Ser. B* 208:135–162.

Baldwin, J., and Chothia, C., 1979. Hemoglobin: the structural changes related to ligand binding and its allosteric mechanism. *J. Mol. Biol.* 129:175–220.

Shulman, R. G., Hopfield, J. J., and Ogawa, S., 1975. Allosteric interpretation of haemoglobin properties. *Quart. Rev. Biophys.* 8:325–420.

Gelin, B. R., Lee, A. W.-M., and Karplus, M., 1983. Hemoglobin tertiary structure change on ligand binding: its role in the cooperative mechanism. *J. Mol. Biol.* 171:489–559.

Friedman, J. M., 1985. Structure, dynamics, and reactivity in hemoglobin. *Science* 228:1273–1280.

Sickle-cell anemia and hemoglobin S

Embury, S. H., 1986. The clinical pathophysiology of sickle-cell disease. *Ann. Rev. Med.* 37:361–376.

Noguchi, C. T., and Schechter, A. N., 1985. Sickle hemoglobin polymerization in solution and in cells. *Ann. Rev. Biophys. Biophys. Chem.* 14:239–263.

Herrick, J. B., 1910. Peculiar elongated and sickle-shaped red blood corpuscles in a case of severe anemia. *Arch. Intern. Med.* 6:517–521.

Pauling, L., Itano, H. A., Singer, S. J., and Wells, I. C., 1949. Sickle cell anemia: a molecular disease. *Science* 110:543–548.

Ingram, V. M., 1957. Gene mutation in human haemoglobin: the chemical difference between normal and sickle cell haemoglobin. *Nature* 180:326–328.

Allison, A. C., 1956. Sickle cells and evolution. *Sci. Amer.* 195(2):87–94. [Available as *Sci. Amer.* Offprint 1065.]

Other genetic disorders of hemoglobin

Embury, S. H., Scharf, S. J., Saiki, R. K., Gholson, M. A., Golbus, M., Arnheim, N., and Erlich, H. A., 1987. Rapid prenatal diagnosis of sickle cell anemia by a new method of DNA analysis. *New Eng. J. Med.* 316:656–661.

Winslow, R. M., and Anderson, W. F., 1983. The hemoglobinopathies. *In* Stanbury, J. B., Wyngaarden, J. B., Fredrickson, D. S., Goldstein, J. L., and Brown, M. S., (eds.), *The Metabolic Basis of Inherited Disease* (5th ed.), pp. 1666–1710. McGraw-Hill.

Kan, Y. W., 1983. The thalassemias. *In* Stanbury, J. B., Wyngaarden, J. B., Fredrickson, D. S., Goldstein, J. L., and Brown, M. S., (eds.), *The Metabolic Basis of Inherited Disease* (5th ed.), pp. 1711–1725.

Weatherall, D. J., and Clegg, J. B., 1982. Thalassemia revisited. *Cell* 29:7–9.

Orkin, S. H., and Kazazian, H. H., Jr., 1984. The mutation and polymorphism of the human β-globin gene and its surrounding DNA. *Ann. Rev. Genet.* 18:131–171.

Honig, G. R., and Adams, J. G., 1986. *Human Hemoglobin Genetics.* Springer-Verlag.

Stamatoyannopoulos, G., Niehaus, A. W., Leder, P., and Majerus, P. W., (eds.), 1986. *Molecular Basis of Blood Diseases.* Saunders.

DNA probe detection of abnormal hemoglobins

Chang, J. C., and Kan, Y. W., 1982. A sensitive new prenatal test for sickle cell anemia. *New Engl. J. Med.* 307:30–32.

Goosens, M., Dumez, Y., Kaplan, L., Lupker, M., Charbet, C., Henrion, R., and Rosa, J., 1983. Prenatal diagnosis of sickle-cell anemia in the first trimester of pregnancy. *New Engl. J. Med.* 309:831–833.

Saiki, R. K., Scharf, S., Faloona, F., Mullis, K. B., Horn, G. T., Erlich, H. A., and Arnheim, N., 1985. Enzymatic amplification of beta-globin genomic sequences and restriction site analysis for diagnosis of sickle cell anemia. *Science* 230:1350–1354.

Bunn, H. F., and Forget, B. G., 1986. *Hemoglobin: Molecular, Genetic and Clinical Aspects.* Saunders.

Relation of exons to structural units of proteins

Go, M., 1981. Correlation of DNA exonic regions with protein structural units in haemoglobin. *Nature* 291:90–92.

Craik, C. S., Buchman, S. R., and Beychok, S., 1981. O_2 binding properties of the product of the central exon of beta-globin gene. *Nature* 291:87–90.

De Sanctis, G., Falcioni, G., Giardina, B., Ascoli, F., and Brunori, M., 1986. Mini-myoglobin: preparation and reaction with oxygen and carbon monoxide. *J. Mol. Biol.* 188:73–76.

PROBLEMS

1. The average volume of a red blood cell is 87 cubic micrometers. The mean concentration of hemoglobin in red cells is 34 g/100 ml.
   (a) What is the weight of the hemoglobin contained in a red cell?
   (b) How many hemoglobin molecules are there in a red cell?
   (c) Could the hemoglobin concentration in red cells be much higher than the observed value? (Hint: Suppose that a red cell contained a crystalline array of hemoglobin molecules 65 Å apart in a cubic lattice.)

2. How much iron is there in the hemoglobin of a 70-kg adult? Assume that the blood volume is 70 ml/kg of body weight and that the hemoglobin content of blood is 16 g/100 ml.

3. The myoglobin content of some human muscles is about 8 g/kg. In sperm whale, the myoglobin content of muscle is about 80 g/kg.
   (a) How much O_2 is bound to myoglobin in human muscle and in that of sperm whale? Assume that the myoglobin is saturated with O_2.
   (b) The amount of oxygen dissolved in tissue water (in equilibrium with venous blood at 37°C) is about 3.5×10^{-5} M. What is the ratio of oxygen bound to myoglobin to that directly dissolved in the water of sperm whale muscle?

4. The equilibrium constant K for the binding of oxygen to myoglobin is 10^{-6} M, where K is defined as
   $$K = \frac{[\text{Mb}][O_2]}{[\text{MbO}_2]}$$
   The rate constant for the combination of O_2 with myoglobin is 2×10^7 M^{-1} s^{-1}.
   (a) What is the rate constant for the dissociation of O_2 from oxymyoglobin?
   (b) What is the mean duration of the oxymyoglobin complex?

5. What is the effect of each of the following treatments on the oxygen affinity of hemoglobin A in vitro?
   (a) Increase in pH from 7.2 to 7.4.
   (b) Increase in pCO$_2$ from 10 to 40 torrs.
   (c) Increase in [BPG] from 2×10^{-4} to 8×10^{-4} M.
   (d) Dissociation of $\alpha_2\beta_2$ into monomer subunits.

6. The erythrocytes of birds and turtles contain a regulatory molecule different from BPG. This substance is also effective in reducing the oxygen affinity of human hemoglobin stripped of BPG. Which of the following substances would you predict to be most effective in this regard?
   (a) Glucose 6-phosphate.
   (b) Inositol hexaphosphate.
   (c) HPO$_4^{2-}$.
   (d) Malonate.
   (e) Arginine.
   (f) Lactate.

7. The pK of an acid depends partly on its environment. Predict the effect of the following environmental changes on the pK of a glutamic acid side chain.
   (a) A lysine side chain is brought into close proximity.
   (b) The terminal carboxyl group of the protein is brought into close proximity.
   (c) The glutamic acid side chain is shifted from the outside of the protein to a nonpolar site inside.

8. The concept of linkage is crucial for the understanding of many biochemical processes. Consider a protein molecule P that can bind A or B or both:

   $$\begin{array}{ccc} \text{P} & \xrightleftharpoons{K_A} & \text{PA} \\ K_B \updownarrow & & \updownarrow K_{AB} \\ \text{PB} & \xrightleftharpoons{K_{BA}} & \text{PAB} \end{array}$$

   The dissociation constants for these equilibria are defined as
   $$K_A = \frac{[P][A]}{[PA]} \quad K_B = \frac{[P][B]}{[PB]}$$
   $$K_{BA} = \frac{[PB][A]}{[PAB]} \quad K_{AB} = \frac{[PA][B]}{[PAB]}$$
   (a) Suppose $K_A = 5 \times 10^{-4}$ M, $K_B = 10^{-3}$ M, and $K_{BA} = 10^{-5}$ M. Is the value of the fourth dissociation constant K_{AB} defined? If so, what is it?
   (b) What is the effect of [A] on the binding of B? What is the effect of [B] on the binding of A?

9. Carbon monoxide combines with hemoglobin to form CO-hemoglobin. Crystals of CO-hemoglobin are isomorphous with those of oxyhemoglobin. Each heme in hemoglobin can bind one carbon monoxide molecule, but O_2 and CO cannot simultaneously bind to the same heme. The binding affinity for CO is about 200 times as great as that for oxygen. Exposure for 1 hour to a CO concentration of 0.1% in inspired air leads to the occupancy by CO of about half of the heme sites in hemoglobin, a proportion that is frequently fatal.

   An interesting problem was posed (and partly solved) by J. S. Haldane and J. G. Priestley in 1935:

   > If the action of CO were simply to diminish the oxygen-carrying power of the hemoglobin, without other modification of its properties, the symptoms of CO poisoning would be very difficult to understand in the light of other knowledge. Thus, a person whose blood is half-saturated with CO is practically helpless, as we

have just seen; but a person whose hemoglobin percentage is simply diminished to half by anemia may be going about his work as usual.

What is the key to this seeming paradox?

10. A protein molecule P reversibly binds a small molecule L. The dissociation constant K for the equilibrium

$$P + L \rightleftharpoons PL$$

is defined as

$$K = \frac{[P][L]}{[PL]}$$

The protein transports the small molecule from a region of high concentration $[L_A]$ to one of low concentration $[L_B]$. Assume that the concentrations of unbound small molecules remain constant. The protein goes back and forth between regions A and B.
  (a) Suppose $[L_A] = 10^{-4}$ M and $[L_B] = 10^{-6}$ M. What value of K yields maximal transport? One way of solving this problem is to write an expression for ΔY, the change in saturation of the ligand-binding site in going from region A to B. Then, take the derivative of ΔY with respect to K.
  (b) Treat oxygen transport by hemoglobin in a similar way. What value of P_{50} would give a maximal ΔY? Assume that the P in the lungs is 100 torrs, whereas P in the tissue capillaries is 20 torrs. Compare your calculated value of P_{50} with the physiological value of 26 torrs.

11. A hemoglobin with an abnormal electrophoretic mobility is detected in a screening program. Fingerprinting after tryptic digestion reveals that the amino acid substitution is in the β chain. The normal amino-terminal tryptic peptide (Val-His-Leu-Thr-Pro-Glu-Glu-Lys) is missing. A new tryptic peptide consisting of six amino acid residues is found. Valine is the amino terminal residue of this peptide.
  (a) Which amino acid substitutions are consistent with these data?
  (b) Which single-base changes in DNA sequence could give these amino acid substitutions? The DNA sequence encoding the normal amino-terminal region is GTGCACCTGACTCCTGAGGAGAAG.
  (c) How should the electrophoretic mobility of this hemoglobin compare with those of Hb A and Hb S at pH 8?

12. Does the appearance of a 1.3-kb fragment following digestion with MstII and Southern blotting with a specific probe (p. 169) prove that the DNA sample contains a sickle-cell gene? Could other mutations give the same result?

13. Some mutations in a hemoglobin gene affect all three of the hemoglobins A, A_2, and F, whereas others affect only one of them. Why?

14. Hemoglobin A inhibits the formation of long fibers of hemoglobin S and the subsequent sickling of the red cell upon deoxygenation. Why does hemoglobin A have this effect?

15. Cyanate was a promising antisickling drug until clinical trials uncovered its toxic side effects, such as damage to peripheral nerves. Cyanate carbamoylates the terminal amino groups of hemoglobin. It behaves as a reactive analog of CO_2.

$$R-NH_2 + {}^-NCO + H^+ \longrightarrow R-\underset{O}{\overset{H}{N}}-\underset{\parallel}{C}-NH_2$$

Terminal amino group   Cyanate   Carbamoylated derivative

How does the change in oxygen affinity from this modification have an antisickling effect?

16. What is the effect of each of the following treatments on the number of H^+ bound to hemoglobin A in vitro?
  (a) Increase in pO_2 from 20 to 100 torrs (at constant pH and pCO_2).
  (b) Reaction of hemoglobin with excess cyanate (at constant pH).

CHAPTER 8

Introduction to Enzymes

Enzymes, the catalysts of biological systems, are remarkable molecular devices that determine the pattern of chemical transformations. They also mediate the transformation of different forms of energy. The most striking characteristics of enzymes are their *catalytic power* and *specificity*. Furthermore, the actions of many enzymes are regulated. Nearly all known enzymes are proteins. The recent discovery of catalytically active RNA molecules, however, indicates that proteins do not have an absolute monopoly on catalysis.

Proteins as a class of macromolecules are highly effective in catalyzing diverse chemical reactions because of their capacity to specifically bind a very wide range of molecules. By utilizing the full repertoire of intermolecular forces, enzymes bring substrates together in an optimal orientation, the prelude to making and breaking chemical bonds. In essence, they catalyze reactions by stabilizing transition states, the highest-energy species in reaction pathways. By doing this selectively, an enzyme determines which one of several potential chemical reactions actually occurs. Enzymes can also act as molecular switches in regulating catalytic activity and transforming energy because of their capacity to couple the actions of separate binding sites.

ENZYMES HAVE IMMENSE CATALYTIC POWER

Enzymes accelerate reactions by factors of at least a million. Indeed, most reactions in biological systems do not occur at perceptible rates in the absence of enzymes. Even a reaction as simple as the hydration of

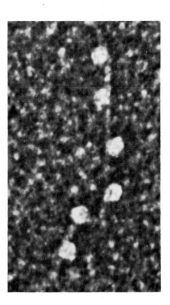

Figure 8-1
Electron micrograph of DNA polymerase I molecules (white spheres) bound to a threadlike synthetic DNA template. [Courtesy of Dr. Jack Griffith.]

carbon dioxide is catalyzed by an enzyme, namely, carbonic anhydrase (p. 38). The transfer of CO_2 from the tissues into the blood and then to the alveolar air would be less complete in the absence of this enzyme. In fact, carbonic anhydrase is one of the fastest enzymes known. Each enzyme molecule can hydrate 10^5 molecules of CO_2 per second. This catalyzed reaction is 10^7 times faster than the uncatalyzed one.

ENZYMES ARE HIGHLY SPECIFIC

Enzymes are highly specific both in the reaction catalyzed and in their choice of reactants, which are called *substrates*. An enzyme usually catalyzes a single chemical reaction or a set of closely related reactions. Side reactions leading to the wasteful formation of by-products rarely occur in enzyme-catalyzed reactions, in contrast with uncatalyzed ones. The degree of specificity for substrate is usually high and sometimes virtually absolute.

Let us consider *proteolytic enzymes* as an example. The reaction catalyzed by these enzymes is the hydrolysis of a peptide bond.

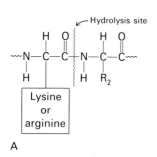

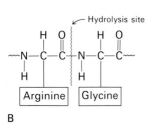

Figure 8-2
Comparison of the specificities of (A) trypsin and (B) thrombin. Trypsin cleaves on the carboxyl side of arginine and lysine residues. Thrombin cleaves Arg-Gly bonds in specific sequences only.

Most proteolytic enzymes also catalyze a different but related reaction, namely, the hydrolysis of an ester bond.

Proteolytic enzymes differ markedly in their degree of substrate specificity. Subtilisin, which comes from certain bacteria, is quite undiscriminating about the nature of the side chains adjacent to the peptide bond to be cleaved. Trypsin, as was mentioned in Chapter 3, is quite specific in that it catalyzes the splitting of peptide bonds on the carboxyl side of lysine and arginine residues only (Figure 8-2A). Thrombin, an enzyme that participates in blood clotting, is even more specific than trypsin. It catalyzes the hydrolysis of Arg-Gly bonds in specific peptide sequences only (Figure 8-2B).

DNA polymerase I, a template-directed enzyme (p. 85), is another highly specific catalyst. The sequence of nucleotides in the DNA strand that is being synthesized is determined by the sequence of nucleotides in another DNA strand that serves as a template. DNA polymerase I is remarkably precise in carrying out the instructions given by the template. The wrong nucleotide is inserted into a new DNA strand less than once in a million times, because DNA polymerase proofreads the nascent product and corrects its mistakes.

THE CATALYTIC ACTIVITIES OF MANY ENZYMES ARE REGULATED

The enzyme that catalyzes the first step in a biosynthetic pathway is usually inhibited by the ultimate product (Figure 8-3). The biosynthesis of isoleucine in bacteria illustrates this type of control, which is called *feedback inhibition*. Threonine is converted into isoleucine in five steps, the first of which is catalyzed by threonine deaminase. This enzyme is inhibited when the concentration of isoleucine reaches a sufficiently high level. Isoleucine inhibits by binding to the enzyme at a regulatory site, which is distinct from the catalytic site. This inhibition is mediated by an *allosteric interaction*, which is reversible. When the level of isoleucine drops sufficiently, threonine deaminase becomes active again, and consequently isoleucine is synthesized once more.

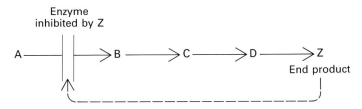

Figure 8-3
Feedback inhibition of the first enzyme in a pathway by reversible binding of the final product.

Figure 8-4
The α-carbon skeleton of calmodulin, a calcium sensor that regulates the activities of many intracellular proteins. The four domains of the protein are shown in different colors. The small circles show bound calcium ions. [Courtesy of Dr. Y. S. Babu and Dr. William J. Cook.]

Enzymes are also controlled by *regulatory proteins*, which can either stimulate or inhibit. The activities of many enzymes are regulated by *calmodulin*, a 17-kd protein that serves as a calcium sensor in nearly all eucaryotic cells (Figure 8-4). The binding of Ca^{2+} to four sites in calmodulin induces the formation of α-helix and other conformational changes that convert it from an inactive to an active form. Activated calmodulin then binds to many enzymes and other proteins in the cell and modifies their activities (p. 989).

Covalent modification is a third mechanism of enzyme regulation. For example, the activities of the enzymes that synthesize and degrade glycogen are regulated by the attachment of a phosphoryl group to a specific serine residue (p. 459). Other enzymes are controlled by the phosphorylation of threonine and tyrosine residues. These modifications are reversed by hydrolysis of the phosphate ester linkage. Specific enzymes catalyze the attachment and removal of phosphoryl and other modifying groups.

Phosphoserine residue   **Phosphothreonine residue**   **Phosphotyrosine residue**

Some enzymes are synthesized in an inactive precursor form, which is activated at a physiologically appropriate time and place. The digestive enzymes exemplify this kind of control, which is called *proteolytic activation*. For example, trypsinogen is synthesized in the pancreas and is activated by peptide-bond cleavage in the small intestine to form the active enzyme trypsin (Figure 8-5). This type of control is also repeatedly used in the sequence of enzymatic reactions that lead to the clotting of blood. The enzymatically inactive precursors of proteolytic enzymes are called *zymogens*.

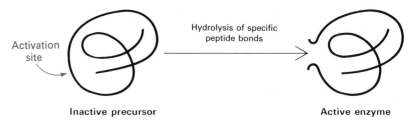

Figure 8-5
Activation of zymogen by hydrolysis of specific peptide bonds.

ENZYMES TRANSFORM DIFFERENT FORMS OF ENERGY

In many biochemical reactions, *the energy of the reactants is converted with high efficiency into a different form.* For example, in photosynthesis, light energy is converted into chemical-bond energy. In mitochondria, the free energy contained in small molecules derived from food is converted into a different currency, the free energy of adenosine triphosphate (ATP). The chemical-bond energy of ATP is then utilized in many different ways. In muscular contraction, the energy of ATP is converted into mechanical energy. Cells and organelles have pumps that utilize ATP to transport molecules and ions against chemical and electrical gradients (Figure 8-6). These transformations of energy are carried out by enzyme molecules that are integral parts of highly organized assemblies.

FREE ENERGY IS THE MOST USEFUL THERMODYNAMIC FUNCTION IN BIOCHEMISTRY

Let us review some key thermodynamic relations. In thermodynamics, a *system* is the matter within a defined region. The matter in the rest of the universe is called the *surroundings*. *The first law of thermodynamics states that the total energy of a system and its surroundings is a constant.* In other words, energy is conserved. The mathematical expression of the first law is

$$\Delta E = E_B - E_A = Q - W \tag{1}$$

in which E_A is the energy of a system at the start of a process and E_B at the end of the process, Q is the heat absorbed by the system, and W is the work done by the system. An important feature of equation 1 is that *the change in energy of a system depends only on the initial and final states and not on the path of the transformation.*

The first law of thermodynamics cannot be used to predict whether a reaction can occur spontaneously. Some reactions do occur spontaneously although ΔE is positive (the energy of the system increases). In such cases, the system absorbs heat from its surroundings. It is evident that a function different from ΔE is required. One such function is the *entropy* (S), which is a measure of the *degree of randomness or disorder of a*

Figure 8-6
Electron micrograph of sodium-potassium pump molecules in a plasma membrane. These densely packed enzyme molecules catalyze the ATP-driven flux of Na^+ and K^+ out of and into cells. [Courtesy of Dr. Guido Zampighi.]

system. The entropy of a system increases (Δ*S* is positive) when it becomes more disordered (Figure 8-7). *The second law of thermodynamics states that a process can occur spontaneously only if the sum of the entropies of the system and its surroundings increases.*

$$(\Delta S_{\text{system}} + \Delta S_{\text{surroundings}}) > 0 \text{ for a spontaneous process} \quad (2)$$

Note that the entropy of a system can decrease during a spontaneous process, provided that the entropy of the surroundings increases so that their sum is positive. For example, the formation of a highly ordered biological structure is thermodynamically feasible because the decrease in the entropy of such a system is more than offset by an increase in the entropy of its surroundings.

One difficulty in using entropy as a criterion of whether a biochemical process can occur spontaneously is that the entropy changes of chemical reactions are not readily measured. Furthermore, the criterion of spontaneity given in equation 2 requires that both the entropy change of the surroundings and that of the system of interest be known. These difficulties are obviated by using a different thermodynamic function called the *free energy*, which is denoted by the symbol *G* (or *F*, in the older literature). In 1878, Josiah Willard Gibbs created the free-energy function by combining the first and second laws of thermodynamics. The basic equation is

$$\Delta G = \Delta H - T \Delta S \quad (3)$$

in which Δ*G* is the change in free energy of a system undergoing a transformation at constant pressure (*P*) and temperature (*T*), Δ*H* is the change in enthalpy of this system, and Δ*S* is the change in entropy of this system. Note that the properties of the surroundings do not enter into this equation. The enthalpy change is given by

$$\Delta H = \Delta E + P \Delta V \quad (4)$$

The volume change, Δ*V*, is small for nearly all biochemical reactions, and so Δ*H* is nearly equal to Δ*E*. Hence,

$$\Delta G \cong \Delta E - T \Delta S \quad (5)$$

Thus, the Δ*G* of a reaction depends both on the change in internal energy and on the change in entropy of the system.

The change in free energy (Δ*G*) of a reaction, in contrast with the change in internal energy (Δ*E*) of a reaction, is a valuable criterion of whether it can occur spontaneously.

1. *A reaction can occur spontaneously only if Δ*G* is negative.*

2. *A system is at equilibrium and no net change can take place if Δ*G* is zero.*

3. *A reaction cannot occur spontaneously if Δ*G* is positive.* An input of free energy is required to drive such a reaction.

Two additional points need to be emphasized here. First, the Δ*G* of a reaction depends only on the free energy of the products (the final state) minus that of the reactants (the initial state). *The ΔG of a reaction is independent of the path (or molecular mechanism) of the transformation.* The mechanism of a reaction has no effect on Δ*G*. For example, the Δ*G* for the oxidation of glucose to CO_2 and H_2O is the same whether it occurs

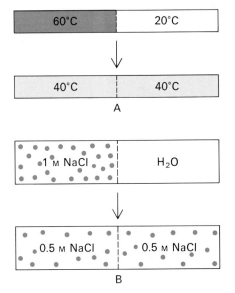

Figure 8-7
Processes that are driven by an increase in the entropy of a system: (A) diffusion of heat; and (B) diffusion of a solute.

by combustion in vitro or by a series of many enzyme-catalyzed steps in a cell. Second, ΔG *provides no information about the rate of a reaction.* A negative ΔG indicates that a reaction can occur spontaneously, but it does not signify whether it will proceed at a perceptible rate. As will be discussed shortly (p. 184), the rate of a reaction depends on the *free energy of activation* ($\Delta G^{\ddagger}$), which is unrelated to ΔG.

STANDARD FREE-ENERGY CHANGE OF A REACTION AND ITS RELATION TO THE EQUILIBRIUM CONSTANT

Consider the reaction

$$A + B \rightleftharpoons C + D$$

The ΔG of this reaction is given by

$$\Delta G = \Delta G^\circ + RT \log_e \frac{[C][D]}{[A][B]} \quad (6)$$

in which ΔG° is the *standard free-energy change*, R is the gas constant, T is the absolute temperature, and [A], [B], [C], and [D] are the molar concentrations (more precisely, the activities) of the reactants. ΔG° is the free-energy change for this reaction under standard conditions—that is, when each of the reactants A, B, C, and D is present at a concentration of 1.0 M (for a gas, the standard state is usually chosen to be 1 atmosphere). Thus, the ΔG of a reaction depends on the *nature* of the reactants (expressed in the ΔG° term of equation 6) and on their *concentrations* (expressed in the logarithmic term of equation 6).

A convention has been adopted to simplify free-energy calculations for biochemical reactions. The standard state is defined as having a pH of 7. Consequently, when H^+ is a reactant, its activity has the value 1 (corresponding to a pH of 7) in equations 6 and 9. The activity of water also is taken to be 1 in these equations. The *standard free-energy change at pH 7*, denoted by the symbol $\Delta G^{\circ\prime}$, will be used throughout this book. The *kilocalorie* will be used as the unit of energy.

The relation between the standard free energy and the equilibrium constant of a reaction can be readily derived. At equilibrium, ΔG = 0. Equation 6 then becomes

$$0 = \Delta G^{\circ\prime} + RT \log_e \frac{[C][D]}{[A][B]} \quad (7)$$

and so

$$\Delta G^{\circ\prime} = -RT \log_e \frac{[C][D]}{[A][B]} \quad (8)$$

The equilibrium constant under standard conditions, K'_{eq}, is defined as

$$K'_{eq} = \frac{[C][D]}{[A][B]} \quad (9)$$

Substituting equation 9 into equation 8 gives

$$\Delta G^{\circ\prime} = -RT \log_e K'_{eq} \quad (10)$$
$$\Delta G^{\circ\prime} = -2.303 \, RT \log_{10} K'_{eq} \quad (11)$$

which can be rearranged to give

$$K'_{eq} = 10^{-\Delta G^{\circ\prime}/(2.303RT)} \quad (12)$$

Units of energy—
A *calorie* (cal) is equivalent to the amount of heat required to raise the temperature of 1 gram of water from 14.5°C to 15.5°C.
A *kilocalorie* (kcal) is equal to 1000 cal.
A *joule* (J) is the amount of energy needed to apply a 1 newton force over a distance of 1 meter. A *kilojoule* (kJ) is equal to 1000 J.
1 kcal = 4.184 kJ

Substituting $R = 1.98 \times 10^{-3}$ kcal mol^{-1} deg^{-1} and $T = 298°K$ (corresponding to 25°C) gives

$$K'_{eq} = 10^{-\Delta G°'/1.36} \quad (13)$$

when $\Delta G°'$ is expressed in kcal/mol. Thus, the standard free energy and the equilibrium constant of a reaction are related by a simple expression. For example, an equilibrium constant of 10 corresponds to a standard free-energy change of -1.36 kcal/mol at 25°C (Table 8-1).

Let us calculate $\Delta G°'$ and ΔG for the isomerization of dihydroxyacetone phosphate to glyceraldehyde 3-phosphate as an example. This reaction occurs in glycolysis (p. 352). At equilibrium, the ratio of glyceraldehyde 3-phosphate to dihydroxyacetone phosphate is 0.0475 at 25°C (298°K) and pH 7. Hence, $K'_{eq} = 0.0475$. The standard free-energy change for this reaction is then calculated from equation 11:

$$\begin{aligned}\Delta G°' &= -2.303\ RT \log_{10} K'_{eq} \\ &= -2.303 \times 1.98 \times 10^{-3} \times 298 \times \log_{10}(0.0475) \\ &= +1.8 \text{ kcal/mol}\end{aligned}$$

Now let us calculate ΔG for this reaction when the initial concentration of dihydroxyacetone phosphate is 2×10^{-4} M and the initial concentration of glyceraldehyde 3-phosphate is 3×10^{-6} M. Substituting these values into equation 6 gives

$$\begin{aligned}\Delta G &= 1.8 \text{ kcal/mol} + 2.303\ RT \log_{10}\frac{3 \times 10^{-6} \text{ M}}{2 \times 10^{-4} \text{ M}} \\ &= 1.8 \text{ kcal/mol} - 2.5 \text{ kcal/mol} \\ &= -0.7 \text{ kcal/mol}\end{aligned}$$

This negative value for the ΔG indicates that the isomerization of dihydroxyacetone phosphate to glyceraldehyde 3-phosphate can occur spontaneously when these species are present at the concentrations stated above. Note that ΔG for this reaction is negative although $\Delta G°'$ is positive. *It is important to stress that whether the ΔG for a reaction is larger, smaller, or the same as $\Delta G°'$ depends on the concentrations of the reactants.* The criterion of spontaneity for a reaction is ΔG, not $\Delta G°'$.

ENZYMES CANNOT ALTER REACTION EQUILIBRIA

An enzyme is a catalyst, and consequently it cannot alter the equilibrium of a chemical reaction. This means that an enzyme accelerates the forward and reverse reaction by precisely the same factor. Consider the interconversion of A and B. Suppose that in the absence of enzyme the forward rate constant (k_F) is 10^{-4} s^{-1} and the reverse rate constant (k_R) is 10^{-6} s^{-1}. The equilibrium constant K is given by the ratio of these rate constants:

$$A \underset{10^{-6} \text{ s}^{-1}}{\overset{10^{-4} \text{ s}^{-1}}{\rightleftharpoons}} B$$

$$K = \frac{[B]}{[A]} = \frac{k_F}{k_R} = \frac{10^{-4}}{10^{-6}} = 100$$

The equilibrium concentration of B is 100 times that of A, whether or not enzyme is present. However, it would take more than an hour to approach this equilibrium without enzyme, whereas equilibrium would be attained within a second in the presence of a suitable enzyme. *Enzymes accelerate the attainment of equilibria but do not shift their position.*

CH$_2$OH
|
C=O
|
CH$_2$OPO$_3^{2-}$

Dihydroxyacetone phosphate

⇅

O   H
 \\ //
  C
  |
H—C—OH
  |
  CH$_2$OPO$_3^{2-}$

Glyceraldehyde 3-phosphate

Table 8-1
Relation between $\Delta G°'$ and K'_{eq} (at 25°C)

K'_{eq}	$\Delta G°'$ (kcal/mol)
10^{-5}	6.82
10^{-4}	5.46
10^{-3}	4.09
10^{-2}	2.73
10^{-1}	1.36
1	0
10	-1.36
10^2	-2.73
10^3	-4.09
10^4	-5.46
10^5	-6.82

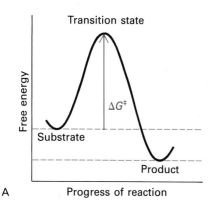

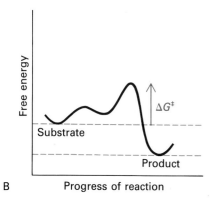

Figure 8-8
Enzymes accelerate reactions by decreasing $\Delta G^{\ddagger}$, the free energy of activation. The free-energy profiles of uncatalyzed (A) and catalyzed (B) reactions are compared.

"I think that enzymes are molecules that are complementary in structure to the activated complexes of the reactions that they catalyze, that is, to the molecular configuration that is intermediate between the reacting substances and the products of reaction for these catalyzed processes. The attraction of the enzyme molecule for the activated complex would thus lead to a decrease in its energy and hence to a decrease in the energy of activation of the reaction and to an increase in the rate of reaction."

LINUS PAULING
Nature 161:707 (1948)

ENZYMES ACCELERATE REACTIONS BY STABILIZING TRANSITION STATES

A chemical reaction of substrate S to form product P goes through a *transition state* $S^{\ddagger}$ that has a higher free energy than either S or P.

$$\underset{\text{Substrate}}{S} \overset{K^{\ddagger}}{\rightleftharpoons} \underset{\substack{\text{Transition} \\ \text{state}}}{S^{\ddagger}} \overset{V}{\longrightarrow} \underset{\text{Product}}{P}$$

The transition state is the most seldom occupied species along the reaction pathway because it has the highest free energy. The *Gibbs free energy of activation*, symbolized by $\Delta G^{\ddagger}$, is equal to the difference in free energy between the transition state and the substrate. The double dagger ($\ddagger$) denotes a thermodynamic quantity of a transition state.

$$\Delta G^{\ddagger} = G_{S^{\ddagger}} - G_S$$

The reaction rate V is proportional to the concentration of $S^{\ddagger}$, which depends on $\Delta G^{\ddagger}$ because it is in equilibrium with S.

$$[S^{\ddagger}] = [S]e^{-\Delta G^{\ddagger}/RT}$$

$$V = \nu[S^{\ddagger}] = \frac{kT}{h}[S]e^{-\Delta G^{\ddagger}/RT}$$

In these equations, k is Boltzmann's constant and h is Planck's constant. The value of kT/h at 25°C is 6.2×10^{12} s^{-1}. Suppose that the free energy of activation is 6.82 kcal/mol. The ratio $[S^{\ddagger}]/[S]$ is then 10^{-5} (see Table 8-1); we have assumed that $[S] = 1$, and so the reaction rate V is 6.2×10^{7} s^{-1}. A decrease of 1.36 kcal/mol in $\Delta G^{\ddagger}$ results in a tenfold faster V.

Enzymes accelerate reactions by decreasing $\Delta G^{\ddagger}$, the activation barrier. The combination of substrate and enzyme creates a new reaction pathway whose transition-state energy is lower than that of the reaction in the absence of enzyme (Figure 8-8). The essence of catalysis is specific binding of the transition state, as will be discussed in the next chapter.

FORMATION OF AN ENZYME-SUBSTRATE COMPLEX IS THE FIRST STEP IN ENZYMATIC CATALYSIS

Much of the catalytic power of enzymes comes from their bringing substrates together in favorable orientations in *enzyme-substrate* (ES) complexes. The substrates are bound to a specific region of the enzyme called the *active site*. Most enzymes are highly selective in their binding of substrates. Indeed, the catalytic specificity of enzymes depends in part on the specificity of binding. Furthermore, the activities of some enzymes are controlled at this stage.

The existence of ES complexes has been shown in a variety of ways:

1. At a constant concentration of enzyme, the reaction rate increases with increasing substrate concentration until a maximal velocity is reached (Figure 8-9). In contrast, uncatalyzed reactions do not show this saturation effect. In 1913, Leonor Michaelis interpreted the *maximal velocity of an enzyme-catalyzed reaction* in terms of the formation of a discrete ES complex. At a sufficiently high substrate concentration, the catalytic sites are filled and so the reaction rate reaches a maximum. Though indirect, this is the oldest and most general evidence for the existence of ES complexes.

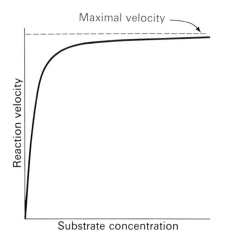

Figure 8-9
Velocity of an enzyme-catalyzed reaction as a function of the substrate concentration.

2. ES complexes have been directly visualized by *electron microscopy*, as in the micrograph of DNA polymerase I bound to its DNA template (see Figure 8-1). *X-ray crystallography* has provided high-resolution images of substrates and substrate analogs bound to the active sites of many enzymes. In the next chapter, we shall take a close look at several of these complexes. Moreover, x-ray studies carried out at low temperatures (to slow reactions down) are providing revealing views of intermediates in enzymatic reactions.

3. The *spectroscopic characteristics* of many enzymes and substrates change upon formation of an ES complex just as the absorption spectrum of deoxyhemoglobin changes markedly when it binds oxygen or when it is oxidized to the ferric state, as described previously (see Figure 7-11, on p. 147). These changes are particularly striking if the enzyme contains a colored prosthetic group. Tryptophan synthetase, a bacterial enzyme that contains a pyridoxal phosphate prosthetic group, affords a nice illustration. This enzyme catalyzes the synthesis of L-tryptophan from L-serine and indole. The addition of L-serine to the enzyme produces a marked increase in the fluorescence of the pyridoxal phosphate group (Figure 8-10). The subsequent addition of indole, the second substrate, quenches this fluorescence to a level lower even than that of the enzyme alone. Thus, fluorescence spectroscopy reveals the existence of an enzyme-serine complex and of an enzyme-serine-indole complex. Other spectroscopic techniques, such as nuclear magnetic resonance and electron spin resonance, also are highly informative about ES interactions.

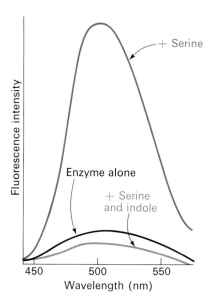

Figure 8-10
Fluorescence intensity of the pyridoxal phosphate group at the active site of tryptophan synthetase changes upon addition of serine and indole, the substrates.

SOME KEY FEATURES OF ACTIVE SITES

The active site of an enzyme is the region that binds the substrates (and the prosthetic group, if any) and contains the residues that directly participate in the making and breaking of bonds. These residues are called the *catalytic groups*. Although enzymes differ widely in structure, specificity, and mode of catalysis, a number of generalizations concerning their active sites can be stated:

1. *The active site takes up a relatively small part of the total volume of an enzyme.* Most of the amino acid residues in an enzyme are not in contact with the substrate. This raises the intriguing question of why enzymes

are so big. Nearly all enzymes are made up of more than 100 amino acid residues, which gives them a mass greater than 10 kd and a diameter of more than 25 Å.

2. *The active site is a three-dimensional entity* formed by groups that come from different parts of the linear amino acid sequence—indeed, residues far apart in the linear sequence may interact more strongly than adjacent residues in the amino acid sequence, as has already been seen for myoglobin and hemoglobin. In lysozyme, an enzyme that will be discussed in more detail in the next chapter, the important groups in the active site are contributed by residues numbered 35, 52, 62, 63, and 101 in the linear sequence of 129 amino acids (Figure 8-11).

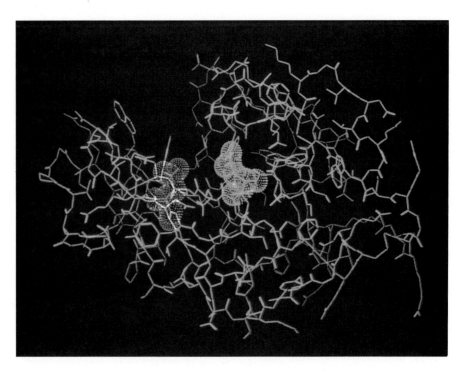

Figure 8-11
Model of lysozyme. The van der Waals surfaces of two catalytically critical residues are shown in color.

3. *Substrates are bound to enzymes by multiple weak attractions.* ES complexes usually have equilibrium constants that range from 10^{-2} to 10^{-8} M, corresponding to free energies of interaction ranging from -3 to -12 kcal/mol. The noncovalent interactions in ES complexes are much weaker than covalent bonds, which have energies between -50 and -110 kcal/mol. As was discussed in Chapter 1 (pp. 7–10), reversible interactions of biomolecules are mediated by electrostatic bonds, hydrogen bonds, van der Waals forces, and hydrophobic interactions. Van der Waals forces become significant in binding only when numerous substrate atoms can simultaneously come close to many enzyme atoms. Hence, the enzyme and substrate should have complementary shapes. The directional character of hydrogen bonds between enzyme and substrate often enforces a high degree of specificity.

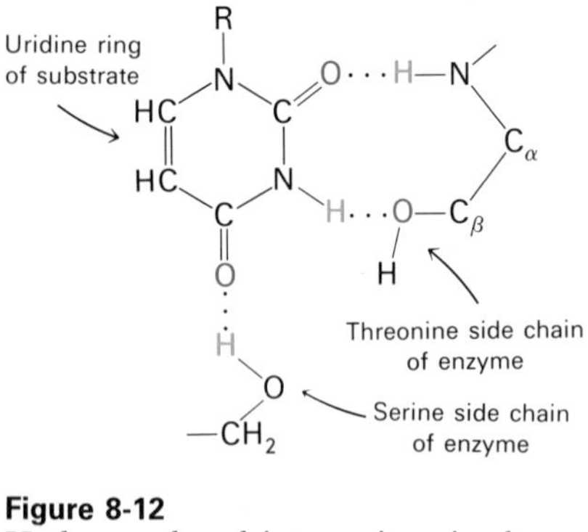

4. *Active sites are clefts or crevices.* In all enzymes of known structure, substrate molecules are bound to a cleft or crevice. Water is usually excluded unless it is a reactant. The nonpolar character of much of the cleft enhances the binding of substrate. However, the cleft may also contain polar residues. It creates a microenvironment in which certain of these residues acquire special properties essential for catalysis. The internal positions of these polar residues are biologically crucial exceptions to the general rule that polar residues are exposed to water.

Figure 8-12
Hydrogen-bond interactions in the binding of a uridine substrate to ribonuclease. [After F. M. Richards, H. W. Wyckoff, and N. Allewell. In *The Neurosciences: Second Study Program*, F. O. Schmidt, ed., (Rockefeller University Press, 1970), p. 970.]

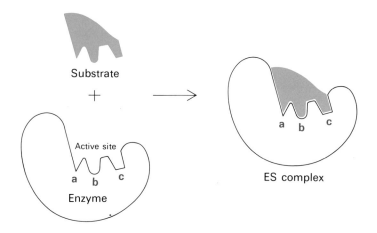

Figure 8-13
Lock-and-key model of the interaction of substrates and enzymes. The active site of the enzyme alone is complementary in shape to that of the substrate.

5. *The specificity of binding depends on the precisely defined arrangement of atoms in an active site.* To fit into the site, a substrate must have a matching shape. Emil Fischer's metaphor of the lock and key (Figure 8-13), expressed in 1890, has proved to be highly stimulating and fruitful. However, it is now evident that the shapes of the active sites of some enzymes are markedly modified by the binding of substrate, as was postulated by Daniel E. Koshland, Jr., in 1958. The active sites of these enzymes have shapes that are complementary to that of the substrate only *after* the substrate is bound. This process of dynamic recognition is called *induced fit* (Figure 8-14).

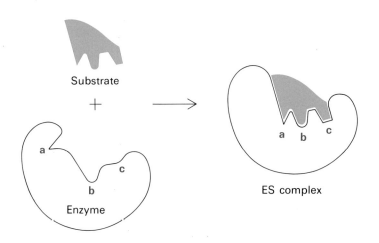

Figure 8-14
Induced-fit model of the interaction of substrates and enzymes. The enzyme changes shape upon binding substrate. The active site has a shape complementary to that of the substrate only after the substrate is bound.

THE MICHAELIS-MENTEN MODEL ACCOUNTS FOR THE KINETIC PROPERTIES OF MANY ENZYMES

For many enzymes, the rate of catalysis, V, varies with the substrate concentration, [S], in a manner shown in Figure 8-15. V is defined as the number of moles of product formed per second. At a fixed concentration of enzyme, V is almost linearly proportional to [S] when [S] is small. At high [S], V is nearly independent of [S]. In 1913, Leonor Michaelis and Maud Menten proposed a simple model to account for these kinetic characteristics. The critical feature in their treatment is that a specific ES complex is a necessary intermediate in catalysis. The model proposed, which is the simplest one that accounts for the kinetic properties of many enzymes, is

$$\mathrm{E} + \mathrm{S} \underset{k_2}{\overset{k_1}{\rightleftharpoons}} \mathrm{ES} \overset{k_3}{\rightarrow} \mathrm{E} + \mathrm{P} \tag{14}$$

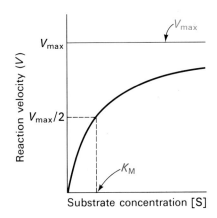

Figure 8-15
A plot of the reaction velocity, V, as a function of the substrate concentration, [S], for an enzyme that obeys Michaelis-Menten kinetics (V_{max} is the maximal velocity and K_M is the Michaelis constant).

An enzyme, E, combines with S to form an ES complex, with a rate constant k_1. The ES complex has two possible fates. It can dissociate to E and S, with a rate constant k_2, or it can proceed to form product P, with a rate constant k_3. It is assumed that almost none of the product reverts to the initial substrate, a condition that holds in the initial stage of a reaction before the concentration of product is appreciable.

We want an expression that relates the rate of catalysis to the concentrations of substrate and enzyme and the rates of the individual steps. The starting point is that the catalytic rate is equal to the product of the concentration of the ES complex and k_3:

$$V = k_3[\text{ES}] \tag{15}$$

Now we need to express [ES] in terms of known quantities. The rates of formation and breakdown of ES are given by

$$\text{Rate of formation of ES} = k_1[\text{E}][\text{S}] \tag{16}$$

$$\text{Rate of breakdown of ES} = (k_2 + k_3)[\text{ES}] \tag{17}$$

We are interested in the catalytic rate under steady-state conditions. In a *steady state*, the concentrations of intermediates stay the same while the concentrations of starting materials and products are changing. This occurs when the rates of formation and breakdown of the ES complex are equal. On setting the right-hand sides of equations 16 and 17 equal,

$$k_1[\text{E}][\text{S}] = (k_2 + k_3)[\text{ES}] \tag{18}$$

By rearranging equation 18,

$$[\text{ES}] = \frac{[\text{E}][\text{S}]}{(k_2 + k_3)/k_1} \tag{19}$$

Equation 19 can be simplified by defining a new constant, K_M, called the *Michaelis constant*:

$$K_M = \frac{k_2 + k_3}{k_1} \tag{20}$$

and substituting it into equation 19, which then becomes

$$[\text{ES}] = \frac{[\text{E}][\text{S}]}{K_M} \tag{21}$$

Now let us examine the numerator of equation 21. The concentration of uncombined substrate, [S], is very nearly equal to the total substrate concentration, provided that the concentration of enzyme is much lower than that of the substrate. The concentration of uncombined enzyme, [E], is equal to the total enzyme concentration, $[\text{E}_T]$, minus the concentration of the ES complex.

$$[\text{E}] = [\text{E}_T] - [\text{ES}] \tag{22}$$

On substituting this expression for [E] in equation 21,

$$[\text{ES}] = ([\text{E}_T] - [\text{ES}])[\text{S}]/K_M \tag{23}$$

Solving equation 23 for [ES] gives

$$[\text{ES}] = [\text{E}_T] \frac{[\text{S}]/K_M}{1 + [\text{S}]/K_M} \tag{24}$$

or

$$[\text{ES}] = [\text{E}_T] \frac{[\text{S}]}{[\text{S}] + K_M} \tag{25}$$

By substituting this expression for [ES] into equation 15, we get

$$V = k_3[\mathrm{E_T}]\frac{[\mathrm{S}]}{[\mathrm{S}] + K_\mathrm{M}} \tag{26}$$

The maximal rate, $V_{\max}$, is attained when the enzyme sites are saturated with substrate—that is, when [S] is much greater than K_M—so that $[\mathrm{S}]/([\mathrm{S}] + K_\mathrm{M})$ approaches 1. Thus,

$$V_{\max} = k_3[\mathrm{E_T}] \tag{27}$$

Substituting equation 27 into equation 26 yields the Michaelis-Menten equation:

$$V = V_{\max}\frac{[\mathrm{S}]}{[\mathrm{S}] + K_\mathrm{M}} \tag{28}$$

This equation accounts for the kinetic data given in Figure 8-15. At very low substrate concentration, when [S] is much less than K_M, $V = [\mathrm{S}]V_{\max}/K_\mathrm{M}$; that is, the rate is directly proportional to the substrate concentration. At high substrate concentration, when [S] is much greater than K_M, $V = V_{\max}$; that is, the rate is maximal, independent of substrate concentration.

The meaning of K_M is evident from equation 28. When $[\mathrm{S}] = K_\mathrm{M}$, then $V = V_{\max}/2$. Thus, K_M *is equal to the substrate concentration at which the reaction rate is half of its maximal value.*

$V_{\max}$ AND K_M CAN BE DETERMINED BY VARYING THE SUBSTRATE CONCENTRATION

The Michaelis constant, K_M, and the maximal rate, $V_{\max}$, can be readily derived from rates of catalysis measured at different substrate concentrations if an enzyme operates according to the simple scheme given in equation 14. It is convenient to transform the Michaelis-Menten equation into one that gives a straight line plot. This can be done by taking the reciprocal of both sides of equation 28 to give

$$\frac{1}{V} = \frac{1}{V_{\max}} + \frac{K_\mathrm{M}}{V_{\max}} \cdot \frac{1}{[\mathrm{S}]} \tag{29}$$

A plot of $1/V$ versus $1/[\mathrm{S}]$, called a *Lineweaver-Burk plot*, yields a straight line with an intercept of $1/V_{\max}$ and a slope of $K_\mathrm{M}/V_{\max}$ (Figure 8-16).

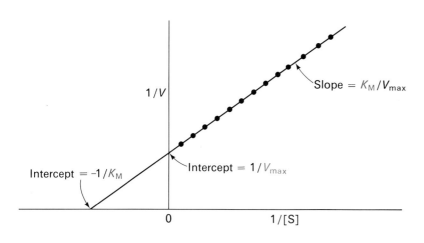

Figure 8-16
A double-reciprocal plot of enzyme kinetics: $1/V$ is plotted as a function of $1/[\mathrm{S}]$. The slope is $K_\mathrm{M}/V_{\max}$, the intercept on the vertical axis is $1/V_{\max}$, and the intercept on the horizontal axis is $-1/K_\mathrm{M}$.

Table 8-2
K_M values of some enzymes

Enzyme	Substrate	K_M (μM)
Chymotrypsin	Acetyl-L-tryptophanamide	5000
Lysozyme	Hexa-*N*-acetylglucosamine	6
β-Galactosidase	Lactose	4000
Threonine deaminase	Threonine	5000
Carbonic anhydrase	CO_2	8000
Penicillinase	Benzylpenicillin	50
Pyruvate carboxylase	Pyruvate	400
	HCO_3^-	1000
	ATP	60
Arginine-tRNA synthetase	Arginine	3
	tRNA	0.4
	ATP	300

SIGNIFICANCE OF K_M AND V_{max} VALUES

The K_M values of enzymes range widely (Table 8-2). For most enzymes, K_M lies between 10^{-1} and 10^{-7} M. The K_M value for an enzyme depends on the particular substrate and also on environmental conditions such as pH, temperature, and ionic strength. The Michaelis constant, K_M, has two meanings. First, K_M is the concentration of substrate at which half the active sites are filled. Once the K_M is known, the fraction of sites filled, f_{ES}, at any substrate concentration can be calculated from

$$f_{ES} = \frac{V}{V_{max}} = \frac{[S]}{[S] + K_M} \tag{30}$$

Second, K_M is related to the rate constants of the individual steps in the catalytic scheme given in equation 14. In equation 20, K_M is defined as $(k_2 + k_3)/k_1$. Consider a limiting case in which k_2 is much greater than k_3. This means that dissociation of the ES complex to E and S is much more rapid than formation of E and product. Under these conditions ($k_2 \gg k_3$),

$$K_M = \frac{k_2}{k_1} \tag{31}$$

The dissociation constant of the ES complex is given by

$$K_{ES} = \frac{[E][S]}{[ES]} = \frac{k_2}{k_1} \tag{32}$$

In other words, K_M *is equal to the dissociation constant of the ES complex if* k_3 *is much smaller than* k_2. When this condition is met, K_M is a measure of the strength of the ES complex: a high K_M indicates weak binding; a low K_M indicates strong binding. It must be stressed that K_M indicates the affinity of the ES complex only when k_2 is much greater than k_3.

The *turnover number* of an enzyme is *the number of substrate molecules converted into product by an enzyme molecule in a unit time when the enzyme is fully saturated with substrate.* It is equal to the kinetic constant k_3. The maximal rate, V_{max}, reveals the turnover number of an enzyme if the concentration of active sites $[E_T]$ is known, because

$$V_{max} = k_3[E_T] \tag{33}$$

For example, a 10^{-6} M solution of carbonic anhydrase catalyzes the formation of 0.6 M H_2CO_3 per second when it is fully saturated with substrate. Hence, k_3 is 6×10^5 s^{-1}. This turnover number is one of the largest known. Each round of catalysis occurs in a time equal to $1/k_3$, which is 1.7 microseconds for carbonic anhydrase. The turnover numbers of most enzymes with their physiological substrates fall in the range from 1 to 10^4 per second (Table 8-3).

KINETIC PERFECTION IN ENZYMATIC CATALYSIS: THE k_{cat}/K_M CRITERION

When the substrate concentration is much greater than K_M, the rate of catalysis is equal to k_3, the turnover number, as described in the preceding section. However, most enzymes are not normally saturated with substrate. Under physiological conditions, the [S]/K_M ratio is typically between 0.01 and 1.0. When [S] $\ll K_M$, the enzymatic rate is much less than k_3 because most of the active sites are unoccupied. Is there a number that characterizes the kinetics of an enzyme under these conditions? Indeed there is, as can be shown by combining equations 15 and 21 to give

$$V = \frac{k_3}{K_M}[\text{E}][\text{S}] \qquad (34)$$

When [S] $\ll K_M$, the concentration of free enzyme, [E], is nearly equal to the total concentration of enzyme [E_T], and so

$$V = \frac{k_3}{K_M}[\text{S}][\text{E}_T] \qquad (35)$$

Thus, when [S] $\ll K_M$, the enzymatic velocity depends on the value of k_3/K_M and on [S].

Are there any physical limits on the value of k_3/K_M? Note that this ratio depends on k_1, k_2, and k_3, as can be shown by substituting for K_M:

$$k_3/K_M = \frac{k_3 k_1}{k_2 + k_3} < k_1 \qquad (36)$$

Thus the ultimate limit on the value of k_3/K_M is set by k_1, the rate of formation of the ES complex. *This rate cannot be faster than the diffusion-controlled encounter of an enzyme and its substrate.* Diffusion limits the value of k_1 so that it cannot be higher than between 10^8 and 10^9 M^{-1} s^{-1}. Hence, the upper limit on k_3/K_M is between 10^8 and 10^9 M^{-1} s^{-1}.

This restriction also pertains to enzymes having more complex reaction pathways than that of equation 14. Their maximal catalytic rate when substrate is saturating, denoted by k_{cat}, depends on several rate constants rather than on k_3 alone. The pertinent parameter for these enzymes is k_{cat}/K_M. In fact, the k_{cat}/K_M ratios of the enzymes acetylcholinesterase, carbonic anhydrase, and triosephosphate isomerase are between 10^8 and 10^9 M^{-1} s^{-1}, which shows that they have attained *kinetic perfection. Their catalytic velocity is restricted only by the rate at which they encounter substrate in the solution.* Any further gain in catalytic rate can come only by decreasing the time for diffusion. Indeed, some series of enzymes are associated into organized assemblies (p. 379) so that the product of one enzyme is very rapidly found by the next enzyme. In effect, products are channeled from one enzyme to the next, much as in an assembly line. Thus, the limit imposed by the rate of diffusion in solution can be partly overcome by confining substrates and products in the limited volume of a multienzyme complex.

Table 8-3
Maximum turnover numbers of some enzymes

Enzyme	Turnover number (per second)
Carbonic anhydrase	600,000
3-Ketosteroid isomerase	280,000
Acetylcholinesterase	25,000
Penicillinase	2,000
Lactate dehydrogenase	1,000
Chymotrypsin	100
DNA polymerase I	15
Tryptophan synthetase	2
Lysozyme	0.5

ENZYMES CAN BE INHIBITED BY SPECIFIC MOLECULES

The inhibition of enzymatic activity by specific small molecules and ions is important because it serves as a major control mechanism in biological systems. Also, many drugs and toxic agents act by inhibiting enzymes. Furthermore, inhibition can be a source of insight into the mechanism of enzyme action: residues critical for catalysis can often be identified by using specific inhibitors.

Enzyme inhibition can be either reversible or irreversible. An *irreversible inhibitor* dissociates very slowly from its target enzyme because it becomes very tightly bound to the enzyme, either covalently or noncovalently. The action of nerve gases on acetylcholinesterase, an enzyme that plays an important role in the transmission of nerve impulses, exemplifies irreversible inhibition. Diisopropylphosphofluoridate (DIPF), one of these agents, reacts with a critical serine residue at the active site on the enzyme to form an inactive diisopropylphosphoryl enzyme (Figure 8-17). Alkylating reagents, such as iodoacetamide, irreversibly inhibit the catalytic activity of some enzymes by modifying cysteine and other side chains (Figure 8-18).

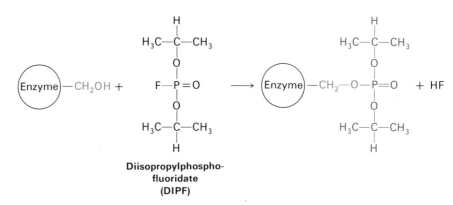

Figure 8-17
Inactivation of chymotrypsin and acetylcholinesterase by diisopropylphosphofluoridate (DIPF).

Figure 8-18
Inactivation of an enzyme with a critical cysteine residue by iodoacetamide.

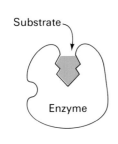

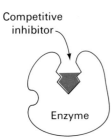

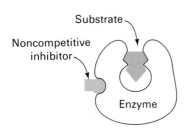

Figure 8-19
Distinction between a competitive inhibitor and a noncompetitive inhibitor: (top) enzyme-substrate complex; (middle) a competitive inhibitor prevents the substrate from binding; (bottom) a noncompetitive inhibitor does not prevent the substrate from binding.

Reversible inhibition, in contrast with irreversible inhibition, is characterized by a rapid dissociation of the enzyme-inhibitor complex. In *competitive inhibition*, the enzyme can bind substrate (forming an ES complex) or inhibitor (EI) but not both (ESI). Many competitive inhibitors resemble the substrate and bind to the active site of the enzyme (Figure 8-19). The substrate is thereby prevented from binding to the same active site. *A competitive inhibitor diminishes the rate of catalysis by reducing the proportion of enzyme molecules bound to a substrate.* A classic example of competitive inhibition is the action of malonate on succinate dehydrogenase, an enzyme that removes two hydrogen atoms from succinate. Malonate differs from succinate in having one rather than two methylene groups. A physiologically important example of competitive inhibition is found in the formation of 2,3-bisphosphoglycerate (BPG, p. 157) from 1,3-bisphosphoglycerate. Bisphosphoglycerate mutase, the enzyme catalyzing this isomerization, is competitively inhibited by even low levels of 2,3-bisphosphoglycerate. In fact, it is not uncommon for an enzyme to be competitively inhibited by its own product because of its structural resemblance to the substrate. Competitive inhibition can be overcome by increasing the concentration of substrate.

In *noncompetitive inhibition,* which is also reversible, the inhibitor and substrate can bind simultaneously to an enzyme molecule. This means that their binding sites do not overlap. A noncompetitive inhibitor acts by decreasing the turnover number of an enzyme rather than by diminishing the proportion of enzyme molecules that are bound to substrate. Noncompetitive inhibition, in contrast with competitive inhibition, cannot be overcome by increasing the substrate concentration. A more complex pattern, called *mixed inhibition,* is produced when an inhibitor both affects the binding of substrate and alters the turnover number of the enzyme.

ALLOSTERIC ENZYMES DO NOT OBEY MICHAELIS-MENTEN KINETICS

The Michaelis-Menten model has greatly affected the development of enzyme chemistry. Its virtues are simplicity and broad applicability. However, the kinetic properties of many enzymes cannot be accounted for by the Michaelis-Menten model. An important group consists of the *allosteric enzymes,* which often display sigmoidal plots (Figure 8-20) of the reaction velocity, V, versus substrate concentration, [S], rather than the hyperbolic plots predicted by the Michaelis-Menten equation (eq. 28). Recall that the oxygen-binding curve of myoglobin is hyperbolic, whereas that of hemoglobin is sigmoidal. The binding of enzymes to substrates is analogous. In allosteric enzymes, the binding of substrate to one active site can affect the properties of other active sites in the same enzyme molecule. A possible outcome of this interaction between subunits is that the binding of substrate becomes cooperative, which would give a sigmoidal plot of V versus [S]. In addition, the activity of allosteric enzymes may be altered by regulatory molecules that are bound to sites other than the catalytic sites, just as oxygen binding to hemoglobin is affected by BPG, H^+, and CO_2.

COMPETITIVE AND NONCOMPETITIVE INHIBITION ARE KINETICALLY DISTINGUISHABLE

Let us return to enzymes that exhibit Michaelis-Menten kinetics. Measurements of the rates of catalysis at different concentrations of substrate and inhibitor serve to distinguish between competitive and noncompetitive inhibition. In *competitive inhibition,* the intercept of the plot of $1/V$ versus $1/[S]$ is the same in the presence and absence of inhibitor, although the slope is different (Figure 8-21). This reflects the fact that V_{max} is not altered by a competitive inhibitor. *The hallmark of competitive inhibition is that it can be overcome by a sufficiently high concentration of substrate.* At a sufficiently high concentration, virtually all the active sites

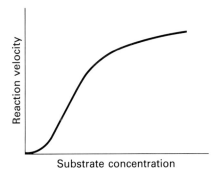

Figure 8-20
Sigmoidal dependence of reaction velocity versus substrate concentration for an allosteric enzyme.

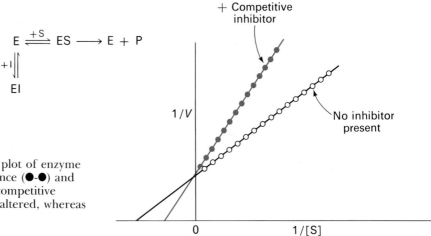

Figure 8-21
A double-reciprocal plot of enzyme kinetics in the presence (●-●) and absence (○-○) of a competitive inhibitor; V_{max} is unaltered, whereas K_M is increased.

are filled by substrate, and the enzyme is fully operative. The increase in the slope of the $1/V$ versus $1/[S]$ plot indicates the strength of binding of competitive inhibitor. In the presence of a competitive inhibitor, equation 29 is replaced by

$$\frac{1}{V} = \frac{1}{V_{max}} + \frac{K_M}{V_{max}}\left(1 + \frac{[I]}{K_i}\right)\left(\frac{1}{[S]}\right) \tag{37}$$

in which $[I]$ is the concentration of inhibitor and K_i is the dissociation constant of the enzyme-inhibitor complex:

$$E + I \rightleftharpoons EI$$

$$K_i = \frac{[E][I]}{[EI]} \tag{38}$$

In other words, the slope of the plot is increased by the factor $(1 + [I]/K_i)$ in the presence of a competitive inhibitor. Consider an enzyme with a K_M of 10^{-4} M. In the absence of inhibitor, $V = V_{max}/2$ when $[S] = 10^{-4}$ M. In the presence of 2×10^{-3} M competitive inhibitor that is bound to the enzyme with a K_i of 10^{-3} M, the apparent K_M will be 3×10^{-4} M. Substitution of these values into equation 29 gives $V = V_{max}/4$.

In *noncompetitive inhibition* (Figure 8-22), V_{max} is decreased to V^I_{max}, and so the intercept on the vertical axis is increased. The new slope,

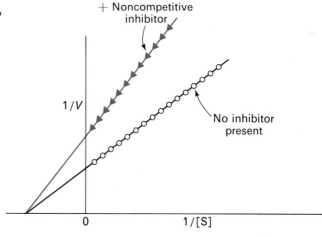

Figure 8-22
A double-reciprocal plot of enzyme kinetics in the presence (▲-▲-▲) and absence (○-○-○) of a noncompetitive inhibitor; K_M is unaltered by the noncompetitive inhibitor, whereas V_{max} is decreased.

which is equal to K_M/V^I_{max}, is larger by the same factor. In contrast with V_{max}, K_M is not affected by this kind of inhibition. *Noncompetitive inhibition cannot be overcome by increasing the substrate concentration.* The maximal velocity in the presence of a noncompetitive inhibitor, V^I_{max}, is given by

$$V^I_{max} = \frac{V_{max}}{1 + [I]/K_i} \tag{39}$$

ETHANOL IS USED THERAPEUTICALLY AS A COMPETITIVE INHIBITOR TO TREAT ETHYLENE GLYCOL POISONING

About fifty deaths occur annually from the ingestion of ethylene glycol, a constituent of permanent-type automobile antifreeze. Ethylene glycol itself is not lethally toxic. Rather, the harm is done by oxalic acid, an oxidation product of ethylene glycol, because the kidneys are severely damaged by the deposition of oxalate crystals. The first committed step in this conversion is the oxidation of ethylene glycol to an aldehyde by alcohol dehydrogenase (Figure 8-23). This reaction can be effectively inhibited by administering a nearly intoxicating dose of ethanol. The basis for this effect is that *ethanol is a competing substrate and so it blocks the oxidation of ethylene glycol to aldehyde products.* The ethylene glycol is then excreted harmlessly. Ethanol is also used as a competing substrate for treating methanol (wood alcohol) poisoning.

Figure 8-23
Formation of oxalic acid from ethylene glycol is inhibited by ethanol.

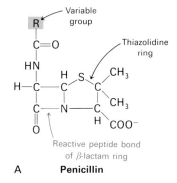

PENICILLIN IRREVERSIBLY INACTIVATES A KEY ENZYME IN BACTERIAL CELL-WALL SYNTHESIS

Penicillin was discovered by Alexander Fleming in 1928, when he observed by chance that bacterial growth was inhibited by a contaminating mold (*Penicillium*). Fleming was encouraged to find that an extract from the mold was not toxic when injected into animals. However, on trying to concentrate and purify the antibiotic, he found that "penicillin is easily destroyed, and to all intents and purposes we failed. We were bacteriologists—not chemists—and our relatively simple procedures were unavailing." Ten years later, Howard Florey, a pathologist, and Ernst Chain, a biochemist, carried out an incisive series of studies that led to the isolation, chemical characterization, and clinical use of this antibiotic.

Penicillin consists of a thiazolidine ring fused to a *β-lactam* ring, to which a variable R group is attached by a peptide bond. In benzyl penicillin, for example, R is a benzyl group (Figure 8-24). This structure can undergo a variety of rearrangements, which accounts for the instability

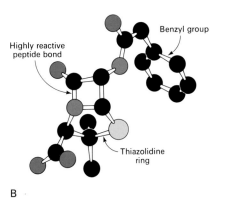

Figure 8-24
(A) Structural formula and (B) model of benzyl penicillin. The reactive site of penicillin is the peptide bond of its β-lactam ring.

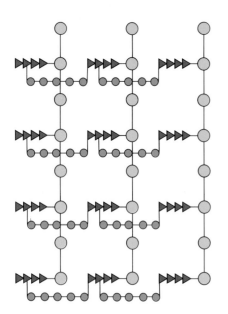

Figure 8-25
Schematic diagram of the peptidoglycan in *Staphylococcus aureus*. The sugars are shown in yellow, the tetrapeptides in red, and the pentaglycine bridges in blue. The cell wall is a single, enormous bag-shaped macromolecule because of extensive cross-linking.

first encountered by Fleming. In particular, the β-lactam ring is very labile. Indeed, this property is closely tied to the antibiotic action of penicillin, as will be evident shortly.

How does penicillin inhibit bacterial growth? In 1957, Joshua Lederberg showed that bacteria ordinarily susceptible to penicillin could be grown in its presence if a hypertonic medium were used. The organisms obtained in this way, called protoplasts, are devoid of a cell wall and consequently lyse when transferred to a normal medium. Hence, it was inferred that penicillin interferes with the synthesis of the bacterial cell wall. The cell-wall macromolecule, called a *peptidoglycan*, consists of linear polysaccharide chains that are cross-linked by short peptides (Figure 8-25). The enormous bag-shaped peptidoglycan confers mechanical support and prevents bacteria from bursting from their high internal osmotic pressure.

In 1965, James Park and Jack Strominger independently deduced that penicillin blocks the last step in cell-wall synthesis, namely the cross-linking of different peptidoglycan strands. In the formation of the cell wall of *Staphylococcus aureus*, the amino group at one end of a pentaglycine chain attacks the peptide bond between two D-alanine residues in another peptide unit (Figure 8-26). A peptide bond is formed between glycine and one of the D-alanine residues, and the other D-alanine residue is released. This cross-linking reaction is catalyzed by *glycopeptide transpeptidase*. Bacterial cell walls are unique in containing D-amino acids, which form cross-links by a mechanism entirely different from that used to synthesize proteins.

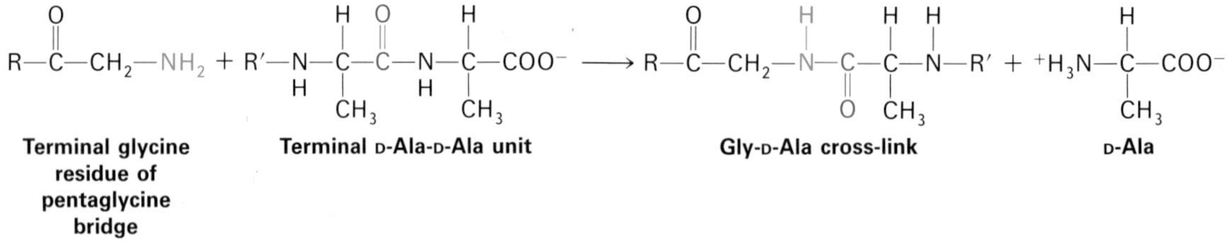

Figure 8-26
The amino group of the pentaglycine bridge in the *S. aureus* cell wall attacks the peptide bond between two D-Ala residues to form a cross-link.

Penicillin inhibits the cross-linking transpeptidase by the Trojan Horse stratagem. The transpeptidase normally forms an *acyl intermediate* with the penultimate D-alanine residue of the D-Ala-D-Ala-peptide (Figure 8-27). This covalent acyl-enzyme intermediate then reacts with the amino group of the terminal glycine in another peptide to form the cross-link. Penicillin is welcomed into the active site of the transpeptidase because it mimics the D-Ala-D-Ala moiety of the normal substrate. Bound penicillin then forms a covalent bond with a serine residue at

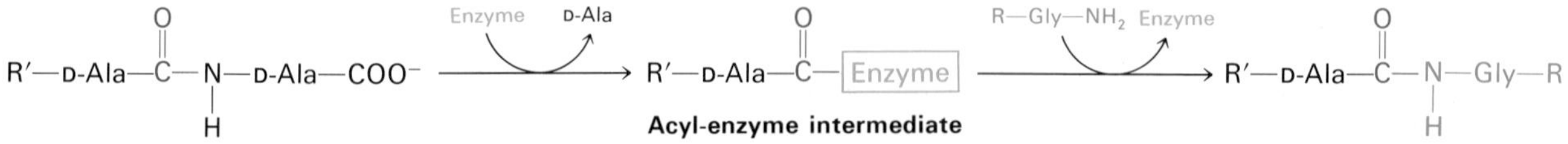

Figure 8-27
An acyl-enzyme intermediate is formed in the transpeptidation reaction.

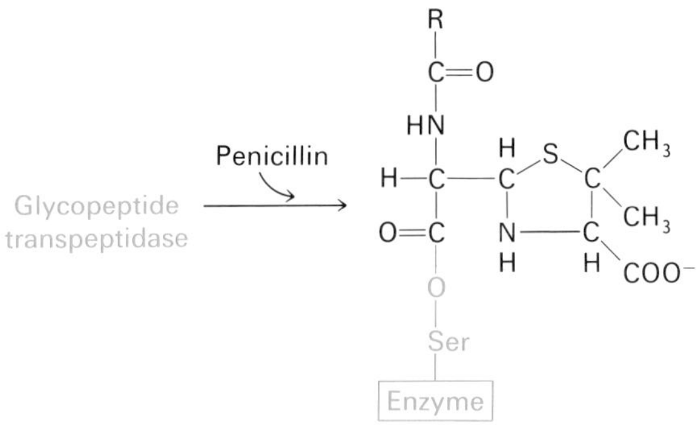

Glycopeptide transpeptidase → Penicillin

[Penicilloyl-enzyme complex structure]

Penicilloyl-enzyme complex
(Enzymatically inactive)

Figure 8-28
Formation of a penicilloyl-enzyme complex, which is indefinitely stable.

the active site of the enzyme (Figure 8-28). *This penicilloyl-enzyme does not react further. Hence, the transpeptidase is irreversibly inhibited.*

Why is penicillin such an effective inhibitor of the transpeptidase? Molecular models show that penicillin resembles acyl-D-Ala-D-Ala, one of the substrates of this enzyme (Figure 8-29). Moreover, the four-membered β-lactam ring of penicillin is strained, which makes it highly reactive. Indeed, the conformation of this part of penicillin is probably very similar to that of the transition state of the normal substrate, a species that interacts strongly with the enzyme. In other words, penicillin is a *transition-state analog,* a striking example of molecular mimicry executed with perfection.

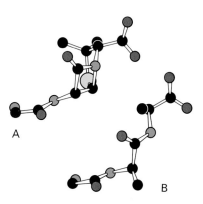

Figure 8-29
The conformation of penicillin in the vicinity of its reactive peptide bond (A) resembles the postulated conformation of the transition state of R—D-Ala—D-Ala (B) in the transpeptidation reaction. [After B. Lee. *J. Mol. Biol.* 61(1971):464.]

SUMMARY

Free energy is the most valuable thermodynamic function for determining whether a reaction can occur and for understanding the energetics of catalysis. Free energy is a measure of the capacity of a system to do useful work at constant temperature and pressure. A reaction can occur spontaneously only if the change in free energy (ΔG) is negative. The ΔG of a reaction is independent of path and depends only on the nature of the reactants and their activities (which can sometimes be approximated by their concentrations). The free-energy change of a reaction that occurs when reactants and products are at unit activity is called the standard free-energy change ($\Delta G°$). Biochemists usually use $\Delta G°'$, the standard free-energy change at pH 7.

The catalysts in biological systems are enzymes, and nearly all of them are proteins. Enzymes are highly specific and have great catalytic power. They enhance reaction rates by factors of at least 10^6. Enzymes do not alter reaction equilibria. Rather, they serve as catalysts by reducing the free energy of activation of chemical reactions. Enzymes accelerate reactions by providing a new reaction pathway in which the transition state (the highest-energy species) has a lower free energy and hence is more accessible than in the uncatalyzed reaction. The first step in catalysis is the formation of an enzyme-substrate complex. Substrates are bound to enzymes at active-site clefts from which water is largely excluded when the substrate is bound. The specificity of enzyme-substrate interactions arises mainly from hydrogen bonding, which is directional, and the shape of the active site, which rejects molecules that do not have a sufficiently complementary shape. The recognition of substrates by enzymes is a dynamic process accompanied by conformational changes at active sites.

The Michaelis-Menten model accounts for the kinetic properties of some enzymes. In this model, an enzyme (E) combines with a substrate (S) to form an enzyme-substrate (ES) complex, which can proceed to form a product (P) or to dissociate into E and S.

$$E + S \underset{k_2}{\overset{k_1}{\rightleftharpoons}} ES \overset{k_3}{\rightarrow} E + P$$

The rate V of formation of product is given by the Michaelis-Menten equation:

$$V = V_{max}\frac{[S]}{[S] + K_M}$$

in which V_{max} is the rate when the enzyme is fully saturated with substrate, and K_M, the Michaelis constant, is the substrate concentration at which the reaction rate is half maximal. The maximal rate, V_{max}, is equal to the product of k_3 and the total concentration of enzyme. The kinetic constant k_3, called the turnover number, is the number of substrate molecules converted into product per unit time at a single catalytic site when the enzyme is fully saturated with substrate. Turnover numbers for most enzymes are between 1 and 10^4 per second.

Enzymes can be inhibited by specific small molecules or ions. In irreversible inhibition, the inhibitor is covalently linked to the enzyme or bound so tightly that its dissociation from the enzyme is very slow. For example, penicillin irreversibly inactivates an essential enzyme in the formation of bacterial cell walls by mimicking the normal substrate in the cross-linking reaction. In contrast, reversible inhibition is characterized by a rapid equilibrium between enzyme and inhibitor. A competitive inhibitor prevents the substrate from binding to the active site. It reduces the reaction velocity by diminishing the proportion of enzyme molecules that are bound to substrate. In noncompetitive inhibition, the inhibitor decreases the turnover number. Competitive inhibition can be distinguished from noncompetitive inhibition by determining whether the inhibition can be overcome by raising the substrate concentration.

The catalytic activity of many enzymes is regulated in vivo. Allosteric interactions, which are defined as interactions between spatially distinct sites, are particularly important in this regard. The enzyme catalyzing the first step in a biosynthetic pathway is usually inhibited by the final product. Enzymes are also controlled by regulatory proteins such as calmodulin, which senses the intracellular Ca^{2+} level. Covalent modifications such as phosphorylation of serine, threonine, and tyrosine side chains are a third means of modulating enzymatic activity. The conversion of an inactive precursor protein into an active enzyme by peptide-bond cleavage, a process termed proteolytic activation, is another recurring regulatory device.

SELECTED READINGS

Where to start

Koshland, D. E., Jr., 1973. Protein shape and biological control. *Sci. Amer.* 229(4):52–64. [An excellent introduction to the importance of conformational flexibility for the specificity and regulation of enzyme action. Available as *Sci. Amer*. Offprint 1280.]

Cech, T. R., 1986. RNA as an enzyme. *Sci. Amer.* 255(5):64–75. [An excellent account of the discovery that RNA, as well as proteins, can be an effective catalyst. Available as *Sci. Amer*. Offprint 1575.]

Books on enzymes

Fersht, A., 1983. *Enzyme Structure and Mechanism* (2nd ed.). W. H. Freeman. [A concise and lucid introduction to enzyme action, with emphasis on physical principles.]

Walsh, C., 1979. *Enzymatic Reaction Mechanisms*. W. H. Freeman. [An excellent account of the chemical basis of enzyme action. This book demonstrates that the large number of enzyme-catalyzed reactions in biological systems can be grouped into a small number of types of chemical reactions.]

Bender, M. L., Bergeron, R. J., and Komiyama, M., 1984. *The Bioorganic Chemistry of Enzymatic Catalysis*. Wiley-Interscience.

Dugas, H., and Penney, C., 1981. *Bioorganic Chemistry: A Chemical Approach to Enzyme Action*. Springer-Verlag.

Dixon, M., and Webb, E. C., 1979. *Enzymes* (3rd ed.). Longmans.

Jencks, W. P., 1969. *Catalysis in Chemistry and Enzymology*. McGraw-Hill.

Hiromi, K., 1979. *Kinetics of Fast Enzyme Reactions*. Halsted Press.

Boyer, P. D., (ed.), 1970. *The Enzymes* (3rd ed.). Academic Press. [This multivolume treatise on enzymes contains a wealth of information. Volumes 1 and 2 (available in a paperback edition) deal with general aspects of enzyme structure, mechanism, and regulation. Volume 3 and subsequent ones contain detailed and authoritative articles on individual enzymes.]

Books on thermodynamics

Edsall, J. T., and Gutfreund, H., 1983. *Biothermodynamics: The Study of Biochemical Processes at Equilibrium*. Wiley.

Klotz, I. M., 1967. *Energy Changes in Biochemical Reactions*. Academic Press. [A concise introduction, full of insight.]

Enzyme kinetics and mechanisms

Fersht, A. R., Leatherbarrow, R. J., and Wells, T. N. C., 1986. Binding energy and catalysis: a lesson from protein engineering of the tyrosyl-tRNA synthetase. *Trends Biochem. Sci.* 11:321–325.

Jencks, W. P., 1975. Binding energy, specificity, and enzymic catalysis: the Circe effect. *Advan. Enzymol.* 43:219–410.

Knowles, J. R., and Albery, W. J., 1976. Evolution of enzyme function and the development of catalytic efficiency. *Biochemistry* 15:5631–5640.

Molecular interactions and binding

Richards, F. M., Wyckoff, H. W., and Allewell, N., 1970. The origin of specificity in binding: a detailed example in a protein-nucleic acid interaction. *In* Schmitt, F. O., (ed.), *The Neurosciences: Second Study Program*, pp. 901–912. Rockefeller University Press.

Davidson, N., 1967. Weak interactions and the structure of biological macromolecules. *In* Quarton, G. C., Melnechuk, T., and Schmitt, F. O., (eds.), *The Neurosciences: A Study Program*, pp. 46–56. Rockefeller University Press.

Control of enzymatic activity

Cheung, W. Y., 1982. Calmodulin. *Sci. Amer.* 246(6):62–70.

Monod, J., Changeux, J.-P., and Jacob, F., 1963. Allosteric proteins and cellular control systems. *J. Mol. Biol.* 6:306–329. [A classic paper that introduced the concept of allosteric interactions.]

Neurath, H., 1985. Proteolytic enzymes, past and present. *Fed. Proc.* 44:2907–2913.

Penicillin and other enzyme inhibitors

Waxman, D. J., and Strominger, J. L., 1983. Penicillin-binding proteins and the mechanism of action of β-lactam antibiotics. *Ann. Rev. Biochem.* 52:825–69.

Abraham, E. P., 1981. The beta-lactam antibiotics. *Sci. Amer.* 244:76–86.

Walsh, C. T., 1984. Suicide substrates, mechanism-based enzyme inactivators: recent developments. *Ann. Rev. Biochem.* 53:493–535.

PROBLEMS

1. The hydrolysis of pyrophosphate to orthophosphate is important in driving forward biosynthetic reactions such as the synthesis of DNA. This hydrolytic reaction is catalyzed in *E. coli* by a pyrophosphatase that has a mass of 120 kd and consists of six identical subunits. For this enzyme, a unit of activity is defined as the amount of enzyme that hydrolyzes 10 μmoles of pyrophosphate in 15 minutes at 37°C under standard assay conditions. The purified enzyme has a V_{max} of 2800 units per milligram of enzyme.

   (a) How many moles of substrate are hydrolyzed per second per milligram of enzyme when the substrate concentration is much greater than K_M?

   (b) How many moles of active site are there in 1 mg of enzyme? Assume that each subunit has one active site.

   (c) What is the turnover number of the enzyme? Compare this value with others mentioned in this chapter.

2. Penicillin is hydrolyzed and thereby rendered inactive by penicillinase (also known as β-lactamase), an enzyme present in some resistant bacteria. The mass of this enzyme in *Staphylococcus aureus* is 29.6 kd. The amount of penicillin hydrolyzed in 1 minute in a 10-ml solution containing 10^{-9} g of purified penicillinase was measured as a function of the concentration of penicil-

lin. Assume that the concentration of penicillin does not change appreciably during the assay.
(a) Plot $1/V$ versus $1/[S]$ for these data. Does penicillinase appear to obey Michaelis-Menten kinetics? If so, what is the value of K_M?
(b) What is the value of V_{max}?
(c) What is the turnover number of penicillinase under these experimental conditions? Assume one active site per enzyme molecule.

[Penicillin]	Amount hydrolyzed (moles)
0.1×10^{-5} M	0.11×10^{-9}
0.3×10^{-5} M	0.25×10^{-9}
0.5×10^{-5} M	0.34×10^{-9}
1.0×10^{-5} M	0.45×10^{-9}
3.0×10^{-5} M	0.58×10^{-9}
5.0×10^{-5} M	0.61×10^{-9}

3. Penicillinase (β-lactamase) hydrolyzes penicillin. Compare penicillinase with glycopeptide transpeptidase.

4. The kinetics of an enzyme are measured as a function of substrate concentration in the presence and absence of 2×10^{-3} M inhibitor (I).
(a) What are the values of V_{max} and K_M in the absence of inhibitor? In its presence?
(b) What type of inhibition is this?
(c) What is the binding constant of this inhibitor?
(d) If $[S] = 1 \times 10^{-5}$ M and $[I] = 2 \times 10^{-3}$ M, what fraction of the enzyme molecules have a bound substrate? A bound inhibitor?
(e) If $[S] = 3 \times 10^{-5}$ M, what fraction of the enzyme molecules have a bound substrate in the presence and absence of 2×10^{-3} M inhibitor? Compare this ratio with the ratio of the reaction velocities under the same conditions.

[S]	Velocity (μmoles/min)	
	No inhibitor	Inhibitor
0.3×10^{-5} M	10.4	4.1
0.5×10^{-5} M	14.5	6.4
1.0×10^{-5} M	22.5	11.3
3.0×10^{-5} M	33.8	22.6
9.0×10^{-5} M	40.5	33.8

5. The kinetics of the enzyme discussed in problem 4 are measured in the presence of a different inhibitor. The concentration of this inhibitor is 10^{-4} M.
(a) What are the values of V_{max} and K_M in the presence of this inhibitor? Compare them with those obtained in problem 4.
(b) What type of inhibition is this?
(c) What is the dissociation constant of this inhibitor?
(d) If $[S] = 3 \times 10^{-5}$ M, what fraction of the enzyme molecules have a bound substrate in the presence and absence of 10^{-4} M inhibitor?

[S]	Velocity (μmoles/min)	
	No inhibitor	Inhibitor
0.3×10^{-5} M	10.4	2.1
0.5×10^{-5} M	14.5	2.9
1.0×10^{-5} M	22.5	4.5
3.0×10^{-5} M	33.8	6.8
9.0×10^{-5} M	40.5	8.1

6. The plot of $1/V$ versus $1/[S]$ is sometimes called a Lineweaver-Burk plot. Another way of expressing the kinetic data is to plot V versus $V/[S]$, which is known as an Eadie-Hofstee plot.
(a) Rearrange the Michaelis-Menten equation to give V as a function of $V/[S]$.
(b) What is the significance of the slope, the vertical intercept, and the horizontal intercept in a plot of V versus $V/[S]$?
(c) Make a sketch of a plot of V versus $V/[S]$ in the absence of an inhibitor, in the presence of a competitive inhibitor, and in the presence of a noncompetitive inhibitor.

7. The hormone progesterone contains two ketone groups. Little is known about the properties of the receptor protein that recognizes progesterone. At pH 7, which amino acid side chains might form hydrogen bonds with progesterone? (Assume that the side chains in the receptor protein have the same pKs as in the amino acids in aqueous solution.)

8. Suppose that two substrates A and B compete for an enzyme. Derive an expression relating the ratio of the rates of utilization of A and B, V_A/V_B, to the concentrations of these substrates and their values of k_3 and K_M. (Hint: Express V_A as a function of k_3/K_M for substrate A, and do the same for V_B.) Is specificity determined by K_M alone?

Appendixes

Answers to Problems

Index

APPENDIX A

Physical Constants and Conversion of Units

Values of physical constants

Physical constant	Symbol	Value
Atomic mass unit (dalton)	amu	1.660×10^{-24} g
Avogadro's number	N	6.022×10^{23} mol^{-1}
Boltzmann's constant	k	1.381×10^{-23} J deg^{-1}
		3.298×10^{-24} cal deg^{-1}
Electron volt	eV	1.602×10^{-19} J
		3.828×10^{-20} cal
Faraday constant	F	9.649×10^{4} C mol^{-1}
		2.306×10^{4} cal volt^{-1} eq^{-1}
Curie	Ci	3.70×10^{10} disintegrations s^{-1}
Gas constant	R	8.315 J mol^{-1} deg^{-1}
		1.987 cal mol^{-1} deg^{-1}
Planck's constant	h	6.626×10^{-34} J s
		1.584×10^{-34} cal s
Speed of light in a vacuum	c	2.998×10^{10} cm s^{-1}

Abbreviations: C, coulomb; cal, calorie; cm, centimeter; deg, degree Kelvin; eq, equivalent; g, gram; J, joule; mol, mole; s, second.

Conversion factors

Physical quantity	Equivalent
Length	1 cm = 10^{-2} m = 10 mm = 10^{4} μm = 10^{7} nm
	1 cm = 10^{8} Å = 0.3937 inch
Mass	1 g = 10^{-3} kg = 10^{3} mg = 10^{6} μg
	1 g = 3.527×10^{-2} ounce (avoirdupoir)
Volume	1 cm^{3} = 10^{-6} m^{3} = 10^{3} mm^{3}
	1 ml = 1 cm^{3} = 10^{-3} l = 10^{3} μl
	1 cm^{3} = 6.1×10^{-2} in^{3} = 3.53×10^{-5} ft^{3}
Temperature	K = °C + 273.15
	°C = (5/9)(°F − 32)
Energy	1 J = 10^{7} erg = 0.239 cal = 1 watt s
Pressure	1 torr = 1 mm Hg (0°C)
	= 1.333×10^{2} newton/m^{2}
	= 1.333×10^{2} pascal
	= 1.316×10^{-3} atmospheres

Mathematical constants

$\pi = 3.14159$
$e = 2.71828$
$\log_e x = 2.303 \log_{10} x$

Standard prefixes

Prefix	Symbol	Factor
kilo	k	10^{3}
hecto	h	10^{2}
deca	da	10^{1}
deci	d	10^{-1}
centi	c	10^{-2}
milli	m	10^{-3}
micro	μ	10^{-6}
nano	n	10^{-9}
pico	p	10^{-12}

APPENDIX B

Atomic Numbers and Weights of the Elements

Element	Symbol	Atomic number	Atomic weight
Actinium	Ac	89	227.03
Aluminum	Al	13	26.98
Americium	Am	95	243.06
Antimony	Sb	51	121.75
Argon	Ar	18	39.95
Arsenic	As	33	74.92
Astatine	At	85	210.99
Barium	Ba	56	137.34
Berkelium	Bk	97	247.07
Beryllium	Be	4	9.01
Bismuth	Bi	83	208.98
Boron	B	5	10.81
Bromine	Br	35	79.90
Cadmium	Cd	48	112.40
Calcium	Ca	20	40.08
Californium	Cf	98	249.07
Carbon	C	6	12.01
Cerium	Ce	58	140.12
Cesium	Cs	55	132.91
Chlorine	Cl	17	35.45
Chromium	Cr	24	52.00
Cobalt	Co	27	58.93
Copper	Cu	29	63.55
Curium	Cm	96	245.07
Dysprosium	Dy	66	162.50
Einsteinium	Es	99	254.09
Erbium	Er	68	167.26
Europium	Eu	63	151.96
Fermium	Fm	100	252.08
Fluorine	F	9	18.99
Francium	Fr	87	223.02
Gadolinium	Gd	64	157.25
Gallium	Ga	31	69.72
Germanium	Ge	32	72.59
Gold	Au	79	196.97
Hafnium	Hf	72	178.49
Helium	He	2	4.00
Holmium	Ho	67	164.93
Hydrogen	H	1	1.01
Indium	In	49	114.82
Iodine	I	53	126.90
Iridium	Ir	77	192.22
Iron	Fe	26	55.85
Khurchatovium	Kh	104	260
Krypton	Kr	36	83.80
Lanthanum	La	57	138.91
Lawrencium	Lr	103	256
Lead	Pb	82	207.20
Lithium	Li	3	6.94
Lutetium	Lu	71	174.97
Magnesium	Mg	12	24.31
Manganese	Mn	25	54.94
Mendelevium	Md	101	255.09
Mercury	Hg	80	200.59
Molybdenum	Mo	42	95.94
Neodymium	Nd	60	144.24
Neon	Ne	10	20.18
Neptunium	Np	93	237.05
Nickel	Ni	28	58.71
Niobium	Nb	41	92.91
Nitrogen	N	7	14.01
Nobelium	No	102	255
Osmium	Os	76	190.20
Oxygen	O	8	16.00
Palladium	Pd	46	106.40
Phosphorus	P	15	30.97
Platinum	Pt	78	195.09
Plutonium	Pu	94	242.06
Polonium	Po	84	208.98
Potassium	K	19	39.10
Praseodymium	Pr	59	140.91
Promethium	Pm	61	145
Protactinium	Pa	91	231.04
Radium	Ra	88	226.03
Radon	Rn	86	222.02
Rhenium	Re	75	186.20
Rhodium	Rh	45	102.91
Rubidium	Rb	37	85.47
Ruthenium	Ru	44	101.07
Samarium	Sm	62	150.40
Scandium	Sc	21	44.96
Selenium	Se	34	78.96
Silicon	Si	14	28.09
Silver	Ag	47	107.87
Sodium	Na	11	22.99
Strontium	Sr	38	87.62
Sulfur	S	16	32.06
Tantalum	Ta	73	180.95
Technetium	Tc	43	98.91
Tellurium	Te	52	127.60
Terbium	Tb	65	158.93
Thallium	Tl	81	204.37
Thorium	Th	90	232.04
Thulium	Tm	69	168.93
Tin	Sn	50	118.69
Titanium	Ti	22	47.90
Tungsten	W	74	183.85
Uranium	U	92	238.03
Vanadium	V	23	50.94
Xenon	Xe	54	131.30
Ytterbium	Yb	70	173.04
Yttrium	Y	39	88.91
Zinc	Zn	30	65.37
Zirconium	Zr	40	91.22

APPENDIX C

pK' Values of Some Acids

Acid	pK' (at 25°C)
Acetic acid	4.76
Acetoacetic acid	3.58
Ammonium ion	9.25
Ascorbic acid, pK'_1	4.10
pK'_2	11.79
Benzoic acid	4.20
n-Butyric acid	4.81
Cacodylic acid	6.19
Carbonic acid, pK'_1	6.35
pK'_2	10.33
Citric acid, pK'_1	3.14
pK'_2	4.77
pK'_3	6.39
Ethylammonium ion	10.81
Formic acid	3.75
Glycine, pK'_1	2.35
pK'_2	9.78
Imidazolium ion	6.95
Lactic acid	3.86
Maleic acid, pK'_1	1.83
pK'_2	6.07
Malic acid, pK'_1	3.40
pK'_2	5.11
Phenol	9.89
Phosphoric acid, pK'_1	2.12
pK'_2	7.21
pK'_3	12.67
Pyridinium ion	5.25
Pyrophosphoric acid, pK'_1	0.85
pK'_2	1.49
pK'_3	5.77
pK'_4	8.22
Succinic acid, pK'_1	4.21
pK'_2	5.64
Trimethylammonium ion	9.79
Tris (hydroxymethyl) aminomethane	8.08
Water	14.0

APPENDIX D

Standard Bond Lengths

Bond	Structure	Length (Å)
C—H	R_2CH_2	1.07
	Aromatic	1.08
	RCH_3	1.10
C—C	Hydrocarbon	1.54
	Aromatic	1.40
C=C	Ethylene	1.33
C≡C	Acetylene	1.20
C—N	RNH_2	1.47
	O=C—N	1.34
C—O	Alcohol	1.43
	Ester	1.36
C=O	Aldehyde	1.22
	Amide	1.24
C—S	R_2S	1.82
N—H	Amide	0.99
O—H	Alcohol	0.97
O—O	O_2	1.21
P—O	Ester	1.56
S—H	Thiol	1.33
S—S	Disulfide	2.05

Answers to Problems

Chapter 2

1. 477 Å (318 residues per strand, 1.5 Å per residue).
2. The methyl group attached to the β carbon of isoleucine sterically interferes with α-helix formation. In leucine, this methyl group is attached to the γ carbon atom, which is farther from the main chain and hence does not interfere.
3. The first mutation destroys activity because valine occupies more space than alanine, and so the protein must take a different shape. The second mutation restores activity because of a compensatory reduction of volume; glycine is smaller than isoleucine.
4. The native conformation of insulin is not the thermodynamically most stable form. Indeed, insulin is formed from proinsulin, a single-chain precursor (p. 994).
5. A segment of the main chain of the protease could hydrogen bond to the main chain of the substrate to form an extended parallel or antiparallel pair of β strands.

Chapter 3

1. (a) Phenyl isothiocyanate.
   (b) Dansyl chloride or dabsyl chloride.
   (c) Urea; β-mercaptoethanol to reduce disulfides.
   (d) Chymotrypsin.
   (e) CNBr.
   (f) Trypsin.
2. 0.01, 0.1, 1, 10, and 100.
3. Each amino acid residue, except the carboxyl-terminal one, gives rise to a hydrazide on reacting with hydrazine. The carboxyl-terminal residue can be identified because it yields a free amino acid.
4. (a) Approximately +1.
   (b) Two peptides.
5. The S-aminoethylcysteine side chain resembles that of lysine. The only difference is a sulfur atom in place of a methylene group.
6. A 1 mg/ml solution of myoglobin (17.8 kd) corresponds to 5.62×10^{-5} M. The absorbance of a 1-cm path length is 0.84, which corresponds to an I_0/I ratio of 6.96. Hence 14.4% of the incident light is transmitted.
7. Tropomyosin is rod shaped, whereas hemoglobin is approximately spherical.
8. 50 kd.
9. Reduction of disulfide bonds by dithiothreitol makes the protein less compact, so that it migrates less rapidly.
10. The positions of disulfide bonds can be determined

by diagonal electrophoresis (p. 57). The disulfide pairing is unaltered by the mutation if the off-diagonal peptides formed from the native and mutant proteins are the same.
11. Electrostatic repulsion between positively charged ϵ-amino groups prevents α-helix formation at pH 7. At pH 10, the side chains become deprotonated, allowing α-helix formation.
12. Poly-L-glutamate is a random coil at pH 7 and becomes α helical below pH 4.5 because the γ-carboxylate groups become protonated.
13. Glycine has the smallest side chain of any amino acid. Its smallness often is critical in allowing polypeptide chains to make tight turns or to approach one another closely.
14. Affinity chromatography on a column containing a covalently attached analog of vasopressin would be an effective purification method. The first step would be to solubilize the receptor by adding a nondenaturing detergent to a membrane preparation. The solubilized receptor would then be added to the affinity column, which would be washed to remove proteins that do not have high affinity for it. The receptor would be eluted by adding vasopressin.
15. The addition of substrate makes the enzyme less compact, which increases its frictional coefficient. The partial specific volume is unlikely to be altered.
16. The centrifugal force F_c corresponds to Ez. The centrifugal field $\omega^2 r$ is analogous to E, and the effective mass m' is analogous to z.
17. (a) A compact particle sediments more rapidly than does an extended one of the same mass and partial specific volume.
    (b) Shape does not affect the position of the molecule in a sedimentation equilibrium experiment.
18. Large molecules emerge first from a gel-filtration column because a smaller volume is accessible to them; they cannot enter the beads. In contrast, a continuous polymer framework impedes the movement of large molecules in gel electrophoresis, causing them to migrate less rapidly than small molecules.
19. Different monoclonal antibodies recognize different determinants (epitopes). Suppose that one antibody is specific for epitope a and the other for epitope b. The Western blot suggests that the 23-kd protein contains a and b, the 57-kd protein contains only a, and the 69-kd protein contains only b.
20. The amino-terminus of myoglobin is flexible. The existence of many different conformations prevents it being seen in the electron density map.
21. A fluorescent-labeled derivative of a bacterial degradation product (e.g., a formylmethionyl peptide) would bind to cells containing the receptor of interest.

3. 5.88×10^3 base pairs.
4. After 1.0 generation, one-half of the molecules would be ^{15}N-^{15}N, the other half ^{14}N-^{14}N. After 2.0 generations, one-quarter of the molecules would be ^{15}N-^{15}N, the other three-quarters ^{14}N-^{14}N. Hybrid ^{14}N-^{15}N molecules would not be observed in conservative replication.
5. The DNA renatured when the heat-killed pneumococci were cooled before they were injected into mice.
6. Non-competent strains may not be able to take up DNA. Alternatively, they may have potent deoxyribonucleases, or they may not be able to integrate fragments of DNA into their genome.
7. In the Hershey-Chase experiment, ^{35}S-labeled T2 viral proteins did not become incorporated into infected cells. The labeled viral proteins were found in the supernatant when infected cells were centrifuged. In contrast, M13 proteins become imbedded in the inner membrane of infected cells; they would appear in the pellet rather than the supernatant after centrifugation. Hershey and Chase would not have been able to separate M13 into genetic and nongenetic parts, as they did for T2.
8. Tritiated thymine.
9. dATP, dGTP, dCTP, and dTTP labeled with ^{32}P in the innermost (α) phosphorus atom.
10. Molecules (a) and (b) would not lead to DNA synthesis because they lack a 3'-OH group (a primer). Molecule (d) has a free 3'-OH at one end of each strand but no template strand beyond. Only (c) would lead to DNA synthesis.
11. A deoxythymidylate oligonucleotide should be used as the primer. The poly rA template specifies the incorporation of dT; hence radioactive dTTP should be used in the assay.
12. The ribonuclease serves to degrade the RNA strand, a necessary step in forming duplex DNA from the RNA-DNA hybrid.
13. (a) Treat one aliquot of the sample with ribonuclease and another with deoxyribonuclease. Test these nuclease-treated samples for infectivity.
    (b) An essential protein enzyme carried by the virus particle seems unlikely because the phenol-treated material was infectious. The nucleic acid contains all the information needed for its replication.
14. Deamination changes the original GC base pair into a GU pair. After one round of replication, one daughter duplex will contain a GC pair, and the other duplex an AU pair. After two rounds of replication, there would be two GC pairs, one AU pair, and one AT pair.

Chapter 4

1. (a) TTGATC; (b) GTTCGA; (c) ACGCGT; and (d) ATGGTA.
2. (a) [T] + [C] = 0.46.
   (b) [T] = 0.30, [C] = 0.24, and [A] + [G] = 0.46.

Chapter 5

1. (a) Deoxyribonucleoside triphosphates versus ribonucleoside triphosphates.
   (b) $5' \to 3'$ for both.
   (c) Semiconserved for DNA polymerase I, conserved for RNA polymerase.

(d) DNA polymerase I needs a primer, whereas RNA polymerase does not.
2. 5′-UAACGGUACGAU-3′
3. The 2′-OH group in RNA acts as an intramolecular catalyst. In the alkaline hydrolysis of RNA, it forms a 2′-3′ cyclic intermediate.
4. Cordycepin terminates RNA synthesis. An RNA chain containing cordycepin lacks a 3′-OH group.
5. Leu-Pro-Ser-Asp-Trp-Met.
6. Poly (Leu-Leu-Thr-Tyr).
7. (a) A codon for lysine cannot be changed to one for aspartate by the mutation of a single nucleotide.
   (b) Arg, Asn, Gln, Glu, Ile, Met, or Thr.
8. A peptide terminating with Lys (UGA is a stop codon); -Asn-Glu-; and -Met-Arg-.
9. Highly abundant amino acid residues have the most codons (e.g., Leu and Ser each have six), whereas the least abundant ones have the fewest (Met and Trp each have only one). Degeneracy allows (a) variation in base composition and (b) decreases the likelihood that a substitution of a base will change the encoded amino acid. If the degeneracy were equally distributed, each of the twenty amino acids would have three codons. Benefits (a) and (b) are maximized by assigning more codons to prevalent amino acids than to less frequently used ones.
10. Phe-Cys-His-Val-Ala-Ala.
11. GUG and GUC are likely to be used more by the alga from the hot springs to increase the melting temperature of its DNA (see p. 82).
12. The genetic code is degenerate. Eighteen of the twenty amino acids are specified by more than one codon. Hence, many nucleotide changes (especially in the third base of a codon) do not alter the nature of the encoded amino acid.
13. (a) Green-red color blindness.
    (b) Recombination between similar genes can lead to gene duplication. The duplicate of an essential gene can undergo mutation and diversification without deleterious consequences. New genes nearly always arise by diversification of duplicated genes.

3. Ovalbumin cDNA should be used. *E. coli* lacks the machinery to splice the primary transcript arising from genomic DNA.
4. (a) No, because most human genes are much longer than 4 kb. One would obtain fragments containing only a small part of a complete gene.
   (b) No, chromosome walking depends on having *overlapping* fragments. Exhaustive digestion with a restriction enzyme produces nonoverlapping, short fragments.
5. Southern blotting of an MstII digest would distinguish between the normal and mutant genes. The loss of a restriction site would lead to the replacement of two fragments on the Southern blot by a single longer fragment (see p. 169). Such a finding would not prove that GTG replaced GAG; other sequence changes at the restriction site could yield the same result.
6. Cech replicated the recombinant DNA plasmid in *E. coli*, and then transcribed the DNA in vitro using bacterial RNA polymerase. He then found that this RNA underwent self-splicing in vitro in the complete absence of any proteins from *Tetrahymena*.
7. (a) Only one of the foreign DNA strands encodes an mRNA specifying a functional protein.
   (b) One could analyze RF molecules from the original single virus infection by restriction enzyme mapping. Asymmetric restriction sites in the foreign DNA fragments and phage chromosome would yield different gel patterns according to the orientation of the foreign DNA with the M13 DNA.
   (c) Single-strand circles from two clones containing the same foreign DNA strand will not hybridize to one another. In contrast, the circles will hybridize to form a duplex if the foreign DNA in them are complements of one another. S1 nuclease digests unpaired DNA strands but not duplex DNA.
8. Knowledge of the amino acid sequence is essential. It would be helpful to know which bonds are highly susceptible to proteolysis, and which residues are critical for the biological function of the peptide.

Chapter 6

1. 5′-GGCATAC-3′.

2.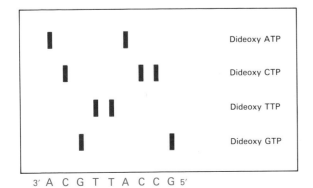

Chapter 7

1. (a) 2.96×10^{-11} g.
   (b) 2.71×10^8 molecules.
   (c) No. There would be 3.22×10^8 hemoglobin molecules in a red cell if they were packed in a cubic crystalline array. Hence, the actual packing density is about 84% of the maximum possible.
2. 2.65 g (or 4.75×10^{-2} moles) of Fe.
3. (a) In humans, 1.44×10^{-2} g (4.49×10^{-4} moles) of O_2 per kg of muscle. In sperm whale, 0.144 g (4.49×10^{-3} moles) of O_2 per kg.
   (b) 128.
4. (a) $k_{off} = k_{on}K = 20$ s^{-1}.
   (b) Mean duration is 0.05 s (the reciprocal of k_{off}).
5. (a) Increased, (b) decreased, (c) decreased, and (d) increased oxygen affinity.

6. Inositol hexaphosphate.
7. The pK is (a) lowered, (b) raised, and (c) raised.
8. (a) Yes. $K_{AB} = K_{BA}(K_B/K_A) = 2 \times 10^{-5}$ M.
   (b) The presence of A enhances the binding of B; hence, the presence of B enhances the binding of A.
9. Carbon monoxide bound to one heme alters the oxygen affinity of the other hemes in the same hemoglobin molecule. Specifically, CO increases the oxygen affinity of hemoglobin and thereby decreases the amount of O_2 released in actively metabolizing tissues. Carbon monoxide stabilizes the quaternary structure characteristic of oxyhemoglobin. In other words, CO mimics O_2 as an allosteric effector.
10. (a) For maximal transport, $K = 10^{-5}$ M. In general, maximal transport is achieved when $K = ([L_A][L_B])^{0.5}$.
    (b) For maximal transport, $P_{50} = 44.7$ torrs, which is considerably higher than the physiological value of 26 torrs. However, it must be stressed that this calculation ignores cooperative binding and the Bohr effect.
11. (a) Lys or Arg at position 6.
    (b) GAG (Glu) to AAG (Lys).
    (c) This mutant hemoglobin moves more rapidly toward the anode than does Hb A and Hb S because it is more positively charged.
12. No. The target site for MstII is CCTNAGG, which encodes Pro-Glu-Glu in hemoglobin A. Other mutations of this heptanucleotide sequence could lead to the loss of the 1.3-kb fragment in the Southern blot—for example, a mutation of CCTGAGG to CTTGAGG (Pro-Glu-Glu to Leu-Glu-Glu).
13. Mutations in the α gene affect all three hemoglobins because their subunit structures are $\alpha_2\beta_2$, $\alpha_2\delta_2$, and $\alpha_2\gamma_2$. Mutations in the β, δ, or γ genes affect only one of them.
14. Deoxy Hb A contains a complementary site, and so it can add on to a fiber of deoxy Hb S. The fiber cannot then grow further because the terminal deoxy Hb A molecule lacks a sticky patch.
15. Carbamoylation of hemoglobin increases its oxygen affinity. Oxygenated Hb S does not sickle.
16. (a and b) Protons are released from hemoglobin because of increased oxygenation.

Chapter 8

1. (a) 31.1 μmoles.
   (b) 0.05 μmoles.
   (c) 622 s^{-1}.
2. Yes. $K_M = 5.2 \times 10^{-6}$ M.
   (b) $V_{max} = 6.84 \times 10^{-10}$ moles/min.
   (c) 337 s^{-1}.
3. Penicillinase, like glycopeptide transpeptidase, forms an acyl-enzyme intermediate with its substrate but transfers it to water rather than to the terminal glycine of the pentaglycine bridge.
4. (a) In the absence of inhibitor, V_{max} is 47.6 μmole/min and K_M is 1.1×10^{-5} M. In the presence of inhibitor, V_{max} is the same, and the apparent K_M is 3.1×10^{-5} M.
   (b) Competitive.
   (c) 1.1×10^{-3} M.
   (d) f_{ES} is 0.243 and f_{EI} is 0.488.
   (e) f_{ES} is 0.73 in the absence of inhibitor and 0.49 in the presence of 2×10^{-3} M inhibitor. The ratio of these values, 1.49, is the same as the ratio of the reaction velocities under these conditions.
5. (a) V_{max} is 9.5 μmole/min. K_M is 1.1×10^{-5} M, the same as without inhibitor.
   (b) Noncompetitive.
   (c) 2.5×10^{-5} M.
   (d) 0.73, in the presence or absence of this noncompetitive inhibitor.
6. (a) $V = V_{max} - (V/[S]) K_M$.
   (b) Slope = $-K_M$, y-intercept = V_{max}, x-intercept = V_{max}/K_M.
   (c)

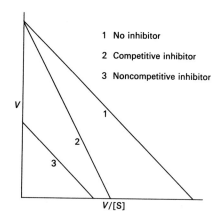

7. Potential hydrogen-bond donors at pH 7 are the side chains of the following residues: arginine, asparagine, glutamine, histidine, lysine, serine, threonine, tryptophan, and tyrosine.
8. The rates of utilization of A and B are given by

$$V_A = \left(\frac{k_3}{K_M}\right)_A [E][A]$$

and

$$V_B = \left(\frac{k_3}{K_M}\right)_B [E][B]$$

Hence, the ratio of these rates is

$$V_A/V_B = \left(\frac{k_3}{K_M}\right)_A [A] \bigg/ \left(\frac{k_3}{K_M}\right)_B [B]$$

Thus, an enzyme discriminates between competing substrates on the basis of their values of k_3/K_M rather than of K_M alone.

INDEX

References to structural formulas are given in **boldface** type.

absolute configuration of amino acids, 17
absorbance, 69
absorption coefficient, 69
absorption spectra
  of DNA, 82
  of myoglobin and hemoglobin, 147
acetate, **4**
acids, 41
acid-base concepts, 41–42
acquired immune deficiency syndrome, 58
acrylamide, **44**
ACTH, 69
actin filaments, 64
active site, 184
  See also individual enzymes
acyl-enzyme intermediate, 196
adaptor molecule, 99
adenine, **72**, **92**
adenine-thymine base pair, 77
adrenocorticotropin, 69
affinity chromatography, 48
Agrobacterium tumefaciens, 135
AIDS, 58
alanine, **18**
alcohol dehydrogenase, 195
Allison, A., 168
allosteric enzymes, 193

allosteric interactions
  in hemoglobin, 154–163
allosteric proteins
  enzymes, 193
  hemoglobin, 154–163
alpha helix, 26
  coiled coil of, 27
alternative splicing, 113
amber (UAG codon), 107
amino acid sequences, 23
  from DNA sequences, 57–58
  determination of, 50–58
  significance of, 23, 58
  See also individual proteins
amino acids, 17–21
  abbreviations of, 21
  absolute configuration, 17
  acidic, 20
  aliphatic, **18**
  aromatic, **18**
  basic, 20
  hydroxyl-bearing, 19
  in secondary structures, 37
  ionization of, 17
  D and L isomers of, 17
  L-isomer, 17
  modified, 24
  pK values of, 42
  sulfur-containing, **19**
ampicillin, 127

Anfinsen, C., 32
angstrom (Å), 27
Angstrom, A. J., 27
antibiotics
  penicillin, 195
antibodies. *See* immunoglobulins
anticodon, 99
antidiuretic hormone, 65
antifreeze poisoning, 195
antigen, 62
antigenic determinant, 62
antiserum, 62
apoprotein, 144
Arber, W., 118
arginine, **20**
8-arginine vasopressin, **65**
asparagine, **20**
aspartate, **20**
Astbury, W. T., 27
autoradiography, 45
Avery, O., 74

bacterial cell wall
  synthesis of, 195
bacteriophage, 128
  M13, 129
  φX174, 85
  T2, 75
Bailey, K., 27

balanced polymorphism, 168
band centrifugation, 50
Barcroft, J., 156
bases
　in nucleotides, 72–73
　as proton acceptors, 41
Benesch, Reinhold, 157
Benesch, Ruth, 157
Benzer, S., 109
benzyl penicillin, 195
Berzelius, J. J., 15
Berg, P., 124
beta pleated sheet, 27–28
beta turn, 28
biochemistry, definition of, 3
1,3-bisphosphoglycerate, **193**
2,3-bisphosphoglycerate, 156–157, **157**, **193**
　binding to hemoglobin, 161–162
Blout, E. R., 37
Bohr effect, 156, 162–163
Bohr, C., 156
Boyer, H., 124
2,3-BPG. *See* 2,3-bisphosphoglycerate
Bragg, L., 151
Brenner, S., 94, 99
buffer, 42
t-butyloxycarbonyl amino acid, **65**

calcium phosphate, 133
calmodulin, 179
　model of, 179
calorie (cal), 182
carbamate, 162
carbamoylation by cyanate, 176
carbon dioxide
　binding of to hemoglobin, 162–163
carbon monoxide, 149
carbonic anhydrase
　catalysis by, 38
　model of, 38
γ-carboxyglutamate, **24**
carboxymethyl (CM) group, **48**
　model of, 12
catalytic power, 177
cDNA, 132–133
CDP-diacylglycerol
　model of, 13
Cech, T., 113
centrifugation, 49
Chain, E., 195
Chargaff, E., 78
Chase, M., 74
chemical cleavage method, 120
chimeric DNA, 125
chromatography, 166
　affinity, 48
　gel-filtration, 47
　ion-exchange, 48
chromosome walking, 131
clostripain, 56
CO₂. *See* carbon dioxide
codon, 99, 100
Cohen, S., 124
cohesive-end method, 126

coiled coil, 27
colinearity, 109
Collman, J., 148
competitive inhibition, 192–195
complementary DNA, 132–133
concanavalin A, 49
conformation, 25
consensus sequence, 98
contact radii, 8
Coomassie blue, 45
cooperative binding
　of oxygen, 154–156
cooperative interactions
　in DNA, 82
coordination position, 146
Corey, R., 25
corticotropin, 69
covalent modification, 179
Crick, F., 76, 99
crown gall, 135
crystallization of myoglobin, 60
cyanate, 176
cyanogen bromide, 55
cysteic acid, **57**
cysteine, **19**
cystine, **23**
cytosine, **72**, **92**

dabsyl chloride, **53**
dalton, 22
Dalton, J., 22
dansyl chloride, **53**
degeneracy of genetic code, 101, 107
denaturation, 33
density-gradient equilibrium sedimentation, 80
deoxyadenosine, **72**
deoxyadenosine 5′-triphosphate, **72**
deoxyadenylate, 73
deoxycytidine, 72
deoxycytidylate, 73
deoxyguanosine, 72
deoxyguanylate, 73
deoxymyoglobin, 146
deoxyribonuclease. *See* DNase I
deoxyribose, **72**
deoxythymidine, 72
deoxythymidylate, 73
1-desamino-8-D-arginine
　vasopressin, **65**
diagonal electrophoresis, 57
dialysis, 47
dicyclohexylcarbodiimide, **65**
dicyclohexylurea, **65**
2′,3′-dideoxy analog, 121, **122**
dideoxy sequencing method, 121
dielectric constants, 10
diethylaminoethyl (DEAE) group, **48**
diffraction, 60
diffusion-limited reaction, 191
dihydroxyacetone phosphate, **183**
diisopropylphosphofluoridate (DIPF), **192**
dimensions, 5
dimethylsulfate, 121

2,3-diphosphoglycerate. *See* 2,3-bisphosphoglycerate
dipolar ion, 17
disulfide bonds, 23, 32
　cleavage of, 57
dithiothreitol, **57**
DNA, **73**
　autoradiograph of, 69
　base pairs in, 7
　chemical synthesis of, 123–124
　chimeric, 125
　circles, 83–84
　cloning of, 124–135
　complementary (cDNA), 132–133
　double helix, 76–78
　electron micrographs of, 13
　gel electrophoresis of, 119–120
　genetic role of, 71–86
　melting of, 80–82
　models of, 3, 77
　polymerases, 84–85
　recombinant, 117–138
　replication of, 78–80, 84–85
　sequencing of, 120–123
　sizes of, 82–83
　solid-phase synthesis of, 123–124
　structure of, 72–73
　supercoiled form of, 84
　transformation by, 73–74
DNA ligase, 126
DNA polymerase I, 84
　electron micrograph of, 177
DNA probes, 131
domains, 31
Doty, P., 95
double helix, 76–78
double-reciprocal plot, 194

Eadie-Hofstee plot, 200
Edman, P., 53
Edman degradation, 53
electron-density maps, 61, 147
electrophoresis, 44
electrophoretic mobility, 165
electroporation, 135
electrostatic bonds, 7
ELISA, 63
embryonic hemoglobins, 150
energetics, 6
energy, 180
enhancer sequences, 98
entropy (S), 180
enzyme, 177–197
　active sites of, 185–187
　catalytic power of, 177
　energy transduction by, 180
　inhibition of, 192–197
　Michaelis-Menten model of, 187–191
　regulation of, 179
　specificity, 178
enzyme-linked immunosorbent assay, 63
enzyme-substrate complex, 184–185
equilibrium constant, 182
erythrocytes, micrograph of, 142, 143

ES complex, 184–185
ethanol
　therapeutic use of, 195
ethidium bromide, 119
ethylene glycol poisoning, 195
exons, 32, 110, 113
　of myoglobin, 149–150

Fasman, G., 37
FBPase 2
feedback inhibition, 179
Fe-protoporphyrin IX, **144**
ferrihemoglobin, 144
ferrimyoglobin, 146
ferritin, 64
ferrohemoglobin, 144
fetal hemoglobin, 150
fingerprinting, 119, 165, 166
Fischer, E., 187
Fleming, A., 195
Florey, H., 195
fluorescamine, **51**
fluorescence microscopy, 64
fluorescence spectroscopy, 185
fluorodinitrobenzene, 52, **53**
fMet, **107**
fMet peptide, **65**
folding units in proteins, 35
formylmethionine, **107**
frameshift mutation, 171
Franklin, R., 76
free energy, 180–183
　of activation, 182
frictional coefficient, 49
fructose bisphosphatase 2

gel electrophoresis, 44
　of DNA, 119–120
　of proteins, 45
gel-filtration chromatography, 47
gene duplication, 112
gene-protein relations, 109
genetic code, 99–109
　degeneracy of, 106
　discovery of, 101–106
　start signals, 107–108
　stop signals, 106–107
　table of, 107
　universality of, 108–109
genetic information, flow of, 91–114
genomic DNA, 130
genomic library, 130
Gibbs, J. W., 181
Gibbs free energy of activation, 184
Gilbert, W., 113, 120
globin, word origin of, 153
globin fold, 153
glutamate, **20**
glutamine, **20**
glyceraldehyde 3-phosphate, **183**
glycine, **18**
glycopeptide transpeptidase, 196
Griffith, F., 74
Gross, E., 55
growth hormone, 134

Grunberg-Manago, M., 102
GTP, 72
guanidine hydrochloride, **32**
guanine, **72**, **92**
guanine-cytosine base pair, 77
guanyl-nucleotide-binding protein.
　See G-proteins

hairpin turns, 28
Haldane, J. S., 156
helix
　α, 26
　DNA double, 76–78
heme, **144**
　in hemoglobin, 160
　model of, 144
　in myoglobin, 147
heme-heme interaction, 156
hemo, word origin of, 153
hemoglobin, 150–171
　adult, 150–171
　allosteric interactions of, 154–163
　binding of BPG to, 161–162
　binding of CO_2 to, 162–163
　Bohr effect in, 156
　cooperative binding of O_2 to, 154–156
　defective synthesis of, 171
　effect of BPG on, 156–157
　embryonic, 150–151
　fetal, 150–151
　gene expression of, 151
　invariant residues of, 153
　model of, 31
　molecular pathology of, 170
　quaternary structure of, 158–162
　salt links in, 159
hemoglobin A, 150
hemoglobin F, 157
hemoglobin Hammersmith, 170
hemoglobin M, 170
hemoglobin S, 164–169
Henderson-Hasselbalch equation, 41
Herrick, J., 163
Herriott, R., 74
Hershey, A., 74
hexokinase
　crystal of, 15
Hill, A., 155
Hill coefficient, 155
Hill plot, 155
histidine, **20**
Hogness box, 98
homoserine lactone, **55**
Hurwitz, J., 96
hybridization of nucleic acids, 95
hybridoma cells, 63
hydrazine, 121
hydrogen bonds, 7
　in DNA, 77
　in enzyme-substrate complexes, 186
　lengths of, 7
　in proteins, 29
hydrophobic attractions, 10
hydroxylamine, 56

4-hydroxyproline, **24**
hyperchromism, 82

ice, 9
immunoassays, 63
immunochemical screening, 133
immunoelectron microscopy, 64
immunoglobulins, 62
　model of, 31
　as reagents, 62–64
induced fit, 216
　model, 187
Ingram, V., 165
inhibition of enzymes, 192–197
insertional inactivation, 127
insulin, **23**
　crystals of, 117
intervening sequence, 110
introns, 110–113
iodoacetamide, **192**
O-iodosobenzoate, 56
ion-exchange chromatography, 48
iron atom, movement of in heme, 159–160
irreversible inhibitor, 192
isoelectric focusing, 46
isoelectric point, 45
isoleucine, **18**

Jacob, F., 93, 94
joule, 182

K_i, 194
K_M, 189–190
　definition of, 189
　determination of, 189
　significance of, 190
　table of values of, 190
Katchalski, E., 37
Kendrew, J., 144
Khorana, H. G., 104
kilobase, 119
kilocalorie, 182
kilodalton, 22
kilojoule, 182
kinetic perfection, 191
kinetics
　of biological processes, 6
　of enzyme-catalyzed reactions, 187–191
Koshland, D. E., Jr., 187

β-lactam ring, 195
β-lactamase, 199
lambda bacteriophage
　in DNA cloning, 128
Lederberg, J., 196
leucine, **18**
leucocytes, 142
Lineweaver-Burk plot, 189
linkage, 175
linkers in DNA cloning, 126–127
lock-and-key model, 187
lysine, **20**

lysozyme,
  model of, 186

M13 phage, 129
MacLeod, C., 74
main chain of proteins, 22
malaria, 168
malonate, **193**
Maloney murine leukemia virus, 134
Marmur, J., 95
mass, determination of, 49
Maxam, A., 120
Maxam-Gilbert method, 120
McCarty, M., 74
melting temperature, 82
Menten, M., 187
β-mercaptoethanol, **32**
Merrifield, R. B., 66
Meselson, M., 79, 94
metallothionein promoter, 134
methanol poisoning, 195
methemoglobin, 170
methionine, **19**
methylenebisacrylamide, **44**
Michaelis, L., 184, 187
Michaelis constant. See K_M
Michaelis-Menten equation, 189
Michaelis-Menten model, 187–191
microinjection, 133
mini-myoglobin, 150
mitochondrion
  ATP synthase of
  genetic code of, 108
mixed disulfide, **32**
mixed inhibition, 193
models, types of molecular, 4
molecular disease, 165
molecular evolution, 113–114, 171
molecular models, 4
moelcular pathology, 170
molecular weight, determination of, 49
monoclonal antibodies, 63
Monod, J., 93
Moore, S., 51
mRNA, 91
  discovery of, 93–95
myo, word origin of, 153
myoglobin, 29–30, 143–150
  amino acid sequence of, 146
  exons of, 149–150
  model of, 145
  oxygen binding by, 146–147

Na$^+$-K$^+$ transporter
  electron micrograph of, 180
Nathans, D., 118
Neel, J., 168
nerve growth factor, 16
NGF, 16
ninhydrin, **50**, 51
Nirenberg, M., 101
2-nitro-5-thiocyanobenzoate, 56
noncompetitive inhibition, 193–194
Northern blotting, 120

nucleation of sickling, 168
nucleoside 5-phosphate, 72
nucleotides, 72

Ochoa, S., 102
ochre (UAA codon), 107
oligonucleotide-directed mutagenesis, 136
opal (UGA codon), 107
oxalic acid, 195
oxygen dissociation curve, 154
oxygen-transporting proteins, 143–171
oxymyoglobin, 146

palindrome, 118
Park, J., 196
Pauling, L., 25, 165, 184
penicillin, **195**
penicillinase, 199
penicilloyl-enzyme complex, **197**
peptide bond, **22**
peptide group, 25
peptides
  chemical synthesis of, 66
  cleavage of, 56
  conformation of, 25
  overlap, 56
  sequence analysis, 196
peptidoglycan, 196
performic acid, **57**
Perutz, M., 144
pH, definition of, 41
phage. See bacteriophage
phase in x-ray diffraction, 61
phenyl isothiocyanate, **53**
phenylalanine, **18**
phenylthiohydantoin, 54
phi angle, 35–36
phosphite triester, **123**
phosphoramidite, **123**
phosphoserine, **24, 179**
  model of, 24
phosphothreonine, **179**
O-phosphotyrosine, **24**, 179
picket-fence iron prophyrin, 148
piperidine, 121
pK
  definition of, 41
  of amino acids, 21
  of proteins, 21
plasmids, 127
  pBR322, 127
  pSC101, 124
  tumor-inducing (Ti), 135
pneumococci, 74
poliovirus, 62
polyacrylamide gel, 44
polyampholytes, 46
polyclonal, 63
polypeptide chain, 22
polypeptides, chemical synthesis of, 64–67
postage-stamp analogy, 160
prediction of conformation, 35–37
Pribnow box, 98

primary structure, 31
probe, 131
proline, **18**
promoter sites, 98
prosthetic group, 29, 144
protein synthesis, 91
  start signals in, 107–108
  stop signals in, 107–108
proteins
  amino acid sequences of, 23
  cleavage of, 55–57
  conformation of, 25–28
  domains of, 111–113
  folding of, 29–37
  functions of, 15–16
  genetically engineered, 136
  hydrogen-bonding of, 29
  levels of structure of, 31
  localization of, 62–64
  methods for study of, 43–67
  modification of, 24
  nonpolar core of, 153
  peptide bonds in, 22
  purification of, 46–49
  specific binding by, 37
  synthesis of, 64–67. See also protein synthesis
  thermodynamic stability of, 34
  word origin of, 15
  x-ray crystallography of, 59–62
proteolytic activation, 179
protoplasts, 135
protoporphyrin IX, **144**
proximal histidine, 160
psi angle, 35–36
PTH-amino acids, 54
purine, **72**
pyrimidine, **72**
pyrrole, 144

quaternary structure, 31
  of hemoglobin, 158

R (relaxed) form
  of hemoglobin, 159
Ramachandran, G. N., 36
Ramachandran plot, 36
random coil, 33
reaction kinetics, 6
recombinant DNA technology, 117–138
  in protein sequencing, 57–58
renaturation, 33
residue, in proteins, 22
resolution, 61
restriction endonucleases (restriction enzymes), 118–120
restriction mapping, 120, 169
restriction-fragment-length polymorphism (RFLP), 120
11-cis retinal, **1029**
  model of, 14
retroviruses, 88, 134
  RNA tumor viruses, 87, 134
reverse transcriptase, 87, 132
reverse turns, 28

reversible inhibition, 192
reversible interactions, 7
reversible oxygenation, 148
rhinovirus 14
ribonuclease A, 30
    amino acid sequence of, 32
    renaturation of, 32–34
ribose, **86**, 92
ribozymes, 113
RNA, **86**
    genetic role of, 86–88
    in evolution, 113–114
    types of, 92
RNA polymerases, 96–97
    DNA-directed, 96
RNA viruses. *See* retroviruses
    retroviruses, 88, 134
RNA world, 113–114
RNA-DNA hybrid, 95
Rosetta stone, 106
rotational symmetry, 118
Rous sarcoma virus, 87
    electron micrograph, 14
Rudall, K. M., 27

salt links, 159
Sanger, F., 23, 52, 121
Sanger dideoxy method, 121
SDS, 44
SDS-polyacrylamide gels, 45
secondary structure, 31
sedimentation coefficient, 49
sedimentation equilibrium, 50
sedimentation velocity, 49
semiconservative replication, 79
sequenators, 54
sequencing
    of DNA, 120–123
    of proteins, 53
serine, **19**
sickle cells, 163, 167–168
sickle-cell anemia, 163–170
    clinical report of, 163
    detection of, 165, 169
    genetics of, 164, 168
    hemoglobin S in, 165
    as molecular defect, 165–166
    protection by against malaria, 168
sickle-cell trait, 164, 168
side chain in proteins, 17, 22
simian virus 40

Sinsheimer, R., 85
site-specific mutagenesis, 136
Smith, H., 118
sodium dodecyl sulfate. *See* SDS
solid-phase DNA synthesis, 123–124
solid-phase immunoassay, 63
solid-phase peptide synthesis, 66
somatotropin, 134
Southern blotting, 119–120, 169
Southern, E. M., 120
sperm whale, 145
Spiegelman, S., 95
splicing of RNA
    alternative, 113
split genes, 111
Stahl, F., 79
standard free-energy change, 182
staphylococcal protease, 56
Stein, W., 51
steroid hormones
stop codons, 106
Strominger, J., 196
succinate, **193**
supercoiled DNA, 84
superhelix, 83
supersecondary structure, 31
Svedberg units, 49
synthetic peptides, 64

T (tense) form
    of hemoglobin, 159
TATA box, 98
tertiary structure, 31
tetracycline, 127
thalassemias, 171
thermal stability
    of DNA, 82
thermodynamics
    laws of, 180
threonine, **19**
thrombin
    specificity of, 178
thymine, **72**
titration curve, 42
tobacco mosaic virus, 86–87
torr, 154
Torricelli, E., 154
transcription, 88, 91
transformation by DNA, 73–74, 133–135
transgenic mice, 134

transition state, 184
    analogs of, 197
translation. *See* protein synthesis
transpeptidation reaction, 196
tRNA, 99
trypsin, 56
    specificity of, 178
    digestion of hemoglobin by, 165
tryptophan, **18**
Tswett, M., 166
tumor-inducing (Ti) plasmids, 135
turnover number of enzymes, 190
two-dimensional electrophoresis, 46
tyrosine, **18**

ultracentrifugation, 49–50
    S values from, 50
uracil, **86**, 92
urea, **32**
    as denaturant, 32

V_{max}, 189–191
valine, **18**
van der Waals bonds, 8
van der Waals contact radii, 8
vasopressin, 65
vectors for cloning, 127–129
viruses
    as vectors, 133–134

water, 9–11
    ionization of, 41
Watson, J., 76
Watson-Crick double helix, 76
Weiss, S., 96
Western blotting, 63, 120
Wilkins, M., 76
Witkop, B., 55
wood alcohol, 195

x-ray crystallography, 59–62

Yanofsky, C., 109

zonal centrifugation, 50
zwitterion, 17
zymogens, 180